Renaissance Craftsmen and Humanistic Scholars

PASSAGEM

ESTUDOS EM CIÊNCIAS CULTURAIS
STUDIES IN CULTURAL SCIENCES
KULTURWISSENSCHAFTLICHE STUDIEN

Herausgegeben von
Peter Hanenberg und Marilia dos Santos Lopes

BAND 10

*Zu Qualitätssicherung und Peer Review
der vorliegenden Publikation*

Die Qualität der in dieser Reihe
erscheinenden Arbeiten wird vor
der Publikation durch beide Herausgeber
der Reihe sowie externe Gutachter geprüft.

*Notes on the quality assurance and
peer review of this publication*

Prior to publication, the quality of the
work published in this series is reviewed
by both editors of the series as well as
by external referees.

Thomas Horst / Marília dos Santos Lopes /
Henrique Leitão (eds.)

Renaissance Craftsmen and Humanistic Scholars

Circulation of Knowledge between Portugal and Germany

Bibliographic Information published by the Deutsche Nationalbibliothek
The Deutsche Nationalbibliothek lists this publication in the Deutsche Nationalbibliografie; detailed bibliographic data is available in the internet at http://dnb.d-nb.de.

The publication of the present volume was made possible thanks to the support of the *Bartholomäus Brüderschaft der Deutschen in Lissabon*, the oldest association of German merchants in Portugal, founded more than 700 years ago.

ISSN 1861-583X
ISBN 978-3-631-68113-8 (Print)
E-ISBN 978-3-653-07237-2 (E-PDF)
E-ISBN 978-3-631-70274-1 (EPUB)
E-ISBN 978-3-631-70275-8 (MOBI)
DOI 10.3726/b10508

© Peter Lang GmbH
Internationaler Verlag der Wissenschaften
Frankfurt am Main 2017
All rights reserved.
Peter Lang Edition is an Imprint of Peter Lang GmbH.

Peter Lang – Frankfurt am Main · Bern · Bruxelles · New York ·
Oxford · Warszawa · Wien

All parts of this publication are protected by copyright. Any utilisation outside the strict limits of the copyright law, without the permission of the publisher, is forbidden and liable to prosecution. This applies in particular to reproductions, translations, microfilming, and storage and processing in electronic retrieval systems.

This publication has been peer reviewed.

www.peterlang.com

Table of Contents

Thomas Horst, Marília dos Santos Lopes and Henrique Leitão
Foreword: Renaissance Craftsmen and Humanistic Scholars 7

Thomas Horst
The Relationship between Portugal and the Holy Roman Empire at
the Beginning of the Early Modern Period: a Brief Introduction 9

Achim Thomas Hack
Friedrich III. und Alfons V., Enea Silvio Piccolomini und
João Fernandes da Silveira. Briefliche Kommunikation zwischen
Portugal und dem Reich in den 1450er-Jahren 37

Jürgen Pohle
Kaiser Maximilian I. und die Rezeption der portugiesischen
Entdeckungen im Nürnberger Kaufmanns- und Gelehrtenkreis am
Ende des 15. Jahrhunderts 57

Marília dos Santos Lopes
Importing Knowledge: Portugal and the Scientific Culture in
Fifteenth and Sixteenth Century's Germany 73

Torsten dos Santos Arnold
Hermann Kellenbenz and the German-Portuguese Economic
Relationships during the Sixteenth Century 91

Yvonne Hendrich
«De insulis et peregrinatione lusitanorum» – Valentim Fernandes als
Vermittler von Informationen zwischen Portugal und Oberdeutschland
zu Beginn des 16. Jahrhunderts 103

Gabriele Kaiser
Leonhard Thurneysser zum Thurn (1531–1596) und sein Nachlass in
der Staatsbibliothek zu Berlin 121

Thomas Horst
A Rediscovered Manuscript about Portuguese Plants and Animals:
Preliminary Observations 133

Yves Schumacher
Basel – Fluchtpunkt der Humanisten und Alchemisten 175

Annemarie Jordan Gschwend
Anthonio Meyting: Artistic Agent, Cultural Intermediary and Diplomat
(1538–1591) 187

Samuel Gessner
Lost Between Centuries: a Celestial Globe (1575) from Augsburg
in the Portuguese Royal Collections 203

Wolfgang Köberer
"The Right Foundation of Seafaring". German-Portuguese Connections
in the Sixteenth Century with Regard to Nautical Science 223

Notes on Contributors 241

Thomas Horst, Marília dos Santos Lopes and Henrique Leitão

Foreword:
Renaissance Craftsmen and
Humanistic Scholars

The study of the relations between Portugal and the German-speaking lands in the fifteenth and sixteenth centuries is a very rich topic that has attracted the interest of scholars for some decades. The maritime expansion of Portugal and Spain created novel conditions that made Iberia the focal point to which converged people from all around Europe. The exciting novelties about the "new worlds" that had been found and the immediate recognition of the enormous potential that now opened to Europeans transformed Lisbon and Seville into cosmopolitan centres bubbling with activity. As is well known, merchants, diplomats, adventurers, naturalists, scholars and craftsmen of all sorts and from many different regions in the German-speaking states travelled to Portugal. Some of them were visitors, but many others settled in Lisbon for long periods and some even permanently. The obvious commercial interest that the very lucrative sea enterprises provided was surely the main stimulus for these events, but it would be simplistic to reduce such complex phenomena to the mere pursuit of profit.

Although studies about the relations between the Holy Roman Empire of the German Nation and the Portuguese kingdom in the period of the maritime discoveries have been many, they have traditionally been pursued along well-established disciplinary boundaries. Thus, economic historians have studied in particular the commercial ventures of the banking and German business companies that operated in Portugal and the other way around; diplomatic and political historians analysed the relations between the Portuguese monarchy and the rulers of German states; and a plethora of others specialists looked into specific areas as diverse as crafts, arts, philosophy and literature.

The collection of different articles in this volume follows this tradition of scholarship. In a sense, therefore, these studies build upon the work of others and follow trails that previous scholars had opened before. But three reasons suggested returning to the topic and organizing an interdisciplinary workshop, which took place in the National Library of Portugal on 20th/21th November 2014, and from where these works and this volume stem.

Firstly, the better understanding of the nature of the networks of communication between Portugal and Germany in the early modern period has shown that they were more rich and complex than previously suspected. In recent years evidence accumulated showing that there was still much to be known and even some large areas were still unexplored. The reader will therefore find in this collection of essays some topics that were very poorly known or that had never been studied before in detail.

Secondly, it was the explicit intention of the organizers of this volume, in order to better capture the nature of what was a complex historical phenomenon, to avoid the excessive constraints that a strictly disciplinary approach entails. It was desired that, as far as possible, the various contributions not only invoked the variety of actors (and their artifacts) present, but also the fact that many of these actors played multiple roles.

Thirdly, the organizers were mostly interested (but not exclusively) in deepening the understanding of what is today usually termed the circulation of knowledge. This, of course, does not mean adopting solely the point of view of intellectual history as it is quite clear today that knowledge travels with people, with artifacts, along commercial lines, and is created and transformed by the intervention of people from all educational and social strata. Hence the general title of the international conference and this book, that simultaneously refers to Renaissance craftsmen and to Humanistic scholars under the broad issue of knowledge circulation.

The studies in this volume bring together texts by some of the most reputed scholars in the field, together with contributions by younger ones and even researchers not affiliated with the academic world. In all cases, however, we were happy and honored to have original contributions by experts in each topic. The collection of diversified essays in this volume does not have the intention to be exhaustive nor to cover the whole field. It is, however, the hope of the editors that this book is not only a supplement to previous scholarship. It rather tries to show the vitality of this interdisciplinary field and the riches that still await researchers interested and willing to probe the history of the close relations between Portugal and Germany in the fifteenth and sixteenth centuries.

Renaissance Craftsmen and Humanistic Scholars. Circulation of Knowledge between Portugal and Germany results from a cooperation between the Interuniversity Centre for the History of Science and Technology (CIUHCT), and the Research Centre for Communication and Culture (CECC), both financed by the National Portuguese Science Foundation FCT. The publication of the present volume was made possible thanks to the support of the *Bartholomäus Brüderschaft der Deutschen in Lissabon*, the oldest association of German merchants in Portugal, founded more than 700 years ago.

Lisbon, July 2016
Thomas Horst, Marília dos Santos Lopes and Henrique Leitão

Thomas Horst

The Relationship between Portugal and the Holy Roman Empire at the Beginning of the Early Modern Period: a Brief Introduction

Abstract: Der vorliegende Beitrag ist als Einführung in die Thematik der deutsch-portugiesischen Kulturbeziehungen im 15./16. Jahrhundert gedacht und hebt exemplarisch einige bedeutende Objekte der materiellen Kultur als Schlaglichter hervor. Darunter sind neben den Grabmälern der aus Portugal stammenden Kaiserin Eleonore von Portugal (1436–1467) und ihrer Kammerzofe Beatrix Lopi († 1453) in Wiener Neustadt auch der im Johanneum in Graz verwahrte Prunkwagen sowie eine Statue in der Hofkirche von Innsbruck hervorzuheben.
Für die Entwicklung der deutsch-portugiesischen Kulturkontakte spielten schließlich nicht nur humanistische Gelehrte, sondern auch in Portugal lebende Mittelsmänner der oberdeutschen Handelsgesellschaften und ihre Vereinigung in der Bartholomäus-Brüderschaft eine besondere Rolle; die dazugehörigen archivalischen Quellen werden bestens mit Reiseberichten und materiellen Objekten, ergänzt. Diese gilt es in Zukunft vermehrt unter wissenshistorischen Aspekten zu untersuchen.

The international workshop "Renaissance Craftsmen and Humanistic Scholars: European Circulation of Knowledge between Portugal and Germany", which we have organized in the National Library of Portugal on 20th/21th November 2014 together with Professora Dr. Marília dos Santos Lopes (Centro de Estudos de Comunicação e Cultura; Universidade Católica Portuguesa) and Professor Dr. Henrique Leitão (Centro Interuniversitário de História das Ciências e da Tecnologia, University of Lisbon) dealt with the circulation of scientific and technological knowledge in the early modern period. At this time the specific cultural relationship and exchange of knowledge beween Portugal and the German-speaking lands played a significant role, which we tried to reconstruct and to analyse, especially with contemporary texts compiled by humanistic scholars (like books, manuscripts, travel journeys etc.) and – as part of the material culture – particular objects, constructed by Renaissance craftsmen (like globes, instruments, tomb stones, carriages, cannons etc.). This article highlights only a few of these objects and gives a short introduction to the interesting topic of cultural exchange of knowledge at the beginning of the early modern period.

The earliest relationship between the Kingdom of Portugal and the Holy Roman Emperor of the German Nation began in the High Middle Ages, when German crusaders helped to recapture the City of Lisbon in 1147.[1] However, the time of the Portuguese *Reconquista* ended in the first half of the thirteenth century. Later, after the Succession crisis (1383–1385), as it is known, the House of Avis established the second dynasty of kings in Portugal (1385–1580) and the "golden era" of the Portuguese discoveries began.

In this age, Infante Pedro, Duke of Coimbra (1392–1449), the brother of the Portuguese king Dom Duarte I (Edward, called the "philosopher", ruled from 1433 to 1438), traveled throughout Europe. Few historians are aware that Pedro even lived at the Habsburg Court, from 1426 to 1428.[2] His arrival in Vienna on 28th March 1426 is well documented in the so-called *Kleine Klosterneuburger Chronik*, a chronicle of the Austrian monastery Klosterneuburg not far from Vienna, where it is written: "im selben jar [...] da kham hergefahrn ein khünigs sun von pordigall, mit seinem volckh, auf 300 guets volckh, er khunt nit teutsch, aber guet lateynisch" ("in the same year [...], a son of the king of Portugal came with 300 good people; he could not speak German, but very well Latin").[3]

Pedro was not only the brother of Infante Henrique ("the Navigator", 1394–1460), who played a significant role for the Portuguese expeditions in these times, but he was also the uncle of Infanta Leonor (Eleanor of Portugal, 1436–1467)[4], who married the Holy Roman Emperor Friedrich III (Frederick III, life dates: 1415–1493) in 1452. Leonor died 15 years later, on 3th September 1467 in Wiener Neustadt, with the young age of 33 years. Her tomb with an impressive tomb slab (fig. 1)[5] can be found in the local Cistercian abbey (Neuklosterkirche zur Heiligen Dreifaltigkeit), where also her chambermaid, the Portuguese noblewoman Beatrix Lopi († 9th April 1453), was buried (fig. 2)[6].

---

1 Cf. for instance Gennrich, 1936: 11.
2 Gomes dos Santos, 1959; Rogers, 1961: 31–58.
3 Cf. the edition by Zeibig, 1851: 250.
4 For a first overview to the topic cf. for instance Hanreich, 198; Hack, 2013; Koller, 1996 and Walsh, 1993. For the date of birth of Leonor cf. Hack, 2012.
5 To the tomb slap cf. Dornik 1966 a; Kohn, 1998: 59–61, Nr. 98, Fig. 39; Hilger, 2013 and Lind, 1869.
6 The features of Beatrix Lopi look very lusitanic, cf. Dornik 1966 b, Gerhartl 1972 and Kohn, 1998: 39, Nr. 66, Fig. 26.

Fig. 1: Tomb slab of Eleanor of Portugal († 3th September 1467) in the "Neuklosterkirche zur Heiligen Dreifaltigkeit", Wiener Neustadt; withdrawn from Kohn, 1998: Fig. 39.

Fig. 2: Tomb slab of Beatrix Lopi († 9th April 1453) in the "Neuklosterkirche zur Heiligen Dreifaltigkeit", Wiener Neustadt; withdrawn from Kohn, 1998: Fig. 26.

One of the most magnificent cultural objects of this time is the upper part of a carriage with the Portuguese coat of arms (fig. 3; constructed probably after Leonor's marriage in 1452), which is preserved today in the "Museum im Palais" in the Steierisches Landesmuseum Joanneum in Graz, Austria.[7]

Leonor's uncle, Infante Pedro, died at the Battle of Alfarrobeira (20th May 1449), where he fought against the troops of his nephew, Dom Afonso V of Portugal (called "the African", 1432–1481, king since 1449, fig. 4).[8] During the reign of Afonso's son Dom João II, "o Príncipe Perfeito" (John II, "the perfect prince", who ruled the country between 1481 and 1495)[9] important arrangements were

---

7   Schramm/Fillitz, 1978: 82 (Nr. 103: „Oberteil eines Prunkwagens") and 218 (fig. 103); Smola 1966.
8   Jaime (1433–1459), a son of Pedro, who also fought in the battle of Alfarrobeira, later made a career in the church and became cardinal in 1456. Pope Pius II made him legat to the Imperial Court at Vienna, but Jaime never went there, because he already died in 1459 in Florence, where he is buried in the basilica San Miniato al Monte, cf. Apfelstadt, 2000.
9   For his biography cf. Horst, 2015. For a general overview about Portugal in the fifteenth century cf. Bernecker/Herbers, 2013: 92–132 („Die »zweite Staatswerdung« – Höhepunkt der portugiesischen Geschichte im Zeitalter der Europäischen Expansion?");

settled with the Catholic kings of Spain to control the new territories overseas (cf. for instance the famous Treaty of Tordesillas in 1494).[10]

Fig. 3: Carriage of Eleanor of Portugal, made after 1452, nearly 3 meters long. Preserved today in the "Museum im Palais" in the "Steierisches Landesmuseum Joanneum" in Graz, Austria.

Fig. 4: King Afonso V of Portugal. Hand-colored sketch (probably drawb around 1470) from the journal of Georg von Ehingen (1428–1508), currently held by the Württembergische Landesbibliothek in Stuttgart.

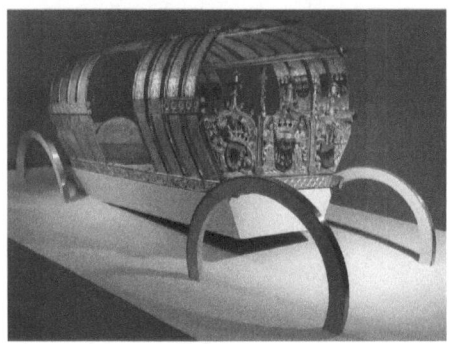

This explains why also João's cousin, King Maximilian I (1459–1519, who was a son of Leonor and Emperor Friedrich III), and his consultant, Konrad Peutinger (1465–1547, an important humanist and town clerk from Augsburg), as well as the German painter Albrecht Dürer (1471–1528)[11], were all very curious about these developments in Western Europe, which had led to the discovery of a New World.[12] An objective remembrance of this relation-

---

Freller, 2012: 171–192 („Vom Rand Europas in das Zentrum der Weltpolitik - Portugal unter Afonso V., João II. und Manoel I."); Jakob, 1969: 72–108 („Die Dynastie Avis [1385–1580]"); Pietschmann, 2011; Schäfer, 1839 as well as the detailed bibliography by Jakob, 1961 and (with newer titles): Ehrhardt, 1994.

10 An important church document for the history of discoveries was also the "Oração de Obediência" of 1485, cf. Hamann, 1971.

11 Cf. for instance his famous woodcut "rhinoceros" from 1515: Cole, 1953 and Clarke, 1986: 16–27 (Chapter 1: "The first Lisbon or ‚Dürer' Rhinoceros of 1515").

12 For the Portuguese expansion and its reflections at the court of Maximilian I, cf. Krendl, 1985; Lopes, 1993; Metzig, 2013; Metzig, 2016; Pohle, 2015 a; Pohle, 2016 and in general (with the edition of contemporary documents) Meyn et al., 1984: 41–150.

ship can be seen in the gothic "Hofkirche" (Court Church) in Innsbruck, Austria, which was built by Emperor Ferdinand I (1503–1564) as a memorial for his grandfather Maximilian I. in 1553–1563. The cenotaph (completed only in 1584)[13] is surrounded by 28 large bronze statues[14], which were created as guards for the tomb between 1502 and 1555. One of these statues (more than 200 cm high and designed in the workshop of Gilg Sesselschreiber around 1509), shows one of Maximilian's Portuguese ancestor: While former historians identified the statue (and standing besides the legendary king Arthur) with Dom Ferdinand I (1345–1383), who was king of Portugal and the Algarve from 1367 until his death, a new detailed analysis of the coat of arms (cf. fig. 5) suggests that Ferdinands grandfather, Dom João I (1357–1433), is portrayed here.[15]

---

13 Maximilian, who even fancied to become pope in 1511 (cf. Wiesflecker, 1963) died in Wiener Neustadt on 12th January 1519 and was buried in the chapel St. Georg in the castle there. He already planned his sepulchral monument in detail during his life time, but it remained unfinished. This is why his grandson Ferdinand I ordered the construction of a new church and monastery in Innsbruck. While the simple tomb remained in Wiener Neustadt, the magnificent marble cenotaph in Innsbruck, which is empty, serves as dynastical memorial, cf. Bange, 1946; Baresel-Brand, 2007: 30–34; Egg, 1988; Haidacher/Diemer, 2004 [on the restauration of the cenotaph]; Ringler, 1958 and Scheicher, 1999.
14 Because of the black colour of these statues, the Franciscan church is known in vernacular language also as "Schwarzmander-Kirche". On the statues cf. the standard work by Oberhammer, 1935. There also existed a bronze statue of "Eleonora Künigin von Portugal, Uxor Friderici terty Rom. Imperatoris Mater Maximiliani I.", but this figure was casted so bad, that it was remelted, cf. ibid.: 96; Hye, 1988: 51 and Ruggenthaler, 2006: 91, Nr. 15 („Dises Pild ist das in dem langen Haar, so gantz schlecht on alle Zier mit plosen Haubt gegossen, und soll in der Zal bleiben, manglen die zway Henndt, Cron, Kertz, Schillt unnd Schrift, steet bey der Kays. Mst. genedigstem entschluß, ob Sy dis Pild gar von newen widerumben wellen giessen lassen oder nit").
15 The coat of arms was put wrongly besides the statue of Elisabeth von Tirol-Görz, cf. Hye, 1988: 53, 54 (fig. 14), 58 and 59 (fig. 24 f.); Oberhammer, 1935: 15 – Cf. also Ringler, 1958: 11 and Ruggenthaler, 2006: 91, Nr. 12 ("Ferdinand Künig von Portgal. An disem Pild ist der Schilt und die Schrift zuveremderem, und anstatt der Cron ain Hertzog Hüetl zumachen").

Fig. 5: The portuguese coat of arms in the "Hofkirche" in Innsbruck can be found wrongly besides the statue of Elisabeth von Tirol-Görz, withdrawn from Hye, 1988: 59, Fig. 25.

Already in the Late Middle Ages, the "Bartholomäus-Brüderschaft" ("confraria de São Bartolomeu")[16], a fraternity devoted to Saint Bartholomew, was founded in Lisbon by German-speaking immigrants. It was a Hanseatic timber merchant, Michael Overstädt (Miguel Sobrevila, probably a counselor of D. Dinis I of Portugal), who possessed a storage yard in Lisbon on the northern bank of the river Tejo. On the side of the present-day Praça do Município he built a chapel, which served as a place for devotions to the Holy Apostle Bartholomew. But the chapel was pulled down only a few years later. In exchange for this, it was incorporated into the larger church of S. Julião, which was completed around 1290/1291. This is the traditional date for the foundation of the "Bartholomäusbruderschaft", which functioned mainly as a guild for the German merchant community in Lisbon.[17] The ecclesiastical fraternity supported the acquisition and maintenance of the Lisbon chapel[18] as well as the salary of its own priest (who ministered already in

---

16 Cf. Ehrhardt, 1990 and Hinsch, 1890. See also Kuder/Ptak, 1984: 7 f. and Strasen/Gândara, 1944: 31–38.
17 Cf. Denk/Schickert, 2010: 16–26; Ehrhardt 1990: 3; Hinsch, 1890: 4; Metzig, 2010: 277, footnote 31 and Pohle, 2000: 146–150. – On the special relations with the Hanse and Hamburg, especially during the early modern times, see Kellenbenz, 1954; Kellenbenz, 1958; Poettering, 2013 and Studemund-Halévy, 2007.
18 Therein were also two relics of Saint Bartholomew, which the fraternity received as a present from the third wife of Dom Manuel I, Eleonor of Austria. The first description of S. Julião can be found in Cardoso, 1666: 322–325, esp. 325.

the German language)[19], but also supported charity-projects in Lisbon, such as the creation of its own cimitery in 1425, and the foundation of the first German hospital in 1495[20]. However, it was during the second half of the fifteenth century[21] when the character of this society changed to become a fraternity of German soldiers ("confraria dos bombardeiros alemães" with "bombardeiros da nómina"[22], cannoneers and Renaissance craftsmen as members[23]), so that the merchants moved into another chapel of S. Julião (devoted to Saint Sebastian). The two fraternities were only reunited in the seventeenth century. The "Bartholomäus-Brüderschaft", which has a long and very interesting history[24], has influenced the cultural life in Lisbon over centuries and still exists today.

---

19 A document from 1551 shows that the German fraternity employed daily a chaplain with the financial income from the houses that the Germans possessed in Lisbon. Alone the earnings of chairity brought altogether 110 cruzados per year, cf. Oliveira, 1551: 5 r: "A confraria de sam Bartolameu, he administrada per alemães tem capella per si com capelão quotediano, tem renda de casas na cidade. E asesmolas valcada anno cento E dez cruzados". – Another text from 1582 relates to the participation of the fraternity in the Corpus Christi procession with green capes and candles, see Velasquez, 1582: 17 v: "La dezima cruz, fue la cofradia de los Flamencos, y como ge[n]te de trato muestran ser caudalosos enla riqueza de sus insignias, tienen la abogacio[n] del Apostol sant Bartholome, sus opas son verdes, y de cera verde traya cada vno vna hacha de quarto paulos enla mano, su cofradia es de mucho numero, que por ser grande su contractacion, habitan en Lisboa, co[n] sus casas de morada, cantidad dellos". This solemn celebration was attended also by the Spanish King Philipp II (r. in Portugal from 1580 to 1598), cf. Denk/Schickert, 2010: 85 f.
20 The famous Nuremberg merchant and cosmographer Martin Behaim died in this hospital, which was also a hostel for the poor and old people, in 1507. However, it did not exist for a long, see Pohle, 2000: 147. Other hospitals were founded in 1799 and 1929, but were also closed soon afterwards, see Ehrhardt, 1990: 8–10.
21 In the first half of the fifteenth Century many letters mention German merchants together with Renaissance craftsmen, see Marques, 1959: 147–150. Later the emigrants were dominated by soldiers, cf. for instance Kellenbenz, 1963.
22 See Denk/Schickert, 2010: 27–41 and in particular Metzig, 2010 ("deutsche Büchsenschützen"; on this profession, which was created in the Late Middle Ages, cf. Leng, 1996).
23 But not all craftsmen were members of this fraternity. In particular Dutch artisans were united in the "Heilig Kreuz und St. Andreas-Bruderschaft" ("Confraria da Santa Cruz e Santo André"), which was founded in 1414, see Pohle, 2000: 149.
24 On the later history of the Bartholomäusbruderschaft cf. Denk/Schickert, 2010 and Mörsdorf, 1957/1958. On the sixteenth century, when the Lisbon inquisition also pursued German emmigrants, see especially Dias, 1995.

In the sixteenth century, the Portuguese empire expanded[25], with the help *inter alia* of German and Flemish mercenaries[26], under the reign of King Dom Manuel I ("the Fortunate", ruled 1495–1521) into Africa[27], India and Asia[28]. In 1514, Manuel gave the famous elephant Hanno[29] as a present to Pope Leo X (1513–1521), and during his reign a new architectural style, called "estilo manuelino" (Manueline style), was established. His third wife, which he married in 1519, was Eleonor of Austria (Eleanor of Castile, 1498–1558, who later became also Queen Consort of France).[30]

---

25 For the iconography of these discoveries cf. the important study by Lopes, 1998 a and Kraus, 2007. A general overview gives Aubin, 1990; Diffie/Winius, 1977 and Poettering, 2014. For an unknown manuscript about these discoveries cf. Nagel, 1970.
26 Cf. for instance the privileges which they got from the Portuguese king: Cassel, 1771; Ferreira, 1969 and Ribeiro, 1922.
27 Hamann, 1968 and Lopes, 2001: 15–17: Even the "Carta Marina Navigatoria", designed by Martin Waldseemüller in 1516, shows the Portuguese king Manuel I riding on a dolphin not far from the Cape of Good Hope, as does the reprint by Lorenz Fries (1530). – To Manuel cf. also Costa, 1970 and Lopes, 1998 b: 17.
28 For the "Estado da India" cf. Feldbauer, 2003 and Malekandathil, 1999.
29 Another indian elephant called Süleyman (Soloman/Soliman) came as a diplomatic gift from the King of Kotte in Ceylon in 1541 first to the Portuguese king Dom João III (John III, 1502–1557) and his wife Catherine of Austria (1507–1578) and later was departed to Vienna in 1551 as wedding present for Archduke Maximilian II, who married his first cousin Mary of Spain (1528–1603, daughter of Emperor Charles V and Isabella of Portugal) in 1548, where he died on 18th December 1553 cf. Jordan Gschwend, 2010; Zollner/Hamberger, 2015 and the novel "A Viagem do Elefante" ("The Elephant's journey") by the Nobel-prize winning Portuguese author José Saramago (1922–2010), cf. Saramogo, 2008.
30 Eleanor of Austria was a daughter of king Philipp I of Castile ("the Handsome", 1478–1506) and Joanna of Castile ("the Mad", 1479–1555), hence she was granddaughter of Emperor Maximilian I, cf. Benavides, 2011: 247–256 – A daughter of Manuel, Isabel of Portugal (1503–1533) married on 10th March 1526 in Seville the Holy Roman Emperor Charles V (1500–1558), who was a brother of Eleanor, cf. Hamann, 1988: 167 f. – A letter of Manuel I to his relative Maximilian I (dated Lisbon, 14th March 1518) is edited by Nagel, 1974, cf. also Lopes, 1998 b: 9, footnote 14. This letter goes back to another letter, which Manuel I sent to the pope Julius II (r. 1503–1513) on 25th September 1507 and which was printed in Abrantes cf. Metzig, 2013: 18 f. and Künast/Zäh, 2003: 157 f., who mention also other letters to the popes; a German copy (printed in Nuremberg in 1508) of another letter is Emanuel [Kunig zu Porthogal], 1508. One of these letters (from Manuel I to pope Leo X) was published also in Vienna in 1513, cf. Emanuel [Regis Portugaliae], 1513. – On the correspondence with the sister of Maximilian I, Kunigunde of Austria (1465–1520, by marriage with Albert IV

Fig. 6: The Poster of the international workshop Renaissance Craftsmen and Humanistic Scholars: European Circulation of Knowledge between Potugal and Germany, hold at the Biblioteca Nacional de Portugal, 20th/21st November 2014 illustrates Jakob Fugger together with his accountant Matthäus Schwartz, cf. Lopes, 1998 a: 35, Fig. I. 12.

Beneath the merchants who settled on the Iberian Peninsula, the German trading houses of the Welser and the Fugger and their factors played an important role: permanent factories of the Upper German merchants, who later financed the discoveries[31], already existed in Portugal from the end of the fifteenth century.[32] The poster of our workshop (cf. fig. 6) portrays Jakob Fugger ("the Rich", 1459–1525) together with his accountant Matthäus Schwartz. This image[33] was painted around 1517 as part of the "Schwartzsches Trachtenbuch" (today in the Herzog-Anton-Ulrich-Museum in Braunschweig, Germany). It represents the two noble men

---

in 1487 Duchess of Bavaria-Munich), who joined the Convent of Püttrich in Munich in 1508 and wrote letters to Manuel in 1519 cf. Kunstmann, 1845: 420 f.
31 Cf. Bernecker, 2000: 202–218; Kellenbenz, 1978 and Werner, 1967.
32 Cf. the groundbreaking study by Pohle, 2000 and various older single case studies by the economic historian Hermann Kellenbenz (for instance: Kellenbenz, 1960; Kellenbenz, 1966 and Kellenbenz, 1970). One of the most important sources is the diary of Lukas Rem (1481–1541), but also the archival documents about Sebastian Kneussel, who was a factor for the Imhof trading house in Lisbon in 1512/13 (cf. Pohle, 2015 b) and the global player Lazarus Nürnberger (1499–1564; cf. Kellenbenz/Walter, 2001) give new perceptions.
33 Lopes, 1998 a: 35, Fig. I. 12.

standing before a bureau, where the correspondence to the various factories was filed. Therein, the Fugger establishment in the City of Lisbon is emphasized in written form – besides the merchant settlements in Rome, Venice, Ofen, Cracow, Milan, Innsbruck, Nuremberg and Antwerp. This image visualizes not only the close economic connections to Portugal, but documents also the transnational exchange in Europe which took place even centuries ago and is not a modern invention.

But how was scientific and technical knowledge transferred in the Renaissance between craftsmen and humanistic scholars? And when did it start exactly? It is not easy to give a precise answer to these questions. But most historians agree that the marriage[34] of Infanta Leonor with the Holy Roman Emperor Friedrich III in the middle of the fifteenth century and the permanent presence of south German trading companies can be seen as a starting point of more intensive connections with the western part of Europe.

Moreover, the first global players like the Nuremberg merchant and cosmographer Martin Behaim (1459–1507)[35], who constructed the first known terrestrial globe around 1493, or Dr. Hieronymus Münzer (Monetarius, † 1508)[36], who met king Dom João II during his voyage and gave a report about Portugal in his famous travelogue of 1494/1495 (Bavarian State Library, Munich, Clm 431, fol. 96 r–275 r)[37], are significant for our topic.

However, Münzer was not the first traveller on the Iberian Peninsula: He stands in a long tradition of travellers[38], who visited Portugal and maybe also transferred their culture, as did, for example, the South Tyrolean troubadour ("min-

---

34 We are well informed about the preparations of this marriage with the report of Nikolaus Lankmann von Falkenstein, who came to Lisbon in 1451 together with Jakob Motz, cf. Hack, 1999 (who is preparing an edition of this interesting travelogue) and Reichert, 2009: Nr. 1, 14–20. – About the festivities in Lisbon on the occasion of this marriage, which was confirmed symbolical ("per procurationem") with Leonor, cf. in particular Hack, 2016 and a recently discovered manuscript in a miscellany (written around 1475) of the Bavarian State Library, Cgm 5482, fol. 99 v–107 r, which gives a German summary.
35 Pohle, 2007 and more in general Willers, 1992. – On the economic relations with Nuremberg in the 15th and 16th centuries cf. Kellenbenz, 1967.
36 Classen, 2003; Herbers, 2000 and Jaspert, 2016. – More general to medieval travellers, who visited the Iberian Peninsula, cf. Reichert, 2001: 90–101 ("Unterwegs zu den Grenzen Europas").
37 In the Lisbon church Santa Maria de Luce (Nossa Senhora da Luz) Münzer has seen also an exotic "Wunderkammer". To Münzers voyage through the Iberian Peninsula cf. Hurtienne, 2009 and Münzer, 2006. – Münzer also wrote a famous letter to the Portuguese king João II in 1493, cf. Metzig, 2013: 21 and Pohle (in this volume): 63.
38 Cf. Marques, 1995.

nesinger") Oswald von Wolkenstein († 1445)[39] already in 1415 on the occasion of the conquest of Ceuta[40]. Further travellers were mainly pilgrims to Santiago de Campostela and visited Portugal (in particular Lisbon and the Kap Finisterre) on their return journeys[41]: We have to mention here Konrad von Scharnachthal, a Patrician of Bern (1445)[42], and Sebastian Ilsung (a Patrician and mayor of Augsburg, † 1468), who reports in the German language about his voyage to Santiago in 1446 (cf. London, British Library, Add. 4326, fol. 1 r–6 r)[43]. Another traveller was the Swabian diplomat and knight Georg von Ehingen (1428–1508), who did a round trip, which led him to the Holy Land and afterwards to the courts of France, Navarra, Castile, Portugal, England and Scotland. In his later autobiography it is

---

39 The Tyrolean poet Oswald von Wolkenstein (around 1376–1445) was also important as a diplomat. He participated as an attendant of Friedrich IV, Duke of Austria and Count of Tyrol ("Friedrich mit der leeren Tasche", 1382–1439) at the Council of Constance (1414–1418). On 16th February 1415 he became a member of the entourage ("Hofgesinde") of the Holy Roman Emperor and King of Hungary, Sigismund of Luxembourg (1368–1437) with a salary of 300 Hungarian Gold ducats per year, cf. Schwab, 1999, 223–227, Nr. 70. His first diplomatic voyage brought him probably to England, Scotland, Ireland and the Iberian Peninsula. If he really visited Lisbon in July, as proposed by Kühn, 2011: 232 f. is uncertain, but he participated (together with other mercenaries) in the conquest of the Moorish city of Ceuta on 21th August 1415 (cf. ibid., 235–238 and Dallapiazza/Molinari, 2011). Afterwards he returned as a rich man via Granada to Perpignan (France, cf. Hartmann, 2015), where he stayed for 50 days and met King Sigismund, who wrote in Paris on 1st April 1416 a letter of consignment for him, cf. Schwab, 1999, 233–235, Nr. 73. In the mid of April of the same year Oswald returned to Constance, cf. Mayr, 1961, 70–81. – Oswald invented the lyric genre of "Reiselieder" (songs which describe adventures), cf. Schallaböck/Müller, 2003. Even if we have no written proof that Oswald really was in Portugal, his song "Durch Barbarei, Arabia […]" (Kl. 44, cf. Ammon, 2007/2008) suggests that he was there, because he mentions therein the Iberian Peninsula ("Durch Arragun, Kastilie, Granaten und Afferen, auss Portigal, Ispanie pis gen dem vinstern steren") – as well as he does in song Kl. 26 "Durch Abenteuer Tal und Berg", I/II, here: I, 6: "auf hölggen gross gen Portigal zu siglen" (written only in 1427).
40 On the Portuguese discovery of Ceuta in 1415, cf. Braga/Braga, 1998: 17–25, Meyn et al., 1984: 50–52 and Schäfer, 1839: 259–291.
41 Cf. Reichert, 2001: 90–97.
42 Konrad von Scharnachthal visited the kingdom of Granada and Portugal, cf. Reichert, 2009: Nr. 3, 28–38, here: 33. – For the relations to Switzerland in general, cf. Fischer, 1960. In 1501 another patrician of Bern, Wolfgang von Laupen († 1519), who was heavily in debt, got a lifelong employment as bombardier in Portugal, cf. Metzig, 2010: 274.
43 The reason for this journey was probably a diplomatic mission for Pope Felix V, cf. Reichert, 2001: 98. Honemann, 1988: 70 and 76 mentions that Ilsung only planned to come to Portugal, but never realized this trip.

reported that in 1457/1458 he helped to defend Ceuta and Granada during the fights with the Muslim armies.[44]

We also know about another journey, the one by the Bohemian noble man Leo von Rožmital († 1485)[45], who made a tour through Western Europe from 1465 to 1467, because his companion Gabriel Tetzel († 1479, a later mayor from Nuremberg) wrote an informative travelogue of their trip. This is why we know that they went from Santiago to Portugal[46]. When they came back to the court in Wiener Neustadt, Leo von Rožmital delivered a letter from Afonso V to his sister Leonor[47]. He played on the lute for her "etlich portugalisch tänz" (Portuguese dances)[48], which pleased not only the queen, but also her young son Maximilian.

Besides a lot of other travels, the voyage of the Silesian nobleman Nikolaus von Popplau through Europe (1483–1486) has to be highlighted, who also visited the cities of Setubal and Lisbon[49].

Another influential person at this time – and very important for the Portuguese-German relationship in general – was the Moravian-German typographer Valentim Fernandes († 1518/1519)[50], who settled down in Lisbon in 1495 and, together with the Germans, played a major overseas role for the transfer of knowledge (cf. for example Bavarian State Library, Cod. Hisp. 27)[51]. One of the Portuguese expeditions to India (1505/1506, led by Francisco de Almeida) was financed by a German trade consortium. For this topic the travellogue of Balthasar Sprenger[52], which was printed three years later with wonderful ethnographic illustrations, is also of great importance. At the same time, a pamphlet with the title *Den rechten*

---

44 Cf. Schmidt, 1997 and Paravicini, 2000: 565–567 and 571–575 (Texts Nr. 2–6, corresponding letters from Afonso V).
45 For Leo von Rožmital and his journey, cf. Paravicini, 2010; Péricard-Méa, 2006 and Stolz, 1988.
46 In 1466 Leo von Rožmital visited Évora, where he stayed for 14 days and met the Portuguese court, which excaped from Lisbon because of the plague, cf. Schmeller, 1844: 174–183, in particular 182: "Der kunig von Portigal was damals geflohen aus Lisbona der haubtstat in Portigal, do er almal hof hält, den sterben in ein stat, heist Ebor". The Portuguese king gave him two horses, two black people, two monkeys and weapons („auch vil leopardenhäut und vil bogen, tarschen, lünzlein und ander heidnische waffen").
47 Paravicini, 2010: 263 f. notices that Rožmital has delivered also a letter of Leonor to her brother in Portugal, which explains the celebratory reception which he got.
48 Schmeller, 1844: 194 f. and Walsh, 1993: 416.
49 For Nikolaus von Popplau cf. Paravicini, 2004 and Reichert, 2009: Nr. 2, 20–28.
50 Hendrich, 2007. For the German printers in Portugal cf. also Ehrhardt, 1996.
51 Massing, 2012.
52 Horst, 2009 and Kunstmann, 1861. Cf. also Bernecker, 2000: 198 f.

weg auß zu fahren von Liszbona gen Kallakuth (cf. fig. 7; *the right way to travel from Lisbon to Calicut*) was printed in Nuremberg[53], together with a small map, which shows Nuremberg and Kalikut as the starting and final points of this enterprise.[54] In the context to this expedition, the geographical image of a globe, which can be found on a triptych of the Dutch painter Quentin (1466–1533), which was painted for Lukas Rem in 1519[55], fits exactly, as does a letter[56] from 26th June 1510, where Valentim Fernandez orders an astrolabe from the Nuremberg merchant Stefan Gabler. On the other hand the Tyrolean craftsmen Gregor Löffler († 1565) casted a canon in 1534, which is today exhibited in the "Museu Militar" in Lisbon.[57]

Fig. 7: Title page of the German pamphlet Den rechten weg auß zu fahren von Liszbona gen Kallakuth *(Nuremberg, ca. 1506). Original in the University Library of Freiburg im Breisgau, J 4672, m. Cf. Online-Ressource, 2013. Freiburger historische Bestände – digital: Drucke des Humanismus und der Reformationszeit (1450–1600). Freiburg im Breisgau, Albert-Ludwigs-Universität, online: http://dl.ub.uni-freiburg.de/diglit/weg1506/0002?sid= 6391129ada036b2e2ebc97e7ebae7083 [seen on 1st May 2016].*

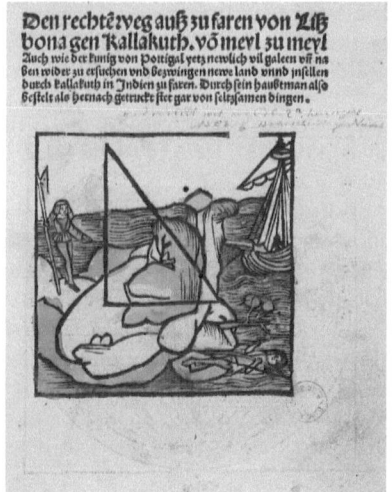

---

53 The Imperial City of Nuremberg was at that time a center for mapmaking and printing, cf. Bernecker, 2000: 194–198.
54 A reproduction of this map can be found in: Horst, 2006: 19, fig. 7. For a description of the pamphlet cf. Horst, 2009: 191 f. and Lopes, 1998 a: 20.
55 Horst, 2006: 14; Horst, 2009: 193 f. and Welser, 1958: 106.
56 Brásio, 1960: 341 and 358.
57 Neuwirth, 2004.

As for the effects of "Shopping in the Renaissance", exotic animals (like parrots), bezoar stones, ivory and many other material objects were traded as part of the contemporary establishment of cabinets of curiosities at European Renaissance courts, international merchants, bankers, collectors and navigators served as cultural mediators.[58] In the sixteenth century more German visitors reported about their journeys to Portugal, so for instance the Nuremberg travel writer Hieronymus Köler the Elder (1507–1574)[59], who stayed more than three months in Lisbon in 1533, or the foot soldier ("Landsknecht") Nikolaus Schmid from Regensburg, who described the discovery of Portugal in 1580 in prosaic verses (Bavarian State Library, Cgm 3008).[60]

There is no place here to discuss all these important highlights[61] for the German-Portuguese relationship in detail. For an introduction to this topic we recommend the reading of the relevant studies by Hermann Kellenbenz (1913–1990)[62], Marion Ehrhardt (1932–2011)[63], António Henrique de Oliveira Marques (1933–2007)[64], Jürgen Pohle and Marília dos Santos Lopes[65], who have all contributed outstandingly to this field of Renaissance culture studies.[66]

---

58 Cf. Pieper, 1999 and especially Jordan Gschwend/Beltz, 2010.
59 Welser, 1874 and Willers, 1992: 837, Nr. 5.10 (Lotte Kuras on his description of the expedition of the Welsers in Venezuela).
60 Hümmerich, 1930. – Vitally important is also a German manuscript at the Austrian National Library, Cod. 9865: The "Aigentliche vnd grundtliche beschreibung / dem vrsprvng vnd anfang des Portugalesischen kriegs" by Johann Holzhammer (a participant of the Spanish crusade against Portugal since 1579) was written in the years between 1583 and 1595. On fol. VI r a colorful illustration shows Philipp II standing on a terrestrial globe in the middle of the ocean, which is controlled by Spanish ships. This figure emblematises his sovereignity over the cosmos as does the coat of arms aside, which shows the coat of arms of Spain and Portugal flanked by the Pillars of Hercules, again a terrestrial globe, and the motto "Plus ultra", cf. Gamillscheg, 1995: 41 [fig.].
61 As an example I want to allude to the politic-diplomatic relations between Portugal and Austria in the eighteenth century cf. Gatzhammer, 1994. It is interesting to see that in this period also Bavarian gaffers imigrated to the Iberian Peninsula. One of these families ("Hahn") still exists (portuguese version: "Gallo"), cf. Paulus, 2010.
62 Kellenbenz, 1968; Kellenbenz, 1970 and Kellenbenz, 1990.
63 Ehrhardt, 1989;
64 Marques, 1960.
65 In particular Lopes, 1998 a.
66 Cf. also the „Portugiesische Forschungen der Görres-Gesellschaft" (20 vols, appeared from 1960 to 1993).

Bibliography

Ammon, Frieder von. 2007/2008. „Wildwüchsiger Gesang vom Ich: Oswalds von Wolkenstein Lied Durch Barbarei, Arabia". *Parapluie* 24, 1–7. [online: http://parapluie.de/archiv/autobiographien/resonanzen/parapluie-autobiographien_resonanzen.pdf].

Apfelstadt, Eric. 2000. "Bishop and Pawn: New Documents fort he Chapel of the Cardinal of Portugal at S. Miniato al Monte, Florence". In: Lowe, Kate J. P. (ed.). *Cultural Links between Portugal and Italy in the Renaissance*. Oxford et al.: Oxford University Press, 183–223.

Aubin, Jean (Ed.). 1990. *La Découverte, le Portugal et l'Europe. Actes du colloque, Paris 26–28 mai 1988*. Paris: Fondation Calouste Gulbenkian.

Bange, Ernst Friedrich. 1946. *Das Grabmal Kaiser Maximilians I. in der Hofkirche zu Innsbruck* (Der Kunstbrief 12). Berlin: Mann.

Baresel-Brand, Andrea. 2007. *Grabdenkmäler nordeuropäischer Fürstenhäuser im Zeitalter der Renaissance 1550–1650*. Kiel: Ludwig.

Benavides, Francisco da Fonseca. 2011. *Reinhas de Portugal. As mulheres que construirem a nação*, 2 vols, (1878/1879). Lisboa: Marcador [reprint].

Bernecker, Walther L. 2000. „Nürnberg und die überseeische Expansion im 16. Jahrhundert". In: Neuhaus, Helmut (Hrsg.). *Nürnberg, Europäische Stadt in Mittelalter und Neuzeit* (Nürnberger Forschungen 29). Nürnberg: Verein für Geschichte der Stadt Nürnberg, 185–218.

Bernecker, Walther L./Herbers, Klaus. 2013. *Geschichte Portugals*. Stuttgart: Kohlhammer.

Braga, Isabel M. R. Mendes Drumond/Braga, Paulo Drumond. 1998. *Ceuta Portuguesa (1415–1656)*. Ceuta: Instituto de Estudios Ceutíes.

Brásio, António. 1960. "Uma carta inédita de Valentim Fernandes". *Boletim da Biblioteca da Universidade de Coimbra* 24, 338–358.

Cardoso, Jorge, 1666. *Agiologio Lusitano dos sanctos, e varoens illustres em virtude do Reino de Portugal, e suas conquistas: consagrado aos gloriosos S. Vicente, e S. Antonio, insigns patronos desta inclyta cidade Lisboa e a seu illustre Cabido Sede Vacante. Tomo III*. Lisboa: Officina Craesbeekiana.

Cassel, Johann Philipp. 1771. *Privilegien und Handelsfreiheiten welche die Könige von Portugal ehedem den deutschen Kaufleuten zu Lissabon ertheilet haben*. Bremen: Witwe und Meier.

Clarke, T. H. 1986. *The Rhinoceros from Dürer to Stubbs: 1515–1799*. London et al.: Sotheby's Publications.

Classen, Albrecht. 2003. „Die iberische Halbinsel aus der Sicht eines humanistischen Nürnberger Gelehrten Hieronymus Münzer: Itinerarium Hispanicum

(1494–1495)". *Mitteilungen des Instituts für Österreichische Geschichtsforschung* 111, 317–340.

Cole, Francis Joseph. 1953. "The History of Albrecht Durer's Rhinoceros in Zoological Literature". In: Underwood, E. Ashworth (ed.). *Science, Medicine and History: Essays on the Evolution of Scientific Thought and Medical Practice, Written in Honour of Charles Singer.* Volume 1. London et al.: Oxford University Press, 337–356 and plates 23–31.

Costa, José Perreira da. 1970. "D. Manuel e a Torre de Tombo". *Portugiesische Forschungen der Goerres-Gesellschaft* 10, 296–303.

Dallapiazza, Michael/Molinari. 2011. „Südfrankreich, die iberische Halbinsel und Nordafrika: zur großen Reise Oswalds von Wolkenstein 1415/1416". In: Müller, Ulrich/Springreth, Margarete (Hrsg.). *Oswald von Wolkenstein. Leben – Werk – Rezeption.* Berlin/New York: De Gruyter, 240–250.

Denk, Thomas/Schickert, Gerhart. 2010. *Die Bartholomäus-Brüderschaft der Deutschen in Lissabon. Entstehung und Wirken, vom späten Mittelalter bis zur Gegenwart – A Irmandade de São Bartolomeu dos Alemães em Lisboa: origem e actividade, do final da Idade Média até à Actualidade: XII–XXI.* Estoril: Sabedoria e Literatura, Lda.

Dias, João José Alves. 1995. „Zur Geschichte der deutschen Kolonie im Portugal des 16. Jahrhunderts. Einige Prozesse der Lissabonner Inquisition". In: Lopes, Marília dos Santos/Hanenberg, Peter/Knefelkamp, Ulrich (Hrsg.). *Portugal und Deutschland auf dem Weg nach Europa. Portugal e Alemanha a caminho para a Europa* (Weltbild und Kulturbegegnung 5). Pfaffenweiler: Centaurus-Verl.-Ges., 27–35.

Diffie, Baily W./Winius, George D. (ed.): *Foundations of the Portuguese Empire 1415–1580* (Europe and the world in the age of expansion 1). Minneapolis: University of Minnesota Press.

Dornik, Hanna. 1966 a. „Grabplatte der Kaiserin Eleonore von Portugal". In: Weninger, Peter (Hrsg.). *Friedrich III. Kaiserresidenz Wiener Neustadt. Ausstellung St. Peter an der Sperr, Wiener Neustadt; 28. Mai bis 30. Oktober 1966* (Katalog des Niederösterreichischen Landesmuseums, Neue Folge 29). Wien: Amt der Niederösterreichischen Landesregierung, 356 f., Nr. 151.

Dornik, Hanna. 1966 b. „Epitaph der Beatrix Lopi". In: Weninger, Peter (Hrsg.). *Friedrich III. Kaiserresidenz Wiener Neustadt. Ausstellung St. Peter an der Sperr, Wiener Neustadt; 28. Mai bis 30. Oktober 1966* (Katalog des Niederösterreichischen Landesmuseums, Neue Folge 29). Wien: Amt der Niederösterreichischen Landesregierung, 356, Nr. 150.

Egg, Erhard. 1988. *Hofkirche in Innsbruck: Grabmal Kaiser Maximilians I.* Ried im Innkreis: Kunstverlag Hofstetter.

Ehrhardt, Marion. 1989. *A Alemanha e os Descobrimentos Portugueses*. Lisboa: Texto Editora.

Ehrhardt, Marion. 1990. *Die Bartholomäus-Brüderschaft der Deutschen in Lissabon. Ein Rückblick – A Irmandade de S. Bartolomeu dos Alemães em Lisboa. Uma Retraspectiva*, Lisboa: Editora Gratica Portuguesa.

Ehrhardt, Marion. 1994. „Auswahlbibliographie zu den deutsch-portugiesischen Kulturbeziehungen". *Zeitschrift für Kulturaustausch* 44/1, 116–124.

Ehrhardt, Marion. 1996. „Frühe deutsche Drucker in Portugal". In: Marques, António Henrique de Oliveira/Opitz, Alfred/Clara, Fernando (Hrsg.). *Portugal – Alemanha – África: do imperialismo colonial ao imperialismo político* (Actas do IV Encontro Luso-alemão). Lissabon: Ed. Colibri, 25–30.

Emanuel [Kunig zu Porthogal]. 1508. *Ein abschrifft eines sandtbriefes So vnserm allerheyligisten vater dem Bapst Julio dem andern gesandt ist, von dem allerdurchleuchtigisten Fursten vnd herren, herren Emanuel Kunig zu Porthogal &c. an dem zwelfften tag des Brachmonds im M.ccccc.viii. jare von wunderbarlichen raysen vnd schieffarten, vnd eroberung landt, stet vnd merckt auch grosser manschlachtung der hayden.* Nürnberg: Georg Stuchs.

Emanuel [Regis Portugaliae]. 1513. *Epistola Potentissimi ac inuictissimi Emanuelis Regis Portugaliæ, & Algarbiorum &c. De victorijs habitis in India & Malacha. Ad. S. in Christo patre[m] & Dominu[m] nostrum Do[minum] Leonem X Pont[ifex] Maximum.* Vienna: per Hieronymu[m] Vietore[m] & Ioannem Singrenium.

Feldbauer, Peter. 2003. *Estado da India. Die Portugiesen in Asien 1498–1620.* Wien: Mandelbaum.

Ferreira, João Albino Pinto. 1969. *Privilégios concedidos pelos Reis de Portugal aos Alemães nos séculos XV e XVI (Documentos Arquivados no Gabinete da História da Cidade).* (Separata do Boletim Cultural 32). Porto: Câmara Municipal.

Fischer, Béat de. 1960. *Dialogue luso-suisse: Essai d'une histoire des relations entre la Suisse et le Portugal du 15ᵉ siècle à la Convention de Stockholm de 1960.* Lisbonne: Ramos Afonso & Moita.

Freller, Thomas. 2012. *Die Geschichte der Iberischen Halbinsel. Mit landeskundlichen und kulturhistorischen Beiträgen von Miguel Vásquez.* Ostfildern: Thorbecke.

Gamillscheg, Ernst. 1995. „Johann Holzhammer. Beschreibung des Portugalesischen Krieges (deutsch)". In: Auer, Alfred/Irblich, Eva (Hrsg). *Natur und Kunst. Handschriften und Alben aus der Ambraser Sammlung Erzherzog Ferdinands II. (1529–1595). Ausstellung des Kunsthistorischen Museums und der Österreichischen Nationalbibliothek, Schloß Ambras, Innsbruck, 23. Juni–24. September 1995.* Wien: Kunsthistorisches Museum, Nr. 2, 40–42.

Gatzhammer, Stefan. 1994. „Politisch-dilpomatische Beziehungen zwischen Portugal und Österreich im 18. Jahrhundert vor dem Hintergrund der Jesuitenfrage". *Mitteilungen des Instituts für österreichische Geschichtsforschung* 102, 359–408.

Gennrich, Paul W. 1936. *Evangelium und Deutschtum in Portugal. Geschichte der Deutschen Evangelischen Gemeinde in Lissabon*. Berlin et al.: de Gruyter.

Gerhartl, Gertrud. 1972. „Das portugiesische Edelfräulein Beatrix Lopi. The tomb stone of the Portuguese Noble Lady Beatrix Lopi". In: Kolarsky, Mathilde (Hrsg.). *Begegnung der Völker in Österreich* (Notring-Jahrbuch 1972). Wien: Notring, 48–50.

Gomes dos Santos, Mauricio Domingos. 1959. "O Infante D. Pedro na Austria-Hungria". *Brotéria* 68, 17–37.

Hack, Achim Thomas. 1999. „Nikolaus Lankmann von Falkenstein". *Bautz. Biographisch-Bibliographisches Kirchenlexikon Zusatz (BBKL)* 16, column 1148–1152.

Hack, Achim Thomas. 2012. „Das Geburtsdatum der Kaiserin Eleonore". *Mitteilungen des Instituts für Österreichische Geschichtsforschung* 120, 146–153.

Hack, Achim Thomas. 2013. „Eine Portugiesin in Österreich um die Mitte des 15. Jahrhunderts. Kultureller Austausch infolge einer kaiserlichen Heirat?". In: Fuchs, Franz/Heinig, Paul-Joachim/Wagendorfer, Martin (Hrsg.). *König und Kanzlist, Kaiser und Papst. Friedrich III. und Enea Silvio Piccolomini in Wiener Neustadt* [Vorträge des interdisziplinären Symposions „Kaiser Friedrich III. (1440–1493) von 8. bis 10. Oktober 2009 in Wiener Neustadt]. Wien et al.: Böhlau, 181–204.

Hack, Achim Thomas. 2016. „Lebende Bilder in Lissabon um die Mitte des 15. Jahrhunderts". In: Dittscheid, Hans-Christoph/Gerstl, Doris/Hespers, Simone (Hrsg.). *Kunst-Kontexte. Festschrift für Heidrun Stein-Kecks*. Petersberg: Michael Imhof Verlag, 91–102.

Haidacher, Christoph/Diemer, Dorthea (Hrsg.). *Maximilian I. Der Kenotaph in der Hofkirche zu Innsbruck*. Innsbruck: Haymon.

Hamann, Brigitte (Hrsg.). 1988. *Die Habsburger. Ein biographisches Lexikon*. München: Piper.

Hamann, Günther. 1968. *Der Eintritt der südlichen Hemisphäre in die europäische Geschichte. Die Erschließung des Afrikaweges nach Asien vom Zeitalter Heinrichs des Seefahrers bis zu Vasco da Gama* (Österreichische Akademie der Wissenschaften, Philosophisch-Historische Klasse, Sitzungsberichte 260; Veröffentlichungen der Kommission für Geschichte der Mathematik und der Naturwissenschaften 6). Wien: Böhlau.

Hamann, Günther. 1971. „Die Oração de Obediência von 1485. Ein kirchenhistorisches Dokument zur Entdeckungsgeschichte". In: Kovács, Elisabeth (Hrsg.). *Festschrift Franz Loidl zum 65. Geburtstag*, vol. 3. Wien: Hollinek, 55–66.

Hanreich, Antonia. 1985. "D. Leonor de Portugal. Esposa do Imperador Frederico III (1436–1467)". In: Scheidl, Ludwig/Palma Caetano, José A. (ed.). *Relações entre a Áustria e Portugal. Testemunhos históricos e culturais*. Coimbra: Livraria Almedina, 3–27.

Hartmann, Sieglinde. 2015. "Oswald von Wolkenstein à Perpignan: le chanteur courtois et son seigneur le roi Sigismond". [unpublished paper, given on 24th September 2015 at the international conference, organized by Catafau, Aymat/Jaspert, Nikolas/Wetzstein, Thomas: *Perpignan 1415: Ein europäisches Gipfeltreffen in der Konzilszeit/Perpignan 1415: un sommet uropéen à l'époque des conciles*].

Hendrich, Yvonne. 2007. *Valentim Fernandes – Ein deutscher Buchdrucker in Portugal um die Wende vom 15. zum 16. Jahrhundert und sein Umkreis* (Mainzer Studien zur Neueren Geschichte 21). Frankfurt am Main: Peter Lang.

Herbers, Klaus. 2000. „ ‚Murcia ist so groß wie Nürnberg' – Nürnberg und Nürnberger auf der Iberischen Halbinsel: Eindrücke und Wechselbeziehungen". In: Neuhaus, Helmut (Hrsg.). *Nürnberg, Europäische Stadt in Mittelalter und Neuzeit* (Nürnberger Forschungen 29). Nürnberg: Verein für Geschichte der Stadt Nürnberg, 151–180.

Hilger, Wolfgang. 2013. „Das Grabdenkmal Kaiserin Eleonores von Portugal in der Neuklosterkirche von Wiener Neustadt". In: Fuchs, Franz/Heinig, Paul-Joachim/Wagendorfer, Martin (Hrsg.). *König und Kanzlist, Kaiser und Papst. Friedrich III. und Enea Silvio Piccolomini in Wiener Neustadt* [Vorträge des interdisziplinären Symposions „Kaiser Friedrich III. (1440–1493) von 8. bis 10. Oktober 2009 in Wiener Neustadt]. Wien et al.: Böhlau, 205–213.

Hinsch, J. D. 1890. „Die Bartolomäus-Brüderschaft der Deutschen in Lissabon". *Hansische Geschichtsblätter* XVII, herausgegeben vom Verein für Hansische Geschichte, Jahrgang 1888, 3–27.

Honemann, Volker 1988. „Sebastian Ilsung als Spanienreisender und Santiagopilger (mit Textedition)". In: Herbers, Klaus (ed.). *Deutsche Jakobspilger und ihre Berichte*. Tübingen: Gunter Narr Verlag, 61–95.

Horst, Thomas. 2006. „Am Anfang war das Gewürz. Vor 500 Jahren kehrte der Allgäuer Balthasar Sprenger von einer Indienfahrt zurück. Er hinterließ einen eindrucksvollen Reisebericht". *Literatur in Bayern* 85 (September 2006), 13–21.

Horst, Thomas. 2009. "The voyage of the Bavarian explorer Balthasar Sprenger to India (1505/1506) at the turning point between the Middle Ages and the Early Modern Times: his travelogue and the contemporary cartography as historical sources". In: Billion, Philipp et al. (Hrsg). *Weltbilder im Mittelalter – Perceptions in the World of the Middle Ages*. Bonn: Bernstein-Verlag, 167–197.

Horst, Thomas. 2015. „Johann II. (1455–1495), König von Portugal". Bautz. *Biographisch-Bibliographisches Kirchenlexikon (BBKL)*, Nachtragsband 36, column 945–972.

Hümmerich, Franz. 1930. *Ein bayerischer Landsknecht über die Eroberung Portugals durch Philipp II. im Jahre 1580* (Miscelânea de Estudos em Honra de D. Carolina Michaelis de Vasconcellos). Coimbra: Imprensa da Universidade.

Hurtienne, René. 2009. „Ein Gelehrter und sein Text. Zur Gesamtedition des Reiseberichts von Dr. Hieronymus Münzer, 1494/95 (Clm 431)". In: Neuhaus, Helmut (Hrsg.). *Erlanger Editionen. Grundlagenforschung durch Quelleneditionen: Berichte und Studien* (Erlanger Studien zur Geschichte 8). Erlangen/Jena: Palm&Enke, 255–272.

Hye, Franz-Heine von. 1988. „Pluriumque Europae Provinciarum Rex et Princeps – Kaiser Maximilians I. genealogisch-heraldische Denkmäler in und um Innsbruck". In: *Staaten, Wappen, Dynastien: XVIII. Internationaler Kongreß für Genealogie und Heraldik in Innsbruck vom 5. bis 9. September 1988* (Veröffentlichungen des Innsbrucker Stadtarchivs, Neue Folge 18). Innsbruck: Stadtmagistrat Innsbruck, 35–63.

Jakob, Ernst Gerhard. 1961. *Deutschland und Portugal. Ihre kulturellen Beziehungen. Rückschau und Ausblick. Eine Bibliographie*. Leiden: Brill.

Jakob, Ernst Gerhard. 1969. *Grundzüge der Geschichte Portugals und seiner Übersee-Provinzen*. Darmstadt: Wissenschaftliche Buchgesellschaft.

Jaspert, Nikolas. 2016. „Hieronymus Münzers deutsche Gastgeber auf der Iberischen Halbinsel. Archivnotizen und Ergänzungen". In: Alraum, Claudia/Holndonner, Andreas/Lehner, Hans-Christian/Scherer, Cornelia/Schlauwitz, Thorsten/Unger, Veronika (Hrsg.). *Zwischen Rom und Santiago. Festschrift für Klaus Herbers zu seinem 65. Geburtstag*. Bochum: Dieter Winkler, 79–98.

Jordan Gschwend, Annemarie. 2010. *The Story of Süleymann: Celebrity elephants and other exotica in Renaissance Portugal*. Zurich: Pachyderm.

Jordan Gschwend, Annemarie/Beltz, Johannes (Hrsg.). 2010. *Elfenbeine aus Ceylon. Luxusgüter für Katharina von Habsburg (1507–1578)*, Ausstellungskatalog. Zürich: Museum Rietberg.

Kellenbenz, Hermann. 1954. *Unternehmerkräfte im Hamburger Portugal- und Spanienhandel 1590–1625* (Veröffentlichungen der Wirtschaftsgeschichtlichen Forschungsstelle e. V. 10). Hamburg: Verlag der Hamburgischen Bücherei.

Kellenbenz, Hermann. 1958. *Sephardim an der unteren Elbe. Ihre wirtschaftliche und politische Bedeutung vom Ende des 16. bis zum Beginn des 18. Jahrhun-*

derts (Beihefte der Vierteljahrschrift für Sozial- und Wirtschaftsgeschichte 42). Wiesbaden: Franz Steiner.

Kellenbenz, Hermann. 1960. "Os mercadores alemães de Lisboa por volta de 1530". *Revista portuguesa de história* 9, 125–140.

Kellenbenz, Hermann. 1963. „Die Unternehmertätigkeit des portugiesischen Prinzen Heinrich und die deutschen Ritter". In: Specht, Karl Gustav (ed.): *Studium sociale. Ergebnisse sozialwissenschaftlicher Forschung der Gegenwart* [Karl Valentin Müller dargebracht]. Köln/Opladen: Westdeutscher Verlag, 751–761.

Kellenbenz, Hermann. 1966. "La participation des capitaux de l'Allemagne méridionale aux entreprises portugaises d'outremer ao tournant du XVe siècle". In: Mollat, Michel and Paul Adam (eds.). *Les aspects internationaux de la décoverte océanique au XVe siècle, Actes du cinquième colloque internacional d'histoire maritime.* Paris: S.E.V.P.E.N., 309–317.

Kellenbenz, Hermann. 1967. „Die Beziehungen Nürnbergs zur Iberischen Halbinsel besonders im 15. und 16. Jahrhundert". Stadtarchiv Nürnberg (Hrsg.). *Beiträge zur Wirtschaftsgeschichte Nürnbergs.* Bd. 1. Nürnberg: Selbstverlag des Stadtrats zu Nürnberg, 456–493.

Kellenbenz, Hermann. 1968. „Jácome Fixer. Deutsche Handelsbeziehungen zu Portugal um 1600". *Portugiesische Forschungen der Görresgesellschaft, Erste Reihe: Aufsätze zur portugiesischen Kulturgeschichte* 8, 251–274.

Kellenbenz, Hermann (ed.), 1970. *Fremde Kaufleute auf der Iberischen Halbinsel* (Kölner Kolloquien zur Internationalen Sozial- und Wirtschaftsgeschichte 1). Köln/Wien: Böhlau Verlag.

Kellenbenz, Hermann. 1978. "The Role of the Great Upper German Families in Financing the Discoveries". *Terrae Incognitae* 10, 45–59.

Kellenbenz, Hermann. 1990. *Die Fugger in Spanien und Portugal bis 1560.* 3 vols. München: Vogel Verlag.

Kellenbenz, Hermann/Walter, Rolf. 2001. *Oberdeutsche Kaufleute in Sevilla und Cadiz. Eine Edition von Notariatsakten aus den dortigen Archiven.* Stuttgart: Steiner.

Kohn, Renate. 1998. *Die Inschriften der Stadt Wiener Neustadt* (Deutsche Inschriften 48; Die Inschriften des Bundeslandes Niederösterreich 2). Wien: Verlag der Österreichischen Akademie der Wissenschaften.

Koller, Erwin. 1996. „Die Verheiratung Eleonores von Portugal mit Kaiser Friedrich III. in zeitgenössischen Berichten". In: Marques, António Henrique de Oliveira/Opitz, Alfred/Clara, Fernando (Hrsg.). *Portugal – Alemanha – África: Do imperialismo colonial ao imperialismo político* (Actas do IV Encontro Luso-alemão). Lissabon: Ed. Colibri, 43–56.

Kraus, Michael. 2007. *Novos Mundos – Neue Welten. Portugal und das Zeitalter der Entdeckungen. Eine Ausstellung des Deutschen Historischen Museums Berlin in Zusammenarbeit mit dem Instituto Camões, Lissabon und der Botschaft von Portugal in Berlin* [24. Oktober 2007 bis 10. Februar 2008]. Berlin: Deutsches Historisches Museum.

Krendl, Peter. 1985. "O imperador Maximiliano I e Portugal". In: Scheidl, Ludwig/ Palma Caetano, José A. (ed.). *Relações entre a Áustria e Portugal. Testemunhos históricos e culturais.* Coimbra: Livraria Almedina, 29–60.

Kuder, Manfred/Ptak, Heinz Peter. 1984. *Deutsch-Portugiesische Kontakte in über 800 Jahren und ihre wechselnde Motivation* (Portugal-Reihe 10). Bammental/ Heidelberg: Klemmerberg.

Kühn, Dieter. 2011. *Ich Wolkenstein. Die Biographie.* Frankfurt am Main: Fischer Taschenbuch Verlag.

Künast, Hans-Jörg/Zäh, Helmut (Hrsg.). 2003. *Die Bibliothek Konrad Peutingers. Edition der historischen Kataloge und Rekonstruktion der Bestände. Bd. 1: Die autographen Kataloge Peutingers, der nicht-juristische Bibliotheksteil.* Tübingen: Niemeyer.

Kunstmann, Friedrich. 1845. „Schreiben des Schwesternhauses zum Pütrich in München an den König Emanuel von Portugal, aus dem Lissaboner Archive mitgetheilt". *Oberbayerisches Archiv für vaterländische Geschichte, herausgegeben von dem Historischen Vereine von und für Oberbayern* 6 (1844), 418–421.

Kunstmann, Friedrich. 1861. *Die Fahrt der ersten Deutschen nach dem portugiesischen Indien.* München: Kaiser.

Leng, Rainer. 1996. „getruwelich dienen mit Buchsenwerk. Ein neuer Beruf im späten Mittelalter: Die Büchsenmeister". In: Rödel, Dieter/Schneider, Joachim Schneider (Hrsg.). *Strukturen der Gesellschaft im Mittelalter. Interdisziplinäre Mediävistik in Würzburg.* Wiesbaden: Reichert, 302–321.

Lind, Karl. 1869. „Der Grabstein der Kaiserin Eleonore (Mit 1 Holzschnitt)". *Mittheilungen der k. k. Central-Commission zur Erforschung und Erhaltung der Baudenkmale* 14, 101–104.

Lopes, Marília dos Santos. 1993. „Die portugiesischen Entdeckungen in deutschen Schriften des 16. Jahrhunderts". *Aufsätze zur portugiesischen Kulturgeschichte* 20 (1988–1992), 132–141.

Lopes, Marília dos Santos. 1998 a. *Wonderful things never yet seen. Iconography of the Discoveries.* Translation from the Portuguese by Clive Gilbert. Lisboa: Quetzal.

Lopes, Marília dos Santos. 1998 b. „Dom Manuel oder das Projekt eines glücklichen Königs". *Periplus. Jahrbuch für Außereuropäische Geschichte* 8, 8–17.

Lopes, Marília dos Santos. 2001. *Da descoberta ao Saber. Os conhecimentos sobre África na Europa dos séculos XVI e XVII.* Viseu: passagem.

Malekandathil, Pius. 1999. *The Germans, the Portuguese and India* (Periplus Parerga 6). Münster: Lit.

Marques, António Henrique de Oliveira. 1959. *Hansa e Portugal na Idade Media.* Lisboa: Universidade de Lisboa.

Marques, António Henrique de Oliveira. 1960. "Relações entre Portugal e a Alemanha no Século XVI". *Revista da Faculdade de Letras de Lisboa*, 36–55.

Marques, António Henrique de Oliveira. 1995. „Deutsche Reisende im Portugal des 15. Jahrhunderts". In: Lopes, Marília dos Santos/Hanenberg, Peter/Knefelkamp, Ulrich (Hrsg.). *Portugal und Deutschland auf dem Weg nach Europa. Portugal e Alemanha a caminho para a Europa* (Weltbild und Kulturbegegnung 5). Pfaffenweiler: Centaurus-Verl.-Ges., 11–26.

Massing, Andreas W. 2012. "V. Fernandes' Five Maps and the early History and Geography of Sam Tomé". *Boletim da Sociedade de Geografia de Lisboa*, Série 130/1-12 (Janeiro-Decembro 2012), 61–74.

Mayr, Norbert. 1961. *Die Reiselieder und Reisen Oswalds von Wolkenstein* (Schlern-Schriften 215). Innsbruck: Wagner.

Metzig, Gregor. 2010. „Kanonen im Wunderland – Deutsche Büchsenschützen im portugiesischen Weltreich (1415–1640)". *Militär und Gesellschaft in der Frühen Neuzeit* 14, 267–298.

Metzig, Gregor. 2013. „Maximilian I. (1486–1519), Portugal und die Expansion nach Übersee". *Jahrbuch für Europäische Überseegeschichte* 11 (2011), 9–44.

Metzig, Gregor. 2016. *Kommunikation und Konfrontation; Diplomatie und Gesandtschaftswesen Kaiser Maximilians I. (1486–1519).* (Bibliothek des Deutschen Historischen Instituts in Rom 130). Berlin et al.: De Gruyter [in print].

Meyn, Matthias et al. (ed.). 1984. *Die großen Entdeckungen* (Dokumente zur Geschichte der europäischen Expansion 2). München: Beck.

Mörsdorf, Klaus. 1957/1958. *A Irmandade de São Bartolomeu dos Alemães em Lisboa. Parecer elaborado a pedido de Mons. Büttner.* Munique/Lisboa: União Gráfica.

Münzer, Jérôme. 2006. *Voyage en Espagne et au Portugal (1494–1495). Introduction, traduction et notes par Michel Tarayre.* Paris: Les Belles Lettres.

Nagel, Rolf. 1970. „Die portugiesischen Entdeckungen in einer rheinischen Handschrift des 16. Jahrhunderts". *Portugiesische Forschungen der Goerres-Gesellschaft* 10, 287–295.

Nagel, Rolf. 1974. „Ein Brief König Manuels I. an Kaiser Maximilian I.". *Portugiesische Forschungen der Goerres-Gesellschaft* 11, 201–205.

Neuwirth, Markus. 2004. „Gregor Löfflers Kanone für Kaiser Karl V. im Museu Militar zu Lissabon". *Mitteilungen des Tiroler Landesmuseums Ferdinandeum* 84, 109–114.

Oberhammer, Vinzenz. 1935. *Die Bronzestandbilder des Maximiliangrabmales in der Hofkirche zu Innsbruck*. Innsbruck: Deutsche Alpenverlags-Gesellschaft.

Oliveira, Cristovão Rodrigues de. 1551. *Summario e[m] que brevemente se contem algvas covsas (assi ecclesiasticas como seculares) qve ha na cidade de Lisboa*. Lixboa: em casa de Germão Galharde.

Paravicini, Werner. 2000. „Georg von Ehingens Reise vollendet". In: Paviot, Jacques/Verger, Jacques (eds.). *Guerre, pouvoir et noblesse au Moyen Âge. Mélanges en l'honneur de Philippe Contamine* (Cultures et civilisations médiévales). Paris: Presses de l'Univ. de Paris-Sorbonne, 547–588.

Paravicini, Werner. 2004. „Der Fremde am Hof. Nikolaus von Papplau auf Europareise 1483–1486". In: Zotz, Thomas (Hrsg.). *Fürstenhöfe und ihre Außenwelt. Aspekte gesellschaftlicher und kultureller Identität im deutschen Spätmittelalter* (Identitäten und Alteritäten 16). Würzburg: Ergon-Verlag.

Paravicini, Werner. 2010. „Leo von Rožmitál unterwegs zu den Höfen Europas (1465–1466)". *Archiv für Kulturgeschichte* 90, 253–308.

Paulus, Georg. 2010. „Bayerische Glasmacher auf der Iberischen Halbinsel. Die um 1740 ausgewanderten Glasmacherfamilien Eder und Hahn". *Blätter des Bayerischen Landesvereins für Familienkunde* 73, 5–39.

Péricard-Méa, Denise. 2006. „Leo von Rozmital, böhmischer Pilger und Botschafter (1465–1467) im werdenden Europa". In: Doležal, Daniel/Kühne, Hartmut (Hrsg.). *Wallfahrten in der europäischen Kultur. Tagungsband Příbram, 26.–29. Mai 2004. Pilgrimage in European culture* (Europäische Wallfahrtsstudien 1). Frankfurt am Main: Peter Lang, 109–119.

Pieper, Renate. 1999. „Papageien und Bezoarsteine. Gesandte als Vermittler von Exotica und Luxuserzeugnissen im Zeitalter Philipps II". In: Edelmayer, Friedrich (Hrsg.): *Hispano-America II: Die Epoche Philipps II. (1556–1598)*. (Studien zur Geschichte und Kultur der Iberischen und Iberoamerikanischen Länder 5). Wien: Verlag für Geschichte und Politik/München: Oldenbourg, 215–224.

Pietschmann, Horst. 2011. „Deutsche und imperiale Interessen zwischen portugiesischer und spanischer Expansion im 15. Jahrhundert". In: Curvelo, Alexandra (Hrsg.). *Portugal und das Heilige Römische Reich (16.–18. Jahrhundert) – Portugal e o Sacro Império (séculos XVI–XVIII)* (Studien zur Geschichte und Kultur der iberischen und iberoamerikanischen Länder 15). Münster: Aschendorff, 15–30.

Poettering, Jorun. 2013. *Handel, Nation und Religion. Kaufleute zwischen Hamburg und Portugal im 17. Jahrhundert*. Göttingen: Vandenhoeck & Ruprecht.

Poettering, Jorun. 2014. „Als die Säulen des Herakles umstürzten. Wissen, Wissenschaft und Herrschaft in der portugiesischen Expansion (15. und 16. Jahrhundert)". *Saeculum. Jahrbuch für Universalgeschichte* 64/2, 257–288.

Pohle, Jürgen. 2000. *Deutschland und die überseeische Expansion Portugals im 15. und 16. Jahrhundert* (Historia profana et ecclesiastica 2). Münster: Lit Verlag.

Pohle, Jürgen. 2007. *Martin Behaim (Martinho da Boémia): Factos, Lendas e Controvérsias* (cadernos do cieg 26). Coimbra: Cieg/MinervaCoimbra.

Pohle, Jürgen. 2015 a. "«Os primeiros alemães a procurar a Índia»: Maximiliano I, Conrad Peutinger e a alta finança alemã estabelecida em Lisboa. The first Germans searching India»: Maximilian I, Conrad Peutinger and the German merchant-bankers established in Lisbon". *Ammentu. Bolletino Storico e Archivistico del Mediterraneo e delle Americhe* 7 (luglio-decembro 2015), 19–28.

Pohle, Jürgen. 2015 b. "Lukas Rem e Sebald Kneussel: Due Agenti Commerciali Tedeschi a Lisbona all'Inicio del Seculo XVI e le loro testimonianze". *Storia Economia* 18/2, 315–329.

Reichert, Folker. 2001. *Erfahrung der Welt. Reisen und Kulturbegegnung im späten Mittelalter*. Stuttgart: Kohlhammer.

Reichert, Folker (Hrsg.). 2009. *Quellen zur Geschichte des Reisens im Spätmittelalter*. Darmstadt: Wissenschaftliche Buchgesellschaft.

Ribeiro, Vítor. 1922. *Privilégios de estrangeiros em Portugal (ingleses, franceses, alemães, flamengos e italianos)*. Memória apresentada á Academia das Ciências de Lisboa (História e Memórias da Academia das Ciências de Lisboa, Nova Série, 2ª Classe: Sciências Morais e Políticas, e Belas Letras XIV). Lisboa: Imprense Nacional.

Ringler, Josef. 1958. *Das Maximiliansgrab in Innsbruck*. Königstein im Taunus: Langewiesche.

Rogers, Francis Millet. 1961. *The Travels of the Infante Dom Pedro of Portugal* (Harvard Studies in Romance Languages 26). Cambridge (Mass.): Harvard University Press.

Ruggenthaler, Oliver OFM. 2006. „Das Maximilian-Mausoleum in der Innsbrucker Hofkirche nach Quellen des Archivs der Tiroler Franziskanerprovinz". *Tiroler Heimat*, Neue Folge 70, 85–97.

Saramogo, José. 2008. *A viagem do Elefante*. Afragide: Caminho.

Schäfer, Heinrich. 1839. *Geschichte von Portugal. Zweiter Band: Vom Erlöschen der echten burgundischen Linie bis zum Schlusse des Mittelalters* (Geschichte der europäischen Staaten II/2). Hamburg/Gotha: Perthes.

Schallaböck, Thomas/Müller, Ulrich. „Gesungene Reiseberichte aus dem 15. Jahrhundert: Die Reiselieder des Oswald von Wolkenstein. Mit einem Anhang: Die letzte Reise Oswalds, oder: Das zweimalige Begräbnis des Wolkensteiners".

In: Erdorff-Kupfer, Xenja von/Giesemann, Gerhard (Hrsg.). *Erkundung und Beschreibung der Welt. Zur Poetik der Reise- und Länderberichte. Vorträge eines interdisziplinären Symposiums vom 19. bis 24. Juni 2000 an der Justus-Liebig-Universität Gießen*, Amsterdam: Rodopi, 163–183.

Scheicher, Elisabeth. 1999. „Kaiser Maximilian plant sein Grabmal". *Jahrbuch des kunsthistorischen Museums Wien*, 81–117.

Schmeller, Johann Andreas (Hrsg.). 1844. *Des böhmischen Herrn Leo's von Rožmital Ritter-, Hof- und Pilger-Reise durch die Abendlande 1465–1467. Beschrieben von zweien seiner Begleiter. Itineris a Leone de Rosmital nobili Bohemo annis 1465–1467 per Germaniam, Angliam, Franciam, Hispaniam, Portugalliam atque Italiam confecti, Commentarii coaevi duo*. Stuttgart: Literarischer Verein.

Schmidt, Susanna. 1997. *Georg von Ehingen, „Reisen nach der Ritterschaft": Stil und Darstellungsmuster einer Ritterbiographie am Übergang vom späten Mittelalter zur frühen Neuzeit*. Bonn: Universität [Dissertation].

Schramm, Percy Ernst/Fillitz, Hermann. 1978. *Ein Beitrag zur Herrschergeschichte von Rudolf I. bis Maximilian I., 1273–1519*. München: Prestel Verlag.

Schwab, Anton (Hrsg.). 1999. *Die Lebenszeugnisse Oswalds von Wolkenstein. Edition und Kommentar*, Band 1: 1382–1419, Nr. 1–92. Wien et al.: Böhlau.

Smola, Gertrud. 1966. „Oberteil eines Prunkwagens". In: Weninger, Peter (Hrsg.). *Friedrich III. Kaiserresidenz Wiener Neustadt. Ausstellung St. Peter an der Sperr, Wiener Neustadt; 28. Mai bis 30. Oktober 1966* (Katalog des Niederösterreichischen Landesmuseums, Neue Folge 29). Wien: Amt der Niederösterreichischen Landesregierung, 357 f., Nr. 153.

Stolz, Michael. 1988. „Die Reise des Leo von Rožmital". In: Herbers, Klaus (ed.). *Deutsche Jakobspilger und ihre Berichte*. Tübingen: Gunter Narr Verlag, 97–121.

Strasen, Ernst August/Gândara, Alfredo. 1944. *Oito Séculos de História Luso-Alemã. Com 320 illustrações no texto, 3 iluminuras, índice de 1800 personagens, e 2 quadros genealógicos anexos*. Berlim: Instituto Ibero-Americano de Berlim.

Studemund-Halévy. 2007. *Portugal in Hamburg*. Herausgegeben von der ZEIT-Stiftung Ebelin und Gerd Bucerius. Hamburg: Ellert &Richter.

Thomas, Lothar. 1944. *Contribuição para a história da filosofia portuguesa. Geschichte der Philosophie in Portugal. Ein Versuch, Erster Band: Die Geschichte der Philosophie in Portugal von den Anfängen bis Ende des 16. Jahrhunderts ausschließlich der Regeneration der Scholastik*. Lisboa: Imprensa Barreiro.

Velasquez, Isidoro. 1582. *La orden que se tuvo en la sole[m]ne procession que hizieron los devotos cofrades del Sa[n]ctissimo Sacrame[n]to de la iglesia de S. Iulian, en la ciudad de Lisboa, celebra[n]do la festiuidad de su cofraria, Domingo, dos de Septie[m]bre de 1582*. Lisboa: en casa de Manuel de Lyra.

Walsh, Katherine. 1993. „Deutschsprachige Korrespondenz der Kaiserin Leonora von Portugal. Bausteine zu einem geistigen Profil der Gemahlin Kaiser Friedrichs III. und zur Erziehung des jungen Maximilian". In: Heinig, Paul-Joachim (Hg.). *Kaiser Friedrich III. (1440–1493) in seiner Zeit. Studien anläßlich des 500. Todestages* (Forschungen zur Kaiser- und Papstgeschichte des Mittelalters 12). Köln u. a.: Böhlau Verlag, 399–445.

Welser, Johann Michael Anton Freiherr von. 1874. „Aus Hieronymus Kölers Aufzeichnungen (Reise nach Spanien 1533)". *Zeitschrift des Historischen Vereins für Schwaben und Neuburg* 1, 321–333.

Welser, Hubert Freiherr von. 1958. „Der Globus des Lukas Rem". *Mitteilungen des Vereins für Geschichte der Stadt Nürnberg* 48, 96–114.

Werner, Theodor Gustav. 1967. „Die Beteiligung der Nürnberger Welser und Augsburger Fugger an der Eroberung des Rio de la Plata und der Gründung von Buenos Aires". In: Stadtarchiv Nürnberg (Hrsg.). *Beiträge zur Wirtschaftsgeschichte Nürnbergs.* Bd. 1. Nürnberg: Selbstverlag des Stadtrats zu Nürnberg, 494–592.

Wiesflecker, Hermann. 1963. „Neue Beiträge zur Frage des Kaiser-Papstplanes Maximilians I. im Jahre 1511". *Mitteilungen des Instituts für Österreichische Geschichtsforschung* 71, 311–332.

Willers, Johannes (Hrsg.). 1992. *Focus Behaim-Globus.* 2 Bde. Nürnberg: Verlag des Germanischen Nationalmuseums.

Zeibig, Hartmann Josef. „Die kleine Klosterneuburger Chronik (1322 bis 1428; Monumenta Claustroneoburgensia 1)". *Archiv zur Kunde österreichischer Geschichtsquellen* 7 (1851): 227–268.

Zollner, Marianne/Hamberger, Edwin. 2015. *Helafant alhier! Die Reise des Elefanten Soliman von Spanien über Mühldorf nach Wien.* Mühldorf am Inn: Stadt Mühldorf am Inn.

Achim Thomas Hack

Friedrich III. und Alfons V., Enea Silvio Piccolomini und João Fernandes da Silveira. Briefliche Kommunikation zwischen Portugal und dem Reich in den 1450er-Jahren

Abstract: This essay reflects on the contacts between Portugal and the German Empire around the mid 15th century, especially on epistolary communications. The most important evidence is the correspondence pursued by humanist Enea Silvio Piccolomini, which is extant in a fragmentary and unique manuscript in the Austrian National Library (ÖNB cvp 3389). It includes two 1453 letters from Piccolomini to João Fernandes da Silveira; unfortunately, no corresponding letters in return have been transmitted. Both parties of the correspondence attest a deep friendship between them; they met in person several times, both before and after 1453 (during the nuptial arrangements 1450 in Naples, in the context of Emperor Frederick III's coronation, apparently also during Piccolomini's pontificate). In addition, the manuscript also contains a letter written by Alfonso V, King of Portugal, and Frederick III's reply, which evidently was written by Enea Silvio Piccolomini. They discuss, amongst other things, the Emperor's Portuguese consort and the state of their marriage; the tone used in these letters is clearly more harsh on both sides. It is telling that none of Eleonore's letters to Portugal are extant.

Lissabon und Wiener Neustadt mögen immer gleich weit voneinander entfernt gelegen haben – die Luftlinie zwischen den beiden Städten beträgt exakt 2.277 Kilometer[1] –, Portugal und das Reich kamen sich bisweilen näher und entfernten sich auch wieder. Diese Art von Nähe existierte immer dann, wenn Austausch und Kommunikation zwischen beiden Ländern vorhanden war. Sie konnte sich auf viele unterschiedliche Gebiete erstrecken – auf Wirtschaft und Politik, aber auch auf Wissenschaft und Philosophie –, sie konnte verschiedene Grade der Intensität annehmen und sie konnte sowohl direkt als auch indirekt (gegebenenfalls sogar über mehrere Zwischenglieder) beschaffen sein.

Es ist hier selbstverständlich nicht der Ort, dieses überaus weite Feld in seiner ganzen Ausdehnung zu kartieren; dazu würden wenige Seiten mit Sicherheit nicht genügen. Vielmehr soll es im Folgenden ausschließlich um die politischen Kontakte und auch bei diesen nur um die auf der höchsten Ebene – das heißt:

---

1 Das entspricht knapp 3.000 Kilometern auf der Straße.

derjenigen der Könige – gehen. Wie sich bald herausstellen wird, gibt es auch hier mehrere Alternativen.

1. Typologie der Kontakte zwischen dem Reich und Portugal

Versucht man, die unterschiedlichen Formen der Kommunikation zwischen Herrschern zu systematisieren, so lassen sich wenigstens drei Formen unterscheiden: persönliche Kontakte, Kontakte durch Gesandte sowie Kontakte mit Hilfe von Briefen. Alle drei sind auch zwischen dem Reich und Portugal bezeugt.

a. Persönliche Kontakte

Die unmittelbarste Form des Kontaktes zwischen Herrschern ist die persönliche Begegnung. Solche ‚Spitzentreffen' hat es das ganze Mittelalter hindurch in nicht unerheblicher Zahl gegeben; für jedes Jahrhundert und für jedes Königreich sind Dutzende bezeugt.[2] Begegnungen zwischen deutschen und portugiesischen Königen sind allerdings nicht darunter.

Fündig wird man dagegen, wenn man auch die Ebene unterhalb der Könige berücksichtigt. So ist der Infant Peter, der drittgeborene Sohn König Johanns I., nicht nur mit König Sigismund zusammengetroffen, sondern längere Zeit in seiner Umgebung bezeugt. Er unterstützte ihn tatkräftig in seinen Kriegen vor allem gegen die Türken (1426–1428).[3] Schon 1418 hatte Sigismund dem Infanten (in Abwesenheit) die Mark Treviso verliehen, Friedrich III. bestätigte 1443 in aller Form diesen Akt.[4]

Bei seiner Ankunft 1426 in Wien wird Peter als *ein khünigs sun von pordigall* beschrieben, der zwar kein Deutsch verstand, aber sehr gut Lateinisch sprach; sein Gefolge beziffert der örtliche Chronist auf dreihundert gut ausgerüstete Krieger.[5] Nach seiner Rückkehr und dem frühen Tod seines älteren Bruders übernahm Peter die vormundschaftliche Regierung für seinen Neffen, den damals sechsjährigen Alfons V. Sie endete am 20. Mai 1449 mit der Schlacht von Alfarrobeira, bei der der Infant nicht nur seine einflussreiche Stellung, sondern auch das Leben verlor.[6]

---

2 Für das Spätmittelalter vgl. Schwedler, 2008, der allerdings nicht auf Portugal eingeht.
3 Gomes dos Santos, 1959; Rogers, 1961: 37–45, 322–335.
4 Vgl. Altmann, 1896–1900: 202 (Nr. 2838), 213 (Nr. 3017); Chmel, 1838: 145 (Nr. 1408), 153 (Nr. 1520), 153 f. (Nr. 1524); Eibl, 1998: 86 f. (Nr. 42); Willich, 1999: 154 f. (Nr. 168). Dazu Rogers, 1961: 16–19; Heinig, 2007: 36 f.
5 Zeibig, 1851: 250.
6 Vgl. Baquero Moreno, 1979–1980; Gomes, 2009: 56–102.

b. Kontakte durch Gesandtschaften

Viel häufiger als eine persönliche Begegnung war der Kontakt durch Gesandte. Schon der oben genannte Infant Peter schickte 1443 mit dem Ritter João Teles sowie dem *Baccalaureus decretorum* Brás Afonso zwei Vertraute, die bei Friedrich III. eine Bestätigung seiner Belehnung mit der Mark Treviso erlangen sollten – und sie hatten damit Erfolg.[7]

Ein ganzes Bündel von Gesandtschaften war einige Jahre später erforderlich, um die Ehe zwischen Friedrich III. und Eleonore, der Nichte Peters und Schwester Alfons' V., anzubahnen. Sie umfassten den Zeitraum von 1448 bis 1452, waren mehrfach ineinander verschränkt und lassen sich vereinfacht folgendermaßen darstellen:[8]

Sommer 1448–Frühjahr 1449: Gesandtschaft Friedrichs III. unter der Leitung von Georg von Volkersdorf und Ulrich Riederer zur Brautschau nach Lissabon.

Herbst 1450–Februar 1451: Gesandtschaft Friedrichs III. unter der Leitung von Enea Silvio Piccolomini, Georg von Volkersdorf und Michael Renz von Pfullendorf nach Neapel zum Abschluss des Heiratsvertrages (Dezember 1450).

Sommer 1450–Frühjahr 1451: Gesandtschaft Alfons' V. unter der Leitung des João Fernandes da Silveira nach Neapel zum Abschluss des Heiratsvertrages (s. oben).

Frühjahr 1451–April 1452: Gesandtschaft des Nikolaus Lankmann und Jakob Motz von Wiener Neustadt nach Lissabon zur Prokurationsheirat mit Eleonore (August 1451); von dort aus Geleit der Braut über Siena (erste Begegnung mit Friedrich) und Rom (Trauung und Kaiserkrönung) nach Neapel (öffentliches Beilager) und schließlich nach Wiener Neustadt (der bevorzugten Residenz Friedrichs III.).

November 1451–Herbst 1452: Geleit der Eleonore durch zahlreiche Portugiesen unter Führung des Markgrafen Alonso von Valença von Lissabon nach Italien, das heißt: Siena, Rom, Neapel, Venedig; nur eine kleine Gruppe folgt ihr von dort aus weiter nach Österreich (darunter auch João Fernandes da Silveira).

---

7 Das am 12. April 1443 in Lissabon ausgestellte Beglaubigungsschreiben für die beiden Gesandten (*Johannes Telez, miles,* sowie *Blasius Alfonsi, in decretis baccalareus*), das Friedrich III. am 27. August 1443 in Wiener Neustadt vidimiert hat, ist bei Chmel, 1837: 128 f. (Nr. 33) abgedruckt. Vgl. auch Faro und Faro, 1961: 268 (Zahlung von 680 Dobra an die beiden Gesandten) sowie die in Anm. 4 angeführten Regesten.

8 Vgl. dazu Pferschy-Maleczek, 1997; Hack, 1999; Winkelbauer, 2007; Hack, 2011. – Bei den zeitlichen Angaben – ohnehin nur grobe Annäherungen – ist im Folgenden immer auch der Rückweg mit einkalkuliert.

c. Kontakte durch Briefverkehr

Eine dritte Möglichkeit – viel weniger aufwändig als die beiden bereits genannten – war der Kontakt durch Briefe. Sie spielte vor allem nach der römischen Heirat und Kaiserkrönung eine große Rolle. An dieser Korrespondenz waren Enea Silvio Piccolomini und João Fernandes da Silveira beteiligt, ja sogar: Friedrich III. und Alfons V. selbst. Diese bisher wenig beachteten Schreiben[9] sollen auf den folgenden Seiten im Zentrum des Interesses stehen.

2. Der Wiener Kodex 3389

Überliefert sind alle hier interessierenden Briefe in einem einzigen Kodex: der Wiener Handschrift 3389. Dabei handelt es sich um ein Papier-Manuskript aus dem 15. Jahrhundert im Quart-Format, das aus drei ursprünglich selbstständigen Teilen besteht: einem Text des Albrecht von Bonstetten über die Schweiz (fol. 1r–38r), dem berühmten Schreiben des Enea Silvio Piccolomini an Sultan Mehmet II. (fol. 40r–79v) sowie einer umfangreichen Briefsammlung, hauptsächlich aus der Feder desselben Verfassers (fol. 80r–193v).[10]

Dieser letzte Teil, um den es im Folgenden ausschließlich gehen soll, ist nur fragmentarisch überliefert: Einzelne Blätter, aber auch ganze Lagen (mindestens drei Quaternionen) müssen verloren gegangen sein. Der Sache nach handelt es sich um eine Abschrift von 186 Briefen, einem Gebet und einem nachträglich gestrichenen Text, die meisten im Autograph des späteren Papstes, mit zahlreichen Korrekturen und in einer Schrift, die in die Mitte der 1450er-Jahre weist. Beinahe alle Briefe wurden von Piccolomini verfasst, die meisten in seinem eigenen Namen, einige jedoch auch im Auftrag Friedrichs III.; hinzu kommen (wie in Briefsammlungen oft üblich) Schreiben, die an ihn bzw. seinen Kaiser gerichtet sind, darunter solche, die von Papst Nikolaus V., dem Dogen Francesco Foscari oder eben Alfons V. von Portugal stammen.

Die Briefe des Wiener Kodex wurden zwischen dem 6. April 1453 und dem 10. Februar 1454 abgefasst und beweisen die überaus rege Korrespondenz des gebürtigen Sienesen: Er verfasste – so muss man schließen – im Durchschnitt mindestens an jedem zweiten Tag einen Brief. Vor allem seit einer wegweisenden Arbeit des Humanismusforschers Georg Voigt (1827–1891) sind die Schreiben

---

9   Die ausführlichste Auseinandersetzung findet sich in einer ungedruckten Wiener Dissertation, vgl. Zierl, 1966: 163–167.
10 Hierzu und zum Folgenden vgl. Voigt, 1865: 332; Weiss, 1897: 87–101; Wolkan, 1918: V–X; Wagendorfer, 2008b: 171–174.

einer interessierten Öffentlichkeit bekannt; einige von ihnen hat dieser selbst, die meisten Anton Weiss (1852–1912), viele erneut Rudolf Wolkan (1860–1927) ediert.[11]

Unter diesen Briefen des Kodex 3389 finden sich vier bzw. fünf Exemplare, die hier näher untersucht werden sollen: ein Schreiben Alfons V. an Friedrich III., die Antwort Friedrichs III. an Alfons V. in zwei Fassungen sowie zwei Briefe des Enea Silvio Piccolomini an João Fernandes da Silveira.[12] Sie hängen nicht nur sehr eng miteinander zusammen, sondern lassen auch auf weitere Schreiben schließen.

3. Zwei königliche Gesandte begegnen sich

Bevor es zu dem intensiven Briefwechsel im Jahre 1453 kam, bestand bereits ein verhältnismäßig enger Kontakt zwischen Enea Silvio Piccolomini und João Fernandes da Silveira. Ja, man kann sogar soweit gehen zu behaupten, dass diese beiden zu den wichtigsten Trägern der deutsch-portugiesischen Beziehungen nach der Mitte des 15. Jahrhunderts gehörten. Wie ist es dazu gekommen? Und noch allgemeiner gefragt: um wen handelt es sich hier überhaupt?

a. Die beiden Protagonisten

Enea Silvio Piccolomini gehört zu denjenigen Personen, die man nicht vorzustellen braucht – und die man auf wenigen Seiten auch gar nicht vorstellen kann. Am 18. Oktober 1405 im Val d'Orcia auf die Welt gekommen – seinen Geburtsort Corsignano hat er später selbst in Pienza umbenannt –, studierte Piccolomini in Siena und Florenz, trat in die Dienste des dortigen Bischofs, Domenico Capranica, und besuchte in dessen Gefolge das Allgemeine Konzil in Basel. Auf diesem nahm er das Amt eines Sekretärs an, wirkte für Papst Felix V., um dann aber, zum *Poeta laureatus* gekrönt, in die Kanzlei Friedrichs III. überzutreten. Hier war er mit unterschiedlichen Aufgaben betraut, unter anderem als Ratgeber, Redner, Diplomat und Stilist. Zugleich betrieb er seine kirchliche Karriere, die ihn zu hohen und höchsten Ämtern führte: 1447 wurde er Bischof von Triest, 1450 Bischof von Siena, 1456 Kardinal und 1458 schließlich Papst. Sein literarisches Œuvre ist riesig und noch längst nicht in allen Facetten beleuchtet; das ist jedoch wenig erstaunlich, wenn man bedenkt, dass kaum eine Textgattung fehlt. Piccolomini

---

11 Trotz der zahlreichen überlieferten Briefe ist Piccolomini als Epistolograph noch kaum erforscht; vgl. jedoch Voigt, 1856–1863, Bd. 2: 277–283; Drücke, 2004; allgemeiner Worstbrock, 1989: 640–646.
12 Zu den beiden Königen vgl. Krieger, 2004: 169–237, und Koller, 2005; sowie Gomes, 2009. Zu den Gesandten s. die beiden folgenden Anmerkungen.

starb am 15. August 1464 in Ancona, wo in diesen Tagen eine Kreuzzugsflotte auslaufen sollte.[13]

Nicht ganz so bekannt ist João Fernandes da Silveira, wobei inzwischen auch er schon Gegenstand einer monographischen Untersuchung geworden ist. Sohn des Juristen Dr. Fernando Afonso da Silveira, der 1430 von Johann I. für seine Verdienste zum Ritter erhoben wurde und seit 1432 als oberster Kanzler des Königreichs Portugal bezeugt ist, und der Catarina Teixeira, Kammerzofe der Infantin Isabella (der späteren Herzogin von Burgund), wurde João Fernandes wohl im zweiten Jahrzehnt des 15. Jahrhunderts geboren, studierte Recht in Lissabon und erwarb schon vor 1443 den juristischen Doktorgrad. Er war zuerst mit Violante Pereira (wahrscheinlich seit 1449) und später mit Maria de Sousa Lobo verheiratet; aus diesen Ehen gingen insgesamt fünf Kinder hervor: vier Söhne und eine Tochter.[14]

Der gelehrte Jurist nahm im Laufe seines Lebens eine ganze Reihe von hohen Ämtern wahr, darunter schon Mitte der 1440er-Jahre das des Kanzlers der Casa do Cível. Vor allem aber trat er als Diplomat hervor, das erste Mal 1448, als er im Namen des soeben zur selbstständigen Regierung gelangten Alfons V. Papst Nikolaus V. die Obödienz leistete. 1456 reiste er zu Calixt III. und anderen italienischen Mächten, um über einen Kreuzzug gegen die Türken zu verhandeln; João Fernandes scheint bis zum Jahre 1460 – also vier Jahre lang – in Italien geblieben zu sein. Nach Portugal zurückgekehrt, wurde er 1463 zum Regedor (Vorsteher) der Casa da Suplicação ernannt.[15]

Höhepunkt seiner Karriere war allerdings die Erhebung zum Baron von Alvito, eine Würde, die 1475 eigens für ihn geschaffen worden ist – bis dahin war er einfacher Herr (Senhor) von Alvito – und die er an seine Nachkommen (bis ins 20. Jahrhundert) vererbte. In der kleinen Stadt ganz im Norden des heutigen Distrikts Beja ließ João Fernandes 1481 den Bau einer Burg beginnen, die bei

---

13 Eine moderne, die zahlreichen Betätigungsfelder Piccolominis abdeckende Biographie liegt leider nicht vor. Vgl. einstweilen (mit jeweils ausführlicher Bibliographie) Voigt, 1856–1863; Worstbrock, 1989; Reinhardt, 2013.

14 Über João Fernandes da Silveira, ohne allerdings die Quellen aus dem Reich zu berücksichtigen, vgl. Faro und Faro, 1961: 264 f.; Rau, 1985: 71 f., 76–79; Gonçalves de Freitas, 2001: 437–442; Pereira Caetano, 2011 (besonders 75–99 und 134–142); ferner Gomes, 2009, mehrfach (s. Register s. v.).

15 João Fernandes hat im Auftrag seines Königs eine ganze Reihe von Gesandtschaften durchgeführt: 1448 nach Rom, 1450/51 nach Neapel, 1451/52 nach Italien und Österreich, 1455 nach Kastilien, 1456–1460 nach Italien (Rom, Florenz, Venedig), 1463 nach Kastilien, 1474 erneut nach Kastilien, 1482 wieder nach Kastilien usw.; vgl. Rau, 1985: 71 f.; Pereira Caetano, 2011: 134–142, und die in der vorausgegangenen Anm. genannte Literatur.

seinem Tod 1484 (oder 1487) noch nicht fertiggestellt war. Sie ist dort bis zum heutigen Tage zu besichtigen.

b. Die erste Begegnung in Neapel

Die erste Begegnung zwischen Enea Silvio Piccolomini und João Fernandes da Silveira fand offenbar Ende 1450 in Neapel statt. Zweck dieses Treffens war es, wie schon weiter oben kurz erwähnt, den Heiratsvertrag zwischen Friedrich III. und Eleonore auszuhandeln. Dazu erhielt der portugiesische Gesandte bereits am 27. Juni 1450 von König Alfons V. eine förmliche Vollmacht, die er offenbar den Verhandlungspartnern überreichte; sie wird bis zum heutigen Tag im österreichischen Haus-, Hof- und Staatsarchiv zu Wien aufbewahrt.[16]

Von Seiten des späteren Kaisers wurde unter anderen Enea Silvio Piccolomini nach Neapel geschickt, der in seiner *Historia Austrialis* über diese Mission ausführlich berichtet. Selbstverständlich erwähnt er auch João Fernandes und charakterisiert ihn dabei als *iuris consultum egregium* und *virum praestabilem*.[17] Der Vertrag wurde in der Gegenwart König Alfons' V. von Neapel, des päpstlichen Legaten, Kardinal Jean le Jeune, sowie von Gesandten zahlreicher italienischer Städte am 10. Dezember 1450 feierlich abgeschlossen.[18]

Enea Silvio kehrte wahrscheinlich im Februar 1451 nach Österreich zurück, João Fernandes muss spätestens im April 1451 ebenfalls in seiner Heimat angekommen sein, denn damals traten die Cortes in Santarém zusammen, um den in Neapel abgeschlossenen Heiratsvertrag zu ratifizieren. Dem Gesandten wurden seine Kosten mit 790 Dobra vergütet.[19]

---

16 Ediert bei Chmel, 1837: 321 (Nr. 153).
17 Knödler und Wagendorfer, 2009: 90, 428. Über die komplizierte Entstehungs- und Editionsgeschichte der *Historia Austrialis* vgl. die Einleitung zur Editio critica sowie Knödler, 2007; Wagendorfer, 2008a; Märtl, 2013.
18 Vgl. den (natürlich stilisierten) Augenzeugenbericht bei Knödler und Wagendorfer, 2009: 90 f., 428 f.; Bellus und Boronkai, 1993–1994: 55 f. Der Vertragstext selbst ist abgedruckt bei Chmel, 1837: 332–338 (Nr. 161); Nascimento, 1992: 62–83 (Nr. 6); João Fernandes da Silveira wird darin als *legum doctor orator etiam et procurator ac mandatarius* bezeichnet.
19 Vgl. Faro und Faro, 1961: 264; Pereira Caetano, 2011: 135 (zum Vergleich: für seinen vierjährigen Aufenthalt in Italien wurden ihm 1462 13.240 Dukaten bezahlt, vgl. Rau, 1985: 69; Pereira Caetano, 2011: 137; weitere Zahlen bei Faro und Faro, 1961). João Fernandes hatte sich aus der Sicht seines Königs so sehr bewährt, dass er später noch mehrfach für Gesandtschaften verwendet wurde, darunter auch 1463 zu den Verhandlungen über eine Ehe Alfons' V. mit Isabella von Kastilien, die aber aus verschiedenen Gründen nicht zustande kam.

c. Erneutes Zusammentreffen in Italien

Allzu lange konnte sich João Fernandes allerdings nicht erholen, denn als im November desselben Jahres Eleonore endlich nach Italien aufbrach (ursprünglich war ein deutlich früheres Datum vorgesehen), war auch der gelehrte Jurist unter ihren Begleitern. Rui de Pina, der Verfasser einer offiziösen Chronik Alfons' V., hebt ihn mit einem Dutzend anderer hochrangiger Persönlichkeiten aus dem sehr zahlreichen Gefolge der Infantin – immerhin mehrere tausend Personen – hervor.[20] Wenig später behauptet er, João Fernandes habe als königlicher Gesandter die stattliche Marmorsäule in Auftrag gegeben, die (vor der Porta Camollia) in Siena an die erste Begegnung Eleonores mit ihrem Bräutigam erinnern sollte (und bis heute erinnert);[21] in Wirklichkeit ging die Initiative vom Rat der italienischen Stadt aus.[22] Von Siena zog der Gesandte weiter nach Rom und nahm dort sicherlich an der Trauung des Paares sowie drei Tage danach an der Kaiserkrönung teil. Wie Rui de Pina ausdrücklich berichtet, wurde der portugiesische Gesandte wenig später vom Kaiser am Altar der Veronica in Sankt Peter zum Ritter geschlagen.[23]

Zur gleichen Zeit, als João Fernandes die Infantin nach Italien begleitete, war Enea Silvio im Gefolge Friedrichs III. ebenfalls südlich der Alpen. Er gehörte sogar zu einer Delegation, die die Braut, wie im Vertrag von 1450 vereinbart, im Hafen von Telamone in Empfang nehmen sollte; allerdings traf ihr Schiff aufgrund widriger Winde in Livorno ein. Dorthin geeilt, begrüßt er Eleonore, geleitete sie in seine Heimatstadt und von dort aus weiter nach Rom. Obwohl er in seinen historiographischen Werken nicht davon berichtet, muss er bei dieser Gelegenheit auch mehrfach mit João Fernandes zusammengetroffen sein.

---

20 Lopes de Almeida, 1977: 762. Rui de Pina hatte auch schon über die Mission nach Neapel 1450 berichtet und in diesem Zusammenhang den Gesandten als *o Doutor Joam Fernandez da Silveira, homem Fydalgo prudente e gram letrado, que despois foy o prymeiro Baram d'Alvito* bezeichnet (ebd.: 759).
21 Vgl. Lopes de Almeida, 1977: 763.
22 Quirin, 1958: 67 mit Anm. 105; dort wird auf den Ratsbeschluss vom 11. März 1452 hingewiesen. Auch Enea Silvio schreibt in seiner *Historia Austrialis* (ab der 2. Redaktion) sowie seinen *Commentarii* die Errichtung der Säule den Sienesen zu, vgl. Knödler und Wagendorfer, 2009: 583; Bellus und Boronkai, 1993–1994: 61. Außerdem steht sie im Zentrum eines Gemäldes Pinturicchios in der Libreria Piccolomini, das die erste Begegnung Friedrichs III. mit Eleonore zeigt. – Über den Aufenthalt des Paares in Siena und dessen Vorbereitung vgl. ausführlich Quirin, 1958
23 Vgl. Lopes de Almeida, 1977: 764. – Zum römischen Ritterschlag 1452 vgl. Hack, 2004 sowie Hack, 2007.

Nach der Kaiserkrönung und einem Besuch in Neapel zog das Kaiserpaar über die Alpen, genauer: nach Wiener Neustadt. Dabei folgte ihnen nicht nur der Sieneser Bischof, sondern auch João Fernandes. Aufgabe des portugiesischen Gesandten war es nämlich, die Erfüllung des Heiratsvertrages – und zwar besonders seiner finanziellen Bestimmungen – sicherzustellen. Soweit sich das erkennen lässt, kam es dabei zu keinen größeren Problemen. So konnte João Fernandes schon am 16. August 1452 gemeinsam mit seinem Kollegen Lopo d'Almeida erklären, „dass Kaiser Friedrich das Heirathsgut, die Widerlage und Morgengabe seiner Gemahlinn Eleonora hinlänglich versichert habe".[24] Zu den übertragenen Besitzungen gehörte unter anderem die Herrschaft Portenau (ital. Pordenone) am Fuße der Vinzentiner Alpen. Um den Gehorsamseid des dortigen Burgkastellans gegenüber Eleonore einzuholen, schickten die beiden portugiesischen Gesandten zwei ihrer Landsleute, Petrus Fynz und Pelagius Varella über die Alpen. Das dafür erforderliche Bevollmächtigungsschreiben wurde am 13. September 1453 im steirischen Leoben ausgestellt.[25]

4. Briefverkehr zwischen Portugal und dem Reich im Jahre 1453

Die Reihe der vier bzw. fünf Briefe, die im Wiener Kodex 3389 überliefert sind, beginnt mit einem Schreiben Alfons' V. an seinen Schwager, Kaiser Friedrich III. Es ist allerdings nicht das erste Mal, dass die beiden Monarchen brieflich miteinander in Verbindung getreten sind. So erwähnt zum Beispiel Enea Silvio in seiner *Historia Australis* einen Brief des Habsburgers (offenbar aus dem Jahr 1449), in dem er dem portugiesischen König Eheverhandlungen in Neapel vorgeschlagen habe. Alfons antwortete ihm ebenfalls mit einem Schreiben und ging darin auf den Vorschlag Friedrichs III. ein.[26] Diese Briefe sind allerdings nicht mehr erhalten.

Das neuerliche Schreiben Alfons' V. an den Habsburger datiert vom 25. März 1453 und wurde in Lissabon abgeschickt.[27] Der Kaiser wird darin in superlativi-

---

24 So zusammenfassend Chmel, 1838: 298 (Nr. 2918), mit Verweis auf das ungedruckte Dokument im Geheimen Haus-Archiv (daraus ein kurzes Zitat im Regest).
25 Vgl. Valentinelli, 1865: 270 f. (Nr. 228).
26 Vgl. Knödler und Wagendorfer, 2009: 89 f., 427. – Der erwähnte Brief Friedrichs III. wird bereits als ein Antwortschreiben (*rescriptum est*) charakterisiert, so dass man auf einen vorausgegangenen Brief Alfons' V. schließen kann.
27 Der Brief wurde bisher nur in Fußnoten abgedruckt, immerhin aber vollständig, vgl. Weiss, 1897: 190; Wolkan, 1918: 585. Wer der Diktator dieses Schreibens ist, wurde bislang noch nicht geklärt.

schen Wendungen als überaus siegreicher Fürst sowie höchst machtvoller Herr bezeichnet und als ruhmreichster Princeps und allergeliebtester Bruder angesprochen. Alfons sendet ihm nicht nur viele Grüße, sondern wünscht ihm auch *de infidelibus gloriose triumphare* – ein vielleicht typisch portugiesischer Wunsch.[28] Gegenstand des Briefes ist die Ehe, näherhin die Ehe zwischen Eleonore und Friedrich III. Alfons V. hat seinen eigenen Worten zufolge gehört, dass der Kaiser seine Gattin außerordentlich schätze und leidenschaftlich liebe, dafür sei er dem Allmächtigen auch sehr dankbar. Diese höflichen Worte zu Beginn des Schreibens können jedoch nicht über eine deutlich negativere Einschätzung des neuen Eheglücks hinwegtäuschen. Denn Alfons gibt unumwunden seiner Hoffnung Ausdruck, dass sich Friedrichs Hinneigung zu seiner Gemahlin noch verstärken werde. Zwar müsse er den Kaiser sicher nicht überreden, doch zwängen ihn die verwandtschaftlichen Bande dazu, Friedrich zu ermahnen und zu bitten (*vos hortari atque rogare*), sein lobenswertes Vorhaben fortzuführen, und wenn er seiner Liebe irgendetwas hinzufügen könne, solle er seine Gattin immer mehr wertschätzen und lieben. Denn nichts sei zumal für einen katholischen Fürsten ehrenhafter und lobenswerter als die vor Gott und den Menschen gleichermaßen verdienstvolle eheliche Liebe. Zum Beweis dafür verweist er auf antike Beispiele (Tiberius und Julia, Mithridates und Hypsikrate) sowie auf die Bibel (Epheser 5, 31). Mit einem erneuten, wortreichen Appell schließt Alfons seinen Brief.

Interessant ist die Tatsache, dass der König dabei die Quelle seiner Erkenntnisse klar benennt: den vortrefflichen Rechtsgelehrten João Fernandes da Silveira, seinen allertreuesten Rat (*relatione eximii legum doctoris, nobilis Johannis Ferdinandi de Silveira, consiliarii nostri fidelissimi*). Dieser war irgendwann nach dem 13. September 1452 aus Österreich in seine Heimat zurückgekehrt und hatte Alfons offenbar persönlich über seine Eindrücke am Kaiserhof berichtet. Worin er genau die Defizite sah, wird nicht explizit benannt. Fest steht allerdings, dass Eleonore auch ein Jahr nach ihrer Vermählung noch kein Kind zur Welt gebracht hatte, ja, dass noch nicht einmal eine Schwangerschaft abzusehen war.[29]

João Fernandes ist der Adressat des nächsten Briefes, der aus der Feder des Enea Silvio stammt; wie sich aus dem kodikologischen Kontext ergibt, muss er in

---

28 Zur Rolle des Heidenkampfes für das Selbstverständnis der portugiesischen Herrscher, zu dieser Zeit im Kult Ferdinands des Standhaften kulminierend, vgl. Rosa, 2007.
29 Eleonore hat ihr erstes Kind am 1. Juni 1454 geboren, vgl. Wolkan, 1918: 489–492 (Nr. 290). Es starb bereits kurz nach der Geburt. Weitere Kinder – drei Knaben und zwei Mädchen – folgten 1455, 1459, 1460, 1465 und 1466. – Zu den allgemeinen Erwartungen vgl. Kohler, 1994.

den ersten Tagen des Juli 1453 abgeschickt worden sein.[30] Der Bischof von Siena unterrichtet darin den vorzüglichen Gelehrten und hochgehrten Freund nach tausend Grüßen über die politischen Verhältnisse in Österreich. Dort herrsche zwar gerade Frieden, doch sei dieser keineswegs sicher. In Ungarn plündere ein Räuber namens Peter Aksamit mit fünftausend Gefolgsleuten das Land, König Ladislaus (Postumus) ziehe ihm mit einem Heer entgegen; zwar habe sich zunächst der erst nach diesem Feldzug in Prag krönen lassen wollen, nun sei aber der Plan geändert worden und er beabsichtige, innerhalb der nächsten vier Wochen die Krone zu empfangen.[31] Der Kaiserin – so schließt er – gehe es gut und ihrem Gemahl ebenso.

Enea Silvio schreibt dem Portugiesen, weil er dessen Vertrauten (*familiarem tuum*) nicht ohne Brief zurückreisen lassen möchte. Was dieser in Graz zu erledigen hatte, erfährt man leider nicht. Trotz der zahlreichen politischen Neuigkeiten handelt es sich bei dem kurzen Schreiben um einen Freundschaftsbrief, versichert sein Verfasser doch gleich zu Beginn: *Que scribam, non sunt alia nisi quia cupio, tibi bene esse teque memoriam mei habere*. Am Ende bittet Enea, der Empfänger möge ihn dem portugiesischen König empfehlen und seine Grüße an Lopo d'Almeida ausrichten, das heißt: an den zweiten Gesandten des Jahres 1452. Und er schließt mit freundschaftlichem Pathos: *Vale et me ama perpetuo, nam tu a me maxime amaris*.

Enea Silvio rechnete offenbar mit einer zügigen Antwort des Adressaten – allerdings vergebens; als er ihm am 23. August 1453 erneut aus Graz schrieb,[32] konfrontierte er ihn gleich zu Beginn seines Briefes mit diesem Versäumnis; wahrscheinlich habe er ihn schon in seiner Heimat – also Italien – vermutet; denn

---

30 Vgl. Voigt, 1865: 403 (Nr. 242); Weiss, 1897: 157 f. (Nr. 49); Wolkan, 1918: 184 f. (Nr. 106). – Die Zeit der Abfassung ergibt sich aus dem vorausgehenden Schreiben an Goro Lolli, das auf den 1. Juli 1453 datiert ist und das vor allem am Ende mit dem Mittelstück des Briefes an João Fernandes wörtlich übereinstimmt, vgl. Wolkan, 1918: 181–184 (Nr. 105).
31 Der hussitische Heerführer Peter Aksamit von Lidéřovic und Kosov, der in den Briefen Piccolominis und in seiner Europa mehrfach erwähnt wird, fiel am 21. Mai 1458 in der Schlacht bei Sárospatak, vgl. zu ihm Palacký, 1857: 331 und 517 f. – Ladislaus wurde am 28. Oktober 1453 in Prag zum König von Böhmen gekrönt.
32 Vgl. Voigt, 1865: 404 (Nr. 265); Wolkan, 1918: 233 f. (Nr. 129). Dieser Brief war schon in der Nürnberger (1481) sowie in den Basler Ausgaben (1551 und 1571) im Druck erschienen. – Zu Lopo d'Almeida vgl. Lapa, 1935; Askins u. a., 2003/2004. Der Genannte ist nicht mit Lopo Velasco zu verwechseln, den Piccolomini auch schon auf dem Basler Konzil kennengelernt hatte, vgl. dazu unten, Anm. 44.

dass er ihn einfach vergessen habe, könne er sich nicht vorstellen. Allein schon diese Bemerkungen, so topisch sie auch sein mögen, beweisen die Intensität der Kommunikation.

Als Grund für seinen langen Verbleib in Deutschland, das er geradezu als Exil empfindet, nennt Piccolomini den Krieg in der Toskana, den der König von Aragon und Neapel – Alfons der Großmütige – dort gegen Florenz führe.[33] Doch hoffe er, bald in seine Diözese zurückkehren zu können, und freue sich besonders auf die Freunde und Verwandten, die er schon seit 22 Jahren vermisse.[34] Er fügt hinzu: *Hec volui tibi velut amico scripsisse, cui omnia consilia communicanda sunt.*

Was die Neuigkeiten am Hofe Friedrichs III. betrifft, verweist Enea Silvio auf den portugiesischen Gesandten.[35] Er selbst berichtet noch über das Wohlbefinden der Kaiserin. Ihre Schönheit werde in ganz Deutschland sehr gepriesen, aber mehr noch ihre Klugheit und Bescheidenheit. Die Heirat einer ihrer Schwestern in dieselbe Region würde er (noch immer) sehr begrüßen, *ut soror sorori consolationi esset* – ein Indiz für Heimweh auch bei Eleonore.[36] Es folgt die übliche Empfehlung an den portugiesischen König sowie Grüße an Lopo d'Almeida, den Enea, wie sich erst hier herausstellt, bereits aus den Zeiten des Basler Konzils kennt.

Erst nach diesen beiden Schreiben an João Fernandes folgt mit Datum vom 28. September 1453 die Antwort Friedrichs III. auf den ungewöhnlich deutlichen Brief Alfons' V.; verfasst hat ihn, wie nicht allein aus der autographen Überlieferung hervorgeht, sein gelehrter Rat und ehemaliger Sekretär, Enea Silvio Piccolomini.[37]

Gleich im Anschluss an das Protokoll, in dem Alfons V. nicht nur als (königlicher) Bruder, sondern auch als überaus geliebter Verwandter bezeichnet wird, nimmt der Verfasser Bezug auf den Bericht des João Fernandes an seinen König. Dabei nennt er diesen, wie immer, *eximius legum doctor*, ergänzt aber *vester et noster consiliarius admodum dilectus*. Es ist der einzige Beleg, dass João Fernandes

---

33 Zum Krieg zwischen Neapel und Florenz (sowie anderen italienischen Mächten), der kurz nach der Abreise Friedrichs III. aus Italien begann und erst mit dem Frieden von Lodi endete, vgl. Ryder, 1990: 285–290.
34 Das heißt seit seiner Abreise nach Basel (Anfang 1432), vgl. Nardi, 1987: 202.
35 Um wen es sich dabei handelt, ist nicht bekannt.
36 Zu den Plänen einer Doppelheirat Friedrichs III. und Ladislaus' mit zwei Töchtern Eduards I. in den 1440er-Jahren vgl. Zierl, 1966: 14–18; Pferschy-Maleczek, 1997: 425; Hack, 2011: 310. Über König Eduard vgl. zuletzt Duarte, 2007.
37 Vgl. Weiss, 1897: 189–192 (Nr. 69); Wolkan, 1918: 584–588 (Nr. IX, Version A).

nicht nur Rat Alfons V., sondern auch Friedrichs III. war – eine seltene Ehre für auswärtige Gesandte.³⁸

Der im Namen des Kaisers verfasste Brief des Sieneser Bischofs ist sehr fein auf das vorausgegangene Schreiben abgestimmt; jeder Teil findet eine genaue Entsprechung. Zunächst preist Friedrich das Glück seiner Ehe in geradezu überschwänglichen Tönen und hebt dabei auch die positiven Eigenschaften seiner Gemahlin wie *modestia, gravitas, lenitas, prudentia* und *morum probitas* gebührend hervor.

Sodann weist er ausführlich auf Vorbilder einer guten Ehe hin, seien sie nun aus der Bibel (Abraham, Isaak, Jakob, Moses) oder aus der antiken Literatur (Alkestis, die Gemahlin des Admetos von Thessalien; Artemisia, die Königin von Karien; Porcia, die Tochter Catos des Jüngeren);³⁹ selbstverständlich wird auch von ihm die berühmte Stelle aus dem Epheser-Brief (5, 31) bemüht.

Vor allem verbittet sich aber Friedrich eine Einmischung in seine Ehe: Ermahnungen benötige er nicht, weder vom Adressaten noch von irgendeiner anderen Person (*neque ad hoc vestre vel alterius cujuspiam exhortatione indigemus*): bei aller Höflichkeit eine sehr deutliche Antwort.

Derselbe Brief an Alfons V. ist im Wiener Kodex 3389 noch ein zweites Mal überliefert, dieses Mal aber nicht im Autograph des Verfassers, sondern von der Hand eines anonymen Schreibers zu Papier gebracht.⁴⁰ In diesem zweiten Brief sind einige Sätze umformuliert; darunter auch die Rüge an Alfons V. Vor allem werden aber die historischen Beispiele stark vermehrt und auch Pompeius, Caesar und Augustus, ja den Karolingern, Ottonen, Saliern und Staufern insgesamt vorbildliche Ehen zugesprochen.

Welches der beiden Schreiben nach Portugal geschickt wurde, ist unbekannt, ebenso der Grund der nachträglichen Überarbeitung. Sollte Alfons V. nicht vor den Kopf gestoßen werden mit der zumindest sehr direkten Formulierung in der ersten Fassung? Oder galt die abgemilderte und um Beispiele vermehrte zweite Fassung einfach als besseres epistolographisches Vorbild?

---

38 Heinig, 1997: 528 f., nennt neben João Fernandes, den er allerdings irrtümlich als Aragonesen bezeichnet, nur noch einen zweiten Rat, der nicht aus dem erweiterten Reichsgebiet stammt.
39 Beispiele für vorbildliche Gattenliebe finden sich zum Beispiel auch in Boccaccios Buch *De mulieribus claris*, vgl. Zaccaria 1970; dazu Müller 1992.
40 Vgl. Weiss, 1897: 197–199 (Nr. 78); Wolkan, 1918: 584–588 (Nr. IX, Version B). Der undatierte Brief ist in der Handschrift zum 3. Oktober eingeordnet. Vgl. dazu schon Zierl, 1966: 166.

5. Briefe Eleonores?

In den Briefen des Jahres 1453 ist viel von Eleonore die Rede, ja sie ist sogar ihr hauptsächlicher Gegenstand. Hat sie aber auch selber Briefe verfasst und an ihre Verwandten im fernen Westen geschickt und umgekehrt solche erhalten? Wie muss man sich ihre kommunikative Situation nach der Etablierung in Österreich vorstellen?

Eine erste Teilantwort lautet: Obwohl nicht wenige Briefe der Eleonore überliefert sind,[41] ist darunter keiner an ihre Geschwister oder einen anderen portugiesischen Adressaten. Das braucht natürlich nicht allzu viel zu bedeuten, wissen wir doch, dass die meisten Briefe des (Spät-)Mittelalters nicht überliefert sind.

Daher verdient hier ein Text Beachtung, der über die ausgedehnte Reise des Leo von Rožmital in den Jahren zwischen 1465 und 1467 quer durch Süd- und Westeuropa und dabei unter anderem auch nach Portugal berichtet. Darin heißt es, der böhmische Adlige habe vor seiner Abreise Eleonore in Wiener Neustadt besucht und von ihr einen Brief an ihren Bruder bekommen. Diesen habe er nicht nur überbracht, sondern von Alfons V. auch ein Gegenschreiben erhalten, das bei der Kaiserin allergrößte Freude auslöste.[42]

Dieser Bericht legt sehr nahe, dass Mitte der 1460er-Jahre und wahrscheinlich auch schon lange davor zwischen Eleonore und ihrem Bruder kein regelmäßiger Kontakt mehr bestand – nicht einmal durch Briefe. Ihre wechselseitige Korrespondenz war offenbar weitgehend von zufälligen Gelegenheiten abhängig.

6. Resümee

Die Suche nach den Kontakten zwischen Portugal und dem Deutschen Reich um die Mitte des 15. Jahrhunderts hat zu dem Briefwechsel des Enea Silvio Piccolomini von 1453/54 geführt, der fragmentarisch und *uno codice* in einer Handschrift der Österreichischen Nationalbibliothek – als cvp 3389 – überliefert ist. Schon dieser Befund zeigt, wie stark unsere Kenntnis Zufällen geschuldet ist, und mahnt zu großer Vorsicht gerade bei negativen Aussagen.

Im Zentrum des Austauschs steht die Korrespondenz zwischen Enea Silvio Piccolomini und João Fernandes da Silveira. Zwar sind nur zwei Schreiben des Sienesen überliefert, allerdings scheinen diese nicht der Beginn des Briefwechsels gewesen zu sein. Außerdem werden umgehende Gegenschreiben aus Portugal erwartet und, wenn sie ausbleiben, auch angemahnt.

---

41 Zum Beispiel die Korrespondenz mit der aus Schottland stammenden Herzogin Eleonore von Tirol, vgl. Walsh, 1993.
42 Vgl. Hack, 2013: 193–195 (mit den einschlägigen Nachweisen).

Obwohl darin sehr unterschiedliche Gegenstände eine Rolle spielen, sind die Schreiben dem Typus der Freundschaftsbriefe zuzuordnen. Die beiden Korrespondenzpartner kannten sich durch ihre diplomatische Tätigkeit im Auftrag ihrer Herrscher; sie waren zuerst Ende 1450 für Verhandlungen über die Ehe Friedrichs III. mit Eleonore in Neapel und erneut im Verlauf des Romzugs 1452 zusammengetroffen. Auch später scheinen sich ihre Wege noch einmal gekreuzt zu haben.[43]

Durch ihre beiden Gesandten waren auch Friedrich III. und Alfons V. in mehr als einer Hinsicht miteinander vernetzt. Den Brief des portugiesischen Herrschers beantwortete Enea Silvio im Namen des Kaisers. Alfons V. bezieht sich in seinem Schreiben auf den mündlichen Bericht des João Fernandes. Nicht zu vergessen in diesem Geflecht ist Lopo d'Almeida, den der spätere Papst regelmäßig grüßen lässt. Er hatte den Romzug als Gesandter des portugiesischen Königs mitgemacht, war Enea Silvio aber schon vom Basler Konzil her bekannt.[44]

In allen Briefen geht es um Kaiserin Eleonore, in den beiden zwischen Friedrich III. und Alfons V. fast ausschließlich. Weshalb sich ihr Bruder so stark für ihre Ehe interessiert, ist vorderhand nicht klar. Allerdings fällt auf, dass von jenem Zeitpunkt an, als sie mit der Geburt eines (ersten) Kindes ihren ‚dynastischen Pflichten' nachzukommen beginnt, soweit heute noch erkennbar, keine Kommunikation mehr stattfindet. Die Kaiserin selbst scheint, was den brieflichen Verkehr mit ihrer alten Heimat betrifft, weitgehend auf seltene und mehr oder weniger zufällige Gelegenheiten angewiesen gewesen zu sein. Aus der portugiesischen Historiographie war sie nach der Kaiserkrönung ohnehin verschwunden.

---

43 So hat João Fernandes im Januar und April 1460 zwei Bullen und einen Geleitbrief Pius' II. erhalten, vgl. Pereira Caetano, 2011: 137; vermutlich haben sich bei dieser Gelegenheit die alten Freunde auch getroffen. – Über den sehr langen Italienaufenthalt des João Fernandes 1456–1460 vgl. Rau, 1985: 75–80 (mit der Edition von Quellen aus Venedig, Genua und Florenz in portugiesischer Übersetzung). Der Aufenthalt wurde von dem königlichen Gesandten auch dazu genutzt, eine Kopie der Weltkarte des Fra Mauro in Auftrag zu geben, vgl. Cattaneo, 2009: 538 (ich danke Herrn Dr. Thomas Horst für diesen Hinweis).

44 Von Basel aus hatte Enea Silvio auch brieflichen Kontakt zu einem anderen Portugiesen, dem Kanonisten Dr. Lupo Velasco, der an der Kurie Eugens IV. tätig war. Den einzigen überlieferten (Antwort-)Brief (vermutlich aus einer längeren Reihe) schickte Piccolomini allerdings erst 1443 aus Wiener Neustadt ab, vgl. Wolkan, 1908: 175–177 (Nr. 72) bzw. van Heck, 2007: 166 f. (Nr. 72); Überbringer dieses Schreibens war Brás Afonso, der Gesandte des Infanten Peter (s. oben, Anm. 7).

Bibliographie

Quellen

Den Vorgaben des Bandes folgend, obwohl für den Historiker sehr ungewohnt, werden die Quellen des 15. Jahrhunderts unter dem Namen des Editors zitiert.

Lopo d'Almeida

Askins, Arthur Lee-Francis u. a. 2003/2004. „A New Set of Cartas de Itália to Afonso V of Portugal from Lopo de Almeida and Luís Gonçalves Malafaia", *Romances Philology* 57, 71–88.

Lapa, Manuel Rodrigues. 1935. *Lopo d'Almeida, Cartas de Itália* (Textos de Literatura Portuguesa 3). Lissabon: Imprensa Nacional.

Giovanni Boccaccio

Zaccaria, Vittorio. 1970. *Boccaccio, De mulieribus claris* (Tutte le opere di Giovanni Boccaccio 10). Mailand (2. Aufl.): Arnoldo Mondadori Editore.

Enea Silvio Piccolomini (Papst Pius II.)

Bellus, Ibolya und Boronkai, Ivan 1993–1994. *Pii secundi pontificis maximi Commentarii*. Budapest: Balassi Kiadó.

van Heck, Adrian. 2007. *Enee Silvii Piccolominei Epistolarium seculare* (Studi e Testi 439). Vatikanstadt: Verlag der Biblioteca Apostolica Vaticana.

Knödler, Julia und Wagendorfer, Martin. 2009. *Eneas Silvius Piccolomini, Historia Australis* (MGH Scriptores rerum Germanicarum 24). Hannover: Hahnsche Buchhandlung.

Weiss, Anton. 1897. *Aeneas Sylvius Piccolomini als Papst Pius II. Sein Leben und Einfluß auf die literarische Cultur Deutschlands. Mit 149 ungedruckten Briefen aus dem Autographen-Codex Nr. 3389 der k. k. Hofbibliothek in Wien*. Graz: Ulrich Moser Verlag.

Wolkan, Rudolf. 1908. *Der Briefwechsel des Eneas Silvius Piccolomini 1: Briefe aus der Laienzeit (1431–1445) 1: Privatbriefe* (Fontes rerum Austriacarum 2, 66). Wien: Verlag der kaiserlichen Akademie der Wissenschaften.

Wolkan, Rudolf. 1918. *Der Briefwechsel des Eneas Silvius Piccolomini 3: Briefe als Bischof von Siena 1: Briefe von seiner Erhebung zum Bischof von Siena bis zum Ausgang des Regensburger Reichstages (23. September 1450–1. Juni 1454)* (Fontes rerum Austriacarum 2, 68). Wien: Verlag der kaiserlichen Akademie der Wissenschaften.

Rui de Pina

Lopes de Almeida, Manuel. 1977. „Rui de Pina, Chronica do Senhor Rey D. Afonso V". In: *Crónicas de Rui de Pina*. Porto: Lello e Irmão, 577–881.

Pseudo-Enekel
Hack, Achim Thomas. 2007. *Ein anonymer Romzugbericht von 1452 (Ps-Enenkel) mit den zugehörigen Personenlisten (Teilnehmerlisten, Ritterschlagslisten, Römische Einzugsordnung)* (Zeitschrift für deutsches Altertum Beiheft 7). Stuttgart: Steiner Verlag.

Klosterneuburger Chronik
Zeibig, Hartmann Josef. 1851. „Die kleine Klosterneuburger Chronik (1322 bis 1428)". *Archiv zur Kunde österreichischer Geschichts-Quellen* 7, 227–268.

Urkunden und Akten
Chmel, Joseph. 1837. *Materialien zur österreichischen Geschichte. Aus Archiven und Bibliotheken*, 1, 2. Wien: Joseph Fink und Sohn.

Nascimento, Aires Augusto. 1992. *Princesas de Portugal. Contratos Matrimonais dos Séculos XV e XVI* (Colecção Medievalia 5). Lissabon: Edições Cosmos.

Valentinelli, Giuseppe. 1865. *Diplomatarium Portusnaonense. Series documentorum ad historia Portusnaonis spectantium quo tempore (1276–1514) domus Austriacae imperio paruit* (Fontes rerum Austriacarum 2, 24). Wien: Verlag der kaiserlichen Akademie der Wissenschaften.

Literatur

Altmann, Wilhelm. *1896–1900. Die Urkunden Sigismunds 1410–1437* (Regesta Imperii 11). Innsbruck: Verlag der Wagner'schen Universitäts-Buchhandlung.

Baquero Moreno, Humberto. 1979–1980. *A Batalha de Alfarrobeira. Antecedentes e significado histórico*. Coimbra: Universidade de Coimbra (zuerst 1973).

Cattaneo, Angelo. 2009. „Orb and Sceptre. Cosmography and World Cartography in Portugal and Italian Cities in the Fifteenth Century". *Archives internationales d'histoire des sciences* 59, 531–553.

Chmel, Joseph. 1838. *Regesta chronologico-diplomatica Friderici IV. Romanorum regis (imperatoris III.)*. Wien: Peter Rohrmann Verlag.

Drücke, Simone. 2004. „Aeneas Silvius Piccolomini als humanistischer Historiograph. Mit einer Edition der frühneuhochdeutschen Übersetzung von Aeneas' Brief an Wilhelm von Stein". In: Staubach, Nikolaus (Hg.). *Rom und das Reich vor der Reformation* (Tradition – Reform – Innovation 7). Frankfurt am Main: Peter Lang Verlag, 271–288.

Duarte, Luís Miguel. 2007. *D. Duarte. Requiem por um rei triste*. Lissabon: Temas e Debates.

Eibl, Elfie-Marita. 1998. *Regesten Kaiser Friedrichs III. (1440–1493). Nach Bibliotheken und Archiven geordnet 9: Sachsen*. Wien u. a.: Böhlau Verlag.

Faro, Maria José und Faro, Jorge. 1961. „Embaixadas enviadas pelos reis de Portugal de 1415 a 1473". In: *Congresso internacional de história dos descobrimentos. Actas III.* Lissabon, 249–270.

Gomes, Saul António. 2009. *D. Afonso V. o Africano.* Lissabon: Temas e Debates.

Gomes dos Santos, Domingos Maurício. 1959. „O Infante D. Pedro na Áustria-Hungria". *Brotéria. Cristianismo e Cultura* 68, 17–37.

Gonçalves de Freitas, Judite Antonieta. 2001. *„Teemos por bem e mandamos". A Burocracia e os seus oficiais em meados de Quattrocentos (1439–1460) 2: Catálogos prosopográficos.* Cascais: Patrimonia.

Hack, Achim Thomas. 1999. *Das Empfangszeremoniell bei mittelalterlichen Papst-Kaiser-Treffen* (Forschungen zur Kaiser- und Papstgeschichte des Mittelalters 18). Köln u. a.: Böhlau Verlag.

Hack, Achim Thomas. 2004. „Der Ritterschlag Friedrichs III. auf der Tiberbrücke 1452. Ein Beitrag zum römischen Krönungszeremoniell des späten Mittelalters". In: Staubach, Nikolaus (Hg.). *Rom und das Reich vor der Reformation* (Tradition – Reform – Innovation 7). Frankfurt am Main: Peter Lang Verlag, 197–236.

Hack, Achim Thomas. 2011. „Eleonore von Portugal". In: Fößel, Amalie (Hg.), *Die Kaiserinnen des Mittelalters.* Regensburg: Friedrich Pustet Verlag, 306–326.

Hack, Achim Thomas. 2013. „Eine Portugiesin in Österreich um die Mitte des 15. Jahrhunderts. Kultureller Austausch infolge einer kaiserlichen Heirat?". In: Fuchs, Franz u. a. (Hg.). *König und Kanzlist, Kaiser und Papst. Friedrich III. und Enea Silvio Piccolomini in Wiener Neustadt* (Forschungen zur Kaiser- und Papstgeschichte des Mittelalters 32). Wien u. a.: Böhlau Verlag, 181–204.

Heinig, Paul-Joachim. 1997. *Kaiser Friedrich III. (1440–1493). Hof, Regierung und Politik,* 1–3 (Forschungen zur Kaiser- und Papstgeschichte des Mittelalters 17/1-3). Weimar u. a.: Böhlau Verlag.

Heinig, Paul-Joachim. 2007. „Konjunkturen des Auswärtigen. ‚State formation' und internationale Beziehungen im 15. Jahrhundert". In: Dünnebeil, Sonja und Ottner, Christine (Hg.). *Außenpolitisches Handeln im ausgehenden Mittelalter. Akteure und Ziele* (Forschungen zur Kaiser- und Papstgeschichte des Mittelalters 27). Wien u. a.: Böhlau Verlag, 21–57.

Kohler, Alfred. 1994. „ ‚Tu felix Austria nube …'. Vom Klischee zur Neubewertung dynastischer Politik in der neueren Geschichte Europas". *Zeitschrift für Historische Forschung* 21, 461–482.

Koller, Heinrich. 2005. *Kaiser Friedrich III.* Darmstadt: Wissenschaftliche Buchgesellschaft.

Knödler, Julia. 2007. „Überlegungen zur Entstehung der ‚Historia Austrialis' ". *Pirckheimer Jahrbuch für Renaissance- und Humanismusforschung* 22, 53–76.

Krieger, Karl-Friedrich. 2004. *Die Habsburger im Mittelalter. Von Rudolf I. bis Friedrich III.* Stuttgart u. a.: Kohlhammer Verlag (zuerst 1994).

Märtl, Claudia. 2013. „Anmerkungen zum Werk des Eneas Silvius Piccolomini (,Historia Austrialis', ,Pentalogus', ,Dialogus')". In: Fuchs, Franz u. a. (Hg.). *König und Kanzlist, Kaiser und Papst. Friedrich III. und Enea Silvio Piccolomini in Wiener Neustadt* (Forschungen zur Kaiser- und Papstgeschichte 32). Wien u. a.: Böhlau Verlag, 1–30.

Müller, Ricarda. 1992. *Ein Frauenbuch des frühen Humanismus* (Palingenesia 40). Stuttgart: Franz Steiner Verlag.

Nardi, Paolo. 1987. Enea Silvio Piccolomini, il cardinale Domenico Capranica e il giurista Tommaso Docci. *Rivista di storia del diritto italiano* 60, 195–203.

Palacký, František. 1857. *Geschichte von Böhmen 4: Das Zeitalter Georgs von Poděbrad 1: Die Zeit von 1439 bis zu K. Ladislaws Tode 1457.* Prag: Friedrich Tempsky.

Pereira Caetano, Pedro Nuno. 2011. *A burocracia régia como veículo para a titulação nobiliárquica. O caso do Dr. João Fernandes da Silveira.* Porto (masch. Master-Arbeit).

Pferschy-Maleczek, Bettina. 1997. „Kaiserin Eleonore". In: Schnith, Karl (Hg.). *Frauen des Mittelalters in Lebensbildern.* Graz u. a.: Verlag Styria, 420–446.

Quirin, Heinz. 1958. „König Friedrich III. in Siena (1452)". In: *Aus Reichstagen des 15. und 16. Jahrhunderts* (Schriftenreihe der Historischen Kommission bei der Bayerischen Akademie der Wissenschaften 5). Göttingen: Vandenhoeck und Ruprecht, 24–79.

Rau, Virgínia. 1985. „Relações diplomáticas de Portugal durante o reinado de D. Afonso V". In: Dies., *Estudos de história medieval.* Lissabon: Editorial Presença, 66–80 [zuerst 1964].

Reinhardt, Volker. 2013. *Pius II. Piccolomini. Der Papst, mit dem die Renaissance begann. Eine Biographie.* München: C. H. Beck Verlag.

Rogers, Francis Millet. 1961. *The Travels of the Infante Dom Pedro of Portugal* (Harvard Studies in Romance Languages 26). Cambridge (Mass.): Harvard University Press.

Rosa, Maria de Lurdes. 2007. „Vom heiligen Grafen zum Morisken-Märtyrer. Funktionen der Sakralität im Kontext der nordafrikanischen Kriege". In: Kraus, Michael und Ottomeyer, Hans (Hg.). *Novos Mundos / Neue Welten. Portugal und das Zeitalter der Entdeckungen.* Dresden: Sandstein Verlag, 89–105.

Ryder, Alan. 1990. *Alfonso the Magnanimous, King of Aragon, Neaples and Sicily, 1396–1458.* Oxford: University Press.

Schwedler, Gerald. 2008. *Herrschertreffen des Spätmittelalters. Formen, Rituale, Wirkungen* (Mittelalter-Forschungen 21). Ostfildern: Jan Thorbecke Verlag.

Voigt, Georg. 1865. „Die Briefe des Aeneas Sylvius vor seiner Erhebung auf den päpstlichen Stuhl, chronologisch geordnet und durch Einfügung von 46 bisher ungedruckten vermehrt, als Vorarbeit zu einer künftigen Ausgabe dieser Briefe". *Archiv für Kunde österreichischer Geschichts-Quellen* 16, 321–424.

Voigt, Georg. 1856–1863. *Enea Silvio de' Piccolomini, als Papst Pius der Zweite, und sein Zeitalter*, 1–3. Berlin: Dietrich Reimer Verlag.

Wagendorfer, Martin. 2008a. „Die Editionsgeschichte der ‚Historia Austrialis' des Eneas Silvius Piccolomini". *Deutsches Archiv für Erforschung des Mittelalters* 64, 65–108 und 597–602.

Wagendorfer, Martin. 2008b. *Die Schrift des Eneas Silvius Piccolomini* (Studi e Testi 441). Vatikanstadt: Verlag der Biblioteca Apostolica Vaticana.

Walsh, Katherine. 1993. „Deutschsprachige Korrespondenz der Kaiserin Leonora von Portugal. Bausteine zu einem geistigen Profil der Gemahlin Kaiser Friedrichs III. und zur Erziehung des jungen Maximilian". In: Heinig, Paul-Joachim (Hg.). *Kaiser Friedrich III. (1440–1493) in seiner Zeit. Studien anläßlich des 500. Todestages* (Forschungen zur Kaiser- und Papstgeschichte des Mittelalters 12). Köln u. a.: Böhlau Verlag, 399–445.

Willich, Thomas. 1999. *Regesten Kaiser Friedrichs III. (1440–1493). Nach Bibliotheken und Archiven geordnet 12: Allgemeine Urkundenreihe (1440–1446)*. Wien u. a.: Böhlau Verlag.

Winkelbauer, Waltraud. 2007. „‚Misit ergo Gergium de Plenavilla'. Die Heiratsvorbereitungen Friedrichs III. im Spiegel von Reisedokumenten des Georg von Volkersdorf". In: Dünnebeil, Sonja und Ottner, Christine (Hg.). *Außenpolitisches Handeln im ausgehenden Mittelalter: Akteure und Ziele* (Forschungen zur Kaiser- und Papstgeschichte des Mittelalters 27). Wien u. a.: Böhlau Verlag, 291–339.

Worstbrock, Franz Josef. 1989. „Piccolomini, Aeneas Silvius". *Verfasser-Lexikon der deutschen Literatur des Mittelalters* 7, 634–669.

Zierl, Antonia. 1966. *Kaiserin Eleonore und ihr Kreis. Eine Biographie (1436–1467)*. Diss. phil. Wien (masch.).

Jürgen Pohle

Kaiser Maximilian I. und die Rezeption der portugiesischen Entdeckungen im Nürnberger Kaufmanns- und Gelehrtenkreis am Ende des 15. Jahrhunderts

Abstract: In the so-called "first age of globalization" the Portuguese overseas expansion influenced decisively the political, economic and cultural relations between Portugal and Germany, as no other event of this age. In the 1490's a more intense intellectual occupation with the maritime expansion of Portugal began in the territory of the Holy Roman Empire. The starting point for the reception of the news about voyages of discovery was Nuremberg, where wealthy merchants and an erudite circle (where important humanists stood out like Hartmann Schedel, Hieronymus Münzer and Conrad Celtis) followed closely and with much curiosity the Portuguese overseas enterprises. Inspired primarily by the information that Martin Behaim had spread throughout his stay in Nuremberg between 1490 and 1493, the German scholars attempted to gain a more accurate picture of the extent of the Portuguese colonial empire and a picture of the world significantly changed. The Holy Roman Emperor Maximilian I entered this debate not so much for humanistic reasons, but mainly for political and dynastic reasons, given its proximity with the House of Avis. This paper tries to shed light on the discussion about the Portuguese discoveries in Nuremberg at the end of the fifteenth century, and on the special role of Martin Behaim and Hieronymus Münzer as mediators and also on the growing interest of Maximilian I in the Portuguese overseas expansion.

Seit dem ausgehenden 15. Jahrhundert gerieten die überseeischen Expeditionen Portugals und das allmählich entstehende portugiesische Kolonialreich zunehmend in den Blickwinkel deutscher Kaufleute und Humanisten. Ausgangspunkt einer intensiveren Beschäftigung mit der portugiesischen Entdeckungsgeschichte war die Reichsstadt Nürnberg. Dort verfolgte nicht nur das dem Handel verschriebene Patriziat, sondern auch ein humanistisch gebildeter Gelehrtenkreis mit großem Interesse die Entwicklung der überseeischen Expansion (Wuttke, 2007). Dazu gehörten bedeutende Kaufmannsfamilien, wie die der Holzschuher, Imhoff, Stromer und Hirschvogel, und so prominente Gelehrte wie Hartmann Schedel (1440–1514), Conrad Celtis (1459–1508) oder Hieronymus Münzer († 1508). Ihr wohl wichtigster Informant war der Nürnberger Kaufmann Martin Behaim (1459–1507), der sich 1484 selbst nach Portugal begeben hatte. 1490 kehrte er in seine fränkische Heimat zurück, wo er bis 1493 verweilte. In diesen drei Jahren entstanden dort drei der wichtigsten Quellen zur Rezeptionsgeschichte der frühen portugiesischen Entdeckungen innerhalb des Heiligen Römischen Reiches, und zwar der sogenannte Behaim-Globus, das Kapitel "Portugalia" in der berühmten Weltchronik des Hartmann Schedel sowie der Brief des Hieronymus Münzer an den portugiesischen König D. João II vom 14. Juli 1493 (Pohle, 2000: 81–96). Das letztgenannte Dokument zeigt, dass sich in der Zwischenzeit auch Kaiser Maximilian I. (reg. 1493/1508–1519) in die in Nürnberg entfachte Debatte über das sich ständig verändernde Weltbild eingeschaltet hatte. Dabei dürften im Falle des Habsburgers nicht nur humanistische sondern vor allem erb- und machtpolitische Gründe eine Rolle gespielt haben. Maximilian war durch seine portugiesische Mutter, D. Leonor (1436–1467), selbst eng mit dem Königshaus der Avis verbunden. Als König D. João II (reg. 1481–1495) im Juli 1491 durch den Tod des Infanten D. Afonso seinen einzigen legitimen Thronfolger verlor, soll Maximilian bereits wenige Monate später bei seinem Cousin um die Wahrung der habsburgischen Erbansprüche nachgesucht haben (Krendl, 2002: 127). In den folgenden Jahren erhielten die diplomatischen Beziehungen zwischen Maximilian und D. João II neue Impulse, die 1494 schließlich zu einem Freundschaftsbündnis der beiden Monarchen führten.

Eine zentrale Rolle innerhalb der sich in dieser Phase intensivierenden deutsch-portugiesischen Beziehungen spielte der in der Forschung höchst umstrittene Martin Behaim. Seit fast 200 Jahren versuchen Historiker auf die sogenannte Behaimfrage, die sich um die Klärung der tatsächlichen Verdienste Behaims in der Entdeckungsgeschichte dreht, eine akzeptable Antwort zu finden, was sich jedoch aufgrund der eher kargen Quellenlage als ein äußerst kompliziertes Unterfangen herausstellte. So kam es dazu, dass Martin Behaim in diversen Geschichtswerken immer wieder unter den großen Seefahrern, Kartographen und Gelehrten seiner Zeit genannt wurde. Zweifellos sind Behaims Leistungen innerhalb der deutschsprachigen, mitunter lokalpatriotischen Geschichtsschreibung früher überbewertet worden. Seit der polemisierenden Studie von Ernest George Ravenstein (1908) stand eigentlich schon zu Beginn des 20. Jahrhunderts fest, dass viele seiner angeblichen Taten einer kritischen Überprüfung nicht standhalten können. Dennoch hat sich das traditionelle Behaim-Bild nur sehr zögerlich gewandelt, was angesichts einer jahrhundertelangen Überlieferung von Legenden und Spekulationen, welche Person und Wirken Behaims unaufhörlich umgaben, auch kaum verwundern kann. In den vergangenen Jahrzehnten hat sich innerhalb der Behaim-Forschung jedoch ganz offensichtlich eine Tendenz abgezeichnet, Behaims Vita mit sachlicher, konstruktiver Kritik zu begegnen, insbesondere hinsichtlich seiner mutmaßlichen Verdienste innerhalb der Entdeckungsgeschichte.[1] Dabei wurde so ziemlich alles in Frage gestellt und zumeist hervorgehoben, "was er alles nicht war". Und dafür gibt es zahlreiche Beispiele (Bräunlein, 1992). Demgegenüber sollte aber nicht vergessen werden, dass Martin Behaim tatsächlich an den überseeischen Unternehmungen der Portugiesen beteiligt war, freilich nicht als bedeutender Seefahrer, sondern eher als Teilnehmer an der einen oder anderen Fahrt in Richtung Westafrika und Azoren.[2] Auch sein Ritterschlag aus dem Jahre 1485 kann wohl kaum in Zweifel gezogen werden.[3] Diese Ehre wurde an der Wende vom 15. zum 16. Jahrhundert schließlich auch anderen deutschen Kaufleuten und "Maurenkämpfern" in Portugal zuteil (Lopes, 2012).[4] Ebenso nachgewiesen sind sein Aufenthalt am portugiesischen Hof (Willers, 1992: 182) und seine Kontakte zu portugiesischen Seefahrern wie Diogo Gomes de Sintra (Nascimento, 2002). Außerdem ist es durchaus möglich und plausibel, dass er astronomische Geräte aus Nürnberg nach Portugal mitbrachte. Schließlich war die fränkische Metropole seinerzeit ein europäisches Zentrum für feinmechanische Arbeiten (Werner, 1965; Bernecker, 2000: 189–191). Wie es um Behaims astronomische Kenntnisse bestellt war, lässt sich hingegen nicht beurteilen. Manche Historiker vermuten, dass er damit am portugiesischen Hof lediglich geprahlt habe. Zuweilen könnte man bezüglich der Bewertung Behaims seitens der jüngeren historischen Forschung das Gefühl gewinnen, dass man mit dem Nürnberger Kaufmann und Abenteurer vergleichsweise strenger umgeht als mit anderen historischen Figuren. Hermann Kellenbenz (1958: 91) brachte es bezüglich der portugiesischen Geschichtsforschung schon Mitte des vorigen Jahrhunderts auf den Punkt, als er bemerkte, dass "(…) die Portugiesen mit ihren Bemühungen,

---

1 Siehe dazu die Arbeiten von Willers (1992), Bräunlein (1992), Knefelkamp (1992; 2007), Pohle (2000; 2007) und Jakob (2007).
2 Aufgrund der Unvereinbarkeit bestimmter Daten, welche auf dem Behaim-Globus und in der Weltchronik des Hartmann Schedel genannt werden, tendierte die kritische Behaim-Forschung zu der Annahme, dass Martin Behaim nicht an einer der Fahrten des Diogo Cão nach Südwestafrika, sondern an einer weniger berühmten Reise beteiligt war. In diesem Zusammenhang wird häufig die Expedition des João Afonso de Aveiro angesprochen, die der Erforschung des Königsreichs Benin diente und insbesondere kommerzielle Zwecke erfüllte.
3 Stadtarchiv Nürnberg [StadtAN], E11/II. FA Behaim, Nr. 570. Es handelt sich hier um eine Handschrift aus der Zeit um 1500, die nicht von Martin Behaim selbst stammt. Vermutlich geht die auf einem losen Zettel stehende Notiz auf ein anderes Mitglied der Familie Behaim zurück. Vgl. *Focus Behaim-Globus*. Tl. 2 (1992: 725 f.).
4 Prominente Beispiele hierfür sind Anton Herwart und Wolfgang Holzschuher, die 1494 bzw. 1503 in Portugal den Ritterschlag erhielten.

Behaims Bedeutung und Rolle zu revidieren, etwas übers Ziel hinausgeschossen (...)" hätten. Ungeachtet aller Spekulationen hinsichtlich seiner tatsächlichen Verdienste bleibt letztendlich festzustellen, dass Martin Behaim einen festen Platz in der Rezeptionsgeschichte der überseeischen Expansion Portugals einnimmt (Ehrhardt, 1989: 24). Angeregt von seinen Informationen, die er Anfang der 90er-Jahre des 15. Jahrhunderts in seiner fränkischen Heimat verstreute, versuchten sich die Gelehrten von der Welt des Entdeckungszeitalters eine Vorstellung zu machen. Dazu trug vor allem der nach Behaim benannte Globus bei.

Der sogenannte "Erdapfel" des Martin Behaim ist der älteste noch erhaltene Erdglobus, dessen Original sich im Germanischen Nationalmuseum in Nürnberg befindet. Er wurde um 1492 fertiggestellt und danach mehrmals restauriert.[5] Da der Globus dem Betrachter in erster Linie das vorkolumbianische Weltbild vermittelt, entsprach er bereits zur Zeit seiner Entstehung nicht mehr dem neuesten Stand der Entdeckungsgeschichte. Sogar die Ergebnisse der portugiesischen Expansion kommen auf dem Globus nur partiell und oftmals ungenau zum Ausdruck (Knefelkamp, 1992), was recht seltsam erscheint, da Behaim den ostatlantischen Raum um die zentrale Azorengruppe aus eigener Erfahrung kannte (Pohle, 2012). Auch seine Beteiligung an einer oder mehreren Fahrten entlang der Westküste Afrikas spiegelt sich auf dem Globus nur unzureichend wider. Behaims maßgebliche Mitarbeit am "Erdapfel" ist von seinen Zeitgenossen ausdrücklich erwähnt worden. So schrieb Hartmann Schedel über den Anteil Behaims an der Entwicklung des Globus: "*Hic globus labore et opera M. B.* [Martin Behaim] *absolutus est*" (*apud* Stauber, 1908: 61). Behaim fungierte demnach als Projektleiter und war wohl für das gesamte inhaltliche Konzept des Globus verantwortlich (Jakob, 2007: 41). Dazu lieferte er eine gedruckte "Mappa mundi", die er dem Nürnberger Rat verkauft hatte (Willers, 1992: 184). Unklar bleibt jedoch, was die inzwischen verschollene Weltkarte eigentlich darstellte. Dennoch geht man in der Behaim-Forschung mehrheitlich davon aus, dass diese Karte eine Hauptquelle für den Globus bildete. Hinzu kamen einige ältere literarische und kartographische Werke, wie die der antiken Autoren Ptolemäus, Plinius und Strabon, der Reisebericht des Marco Polo und die phantastische Reisedarstellung des Johann von Mandeville. Jüngere Analysen der Globusinschriften und des Kartenbildes zeigen, dass auch neuestes Quellenmaterial benutzt wurde, wie zum Beispiel die Karte des Henricus Martellus (Görz, 2007: 83; Knefelkamp, 2007: 75) oder der sogenannte Behaim-Gomes-Bericht.[6]

---

5   Zu Geschichte, Herstellung und Quellen des Behaim-Globus vgl. Willers (1980), Timann (2007) und Görz (2007).
6   "*De prima inuentione Guinee*". Siehe Nascimento (2002). – Bei einem Vergleich des Quellentextes mit den schriftlichen Angaben auf dem Behaim-Globus kam Daniel

Geht man davon aus, dass das Globus-Projekt auf Martin Behaim zurückgeht, stellt sich unwillkürlich die Frage, was ihn dazu angetrieben haben könnte. Nach einer von Götz Freiherr von Pölnitz (1959: 135-136) aufgestellten und von Hermann Kellenbenz (1967: 468) weiterentwickelten These soll es Behaim während seiner Nürnberger Reise von 1490 bis 1493 darum gegangen sein, "(...) das handeltreibende Patriziat der Stadt für ein überseeisches Unternehmen zu gewinnen, wobei der Globus als Beweismittel dienen sollte, um die Kaufleute leichter zu überzeugen". Die These gewinnt an Glaubwürdigkeit, wenn man sich vor Augen führt, dass auf dem Globus insbesondere Gebiete hervorgehoben werden, wo Gewürze wuchsen oder gehandelt wurden (Knefelkamp, 2007: 73).

In der Tat sollte es noch ein weiteres Jahrzehnt dauern, bis sich Nürnberger Handelshäuser in Lissabon niederließen. Dazu gehörten neben den Imhoff auch die Hirschvogel, mit denen die Behaim sowohl verwandtschaftlich wie auch geschäftlich eng verbunden waren. Martin Behaim stand selbst im Dienst der Hirschvogel, und zwar mindestens bis zum Jahre 1484, in dem er seine Reise nach Lissabon antrat. Ob er eventuell sogar in ihrem Auftrag reiste, kann hingegen nicht bewiesen werden. Unabhängig davon bleibt festzuhalten, dass Martin Behaim am Ende des 15. Jahrhunderts nicht der einzige Nürnberger Kaufmann war, der mit Portugal in Kontakt geriet. Es ist überliefert, dass der Gruberfaktor Hans Stromer 1490 in Lissabon verstarb (Biedermann, 1748: Tabula CCCCLXVII) und die Kaufleute Kaspar Fischer und Nikolaus Wolkenstein in den Jahren 1494/95 Hieronymus Münzer auf dessen Reise durch Südwesteuropa begleiteten.[7] An der Wende vom 15. zum 16. Jahrhundert tauchten einige Mitglieder der Familie Holzschuher in Portugal und dem entstehenden portugiesischen Kolonialreich auf. Wolfgang Holzschuher zeichnete sich damals auch im Heidenkampf in Nordafrika aus und wurde in Anerkennung seiner Dienste 1503 vom portugiesischen König zum Ritter geschlagen.[8] Wenige Monate später reiste Peter Holzschuher an Bord einer portugiesischen Flotte nach Indien, wo er im folgenden Jahr verstarb (Pohle, 2000: 202-204).[9] Gleiches Schicksal ereilte Jacob Holzschuher 1504 in Lissabon.[10] Den genannten Personen aus dem Nürnberger Kaufmannskreis

---

López-Cañete Quiles (1995) zu dem Resultat, dass der Behaim-Gomes-Bericht dem "Erdapfel" als Quelle diente. Demzufolge ist der Bericht während Behaims erstem Aufenthalt in Portugal zwischen Ende 1484 und Anfang 1490 entstanden.
7   Zur Reise Münzers und seinem Reisebericht siehe *infra*, Anm. 22.
8   StadtAN, E 49/I Nr. 605. Siehe ebenfalls im Stadtarchiv Nürnberg folgende Signaturen: E 3 Nr. 48, fol. 119v-125; E 49/II Nr. 745 (I; Ib; II; IX).
9   StadtAN, E 49/III Nr. 1, fol. 30.
10 StadtAN, E 49/III Nr. 1, fol. 28.

dürfte es vorangig darum gegangen sein, sich mit den Handelsbedingungen in Portugal vertraut zu machen und sich einen Eindruck von dem in dieser Phase aufblühenden portugiesischen Überseehandel zu verschaffen. Demzufolge darf man wohl davon ausgehen, dass der Behaim-Globus ganz im Sinne der Pölnitzschen These seinen Zweck erfüllt hat.

Eine weitere Quelle der Inspiration für die Nürnberger Kaufleute und Gelehrten hinsichlich der portugiesischen Entdeckungen bildete das Kapitel "Portugalia" in der Weltchronik des Hartmann Schedel (2004: Blatt CCLXXXV). Das auch als "Liber Chronicarum" bekannte Werk war eine Gemeinschaftsproduktion des Nürnberger Humanistenkreises. Neben Hartmann Schedel lieferten auch Conrad Celtis und Hieronymus Münzer Beiträge, die in das aufwendige und bedeutende publizistische Unternehmen einflossen (Rücker, 1980). Es liegen zwei Versionen der Chronik vor, eine lateinische und eine deutsche, welche im Juli und Dezember 1493 in der Werkstatt des Nürnberger Typographen Anton Koberger († 1513) gedruckt wurden. Beide Ausgaben enthalten das oben angesprochene Kapitel über Portugal, welches dem Leser einen kurzen Einblick in die portugiesische Geschichte von den 20er- bis zu den 80er-Jahren des 15. Jahrhunderts unter besonderer Berücksichtigung der überseeischen Expansion Portugals gewährt. Es konnte nachgewiesen werden, dass Hieronymus Münzer zumindest den größten Teil des lateinischen Urtextes verfasst hat.[11] Dieser entspricht inhaltlich im Wesentlichen der deutschen Fassung, enthält aber am Ende noch einen Zusatz, in dem auf den portugiesischen Handel mit afrikanischem Pfeffer eingegangen wird. Außerdem fällt auf, dass bezüglich der hier angesprochenen Reise des Diogo Cão nach Südwestafrika die Zeitangaben voneinander abweichen. Während in der lateinischen Version von einer 26-monatigen Fahrt die Rede ist, hat selbiges Unternehmen nach der deutschen Fassung lediglich 16 Monate gedauert. Unklar bleibt dabei, ob es sich hier um einen einfachen Druckfehler handelt oder aber um eine Korrektur der ursprünglichen Zeitangabe. Auch die Teilnahme Behaims an eben jener Expedition des Diogo Cão bleibt, obwohl sie *expressis verbis* in der Chronik erwähnt wird, höchst umstritten, da das angegebene Abfahrtsjahr 1483 mit den Lebensdaten Behaims nicht vereinbar ist. Es stellt sich hier die Frage, ob das Datum eventuell falsch überliefert oder ob Hieronymus Münzer etwa bei der Niederschrift der Angaben Behaims ein Fehler unterlaufen sein könnte. Beide Möglichkeiten muten jedoch etwas merkwürdig an, wenn man bedenkt, dass zwischen Behaim und Münzer Anfang der 1490er-Jahre ein reger geistiger Austausch stattfand. Ein eindeutiger

---

11 Vgl. dazu Rücker (1980: 190) und *Focus Behaim-Globus*. Tl. 2 (1992: 734).

Beleg dafür ist der Inhalt eines Schreibens, das der Nürnberger Humanist im Sommer 1493 an den portugiesischen König richtete.

Am 14. Juli 1493 unterzeichnete Hieronymus Münzer in Nürnberg einen Brief an D. João II, in dem er den portugiesischen Monarchen im Namen Kaiser Maximilians I. zu einem gemeinsamen überseeischen Unternehmen, welches an die Ostküste Asiens nach Cathay führen sollte, einlud.[12] In diesem Schreiben ging Münzer zunächst auf die Expansionspolitik Portugals seit der Zeit des Infanten D. Henrique ein und lobte dabei den Missionseifer der portugiesischen Krone in Westafrika. Anschließend sprach er direkt die ins Auge gefasste "Westfahrt nach Cathay" an (*apud* Grauert, 1908: 317–319):

> In Erwägung dieser Umstände hat Maximilian, der unbesiegbare König der Römer, der durch seine Mutter selbst ein Portugiese ist, durch meinen, wenn auch noch so schmucklosen Brief, Deine Majestät einladen wollen, das östliche Land des sehr reichen Cathay aufzusuchen. (…) so sage ich, daß der Anfang des bewohnbaren Ostens dem Ende des bewohnbaren Westens sehr nahe liege. (…)
>
> Wenn du diese Expedition aber durchführst, wird man dich wie Gott erheben oder wie einen zweiten Herkules, und du wirst, wenn es dir beliebt, für diese Fahrt auch einen von unserm Könige Maximilian abgesandten Gefährten haben, den Herrn Martin Behaim, ganz besonders um dies durchzuführen, und viele andere kundige Seeleute, welche die Breite des Meeres durchsegeln werden, indem sie ihren Weg von den Habichtsinseln (Azoren) aus nehmen mit ihrem kühnen Unternehmungsgeist, ihrem Zylinder, ihrem Quadranten, Astrolabium und anderen Instrumenten.

Die Textpassage zeigt zum einen, dass der Gedanke, die Küste Ostasiens auf einer verhältnismäßig kurzen Fahrt in westlicher Richtung über den Atlantischen Ozean erreichen zu können, so wie es Toscanelli vermutet hatte, auch im Nürnberger Gelehrtenkreis erwogen wurde. Münzer und Behaim teilten demnach die gleiche strategische Idee, die Kolumbus ein Jahr zuvor zu seiner ersten Westindienfahrt angetrieben hatte (Rötzer, 2005). Es ging darum, von Europa aus eine möglichst schnelle Route zu den Gewürzländern Asiens zu finden.

Der Inhalt des Münzer-Briefs unterstreicht zum anderen das besondere Interesse Maximilians I. an der überseeischen Expansion Portugals. Er verdeutlicht,

---

12 Von dem ursprünglich in lateinischer Sprache abgefassten Brief existiert nur noch eine unvollständige Abschrift Hartmann Schedels (Bayerische Staatsbibliothek München [BSB], 4 Inc.c.a. 424; abgedruckt in: Stauber, 1908: 251). Es hat sich hingegen eine portugiesische Übersetzung des Briefes erhalten, die aus dem frühen 16. Jahrhundert stammt und von dem Dominikanermönch Álvaro da Torre verfasst wurde (BSB, Rar. 204, Beiband 1, fol. 18v–19v; eine Transkription des Textes mit einer deutschen Übersetzung bei Grauert, 1908: 315–319).

dass der Römische König seinen Cousin nicht nur zu einer Westfahrt nach Cathay ermuntern wollte, sondern auch dazu bereit war, die Expedition personell und materiell zu unterstützen, wobei hier ausdrücklich Martin Behaim als "Abgesandter" Maximilians und einige astronomische Geräte erwähnt werden. Außerdem wurde eine rasche Ausführung der Expedition empfohlen, und zwar von den Azoren aus.[13] Der Rat Münzers, sich bei dem geplanten Unternehmen zu beeilen, könnte meines Erachtens ein Hinweis darauf sein, dass zur Zeit der Abfassung des Briefes bereits Nachrichten von der Westindienreise des Kolumbus in Nürnberg vorlagen, wie Walther L. Bernecker (2000: 190 f.) annimmt.

Es ist nicht eindeutig zu ermitteln, wann D. João II den Brief erhielt. José Manuel Garcia (2012: 39) geht davon aus, dass er ihm von Martin Behaim persönlich zugestellt wurde, und zwar unmittelbar nach dessen Rückkehr nach Portugal. Ob der portugiesische König auf das Schreiben geantwortet hat, ist ebenfalls nicht zu belegen. Sicherlich blieb der Brief in Portugal nicht unbeachtet, denn schließlich konnte es der portugiesischen Krone im Vorfeld der Vertragsverhandlungen von Tordesillas nur recht sein, wenn der Römische König seine Hilfe anbot. Nach Hermann Kellenbenz (1969: 78) muss man "(...) den Inhalt dieses Briefs im Zusammenhang sehen mit den Bemühungen der portugiesischen Regierung, in engeren Kontakt zu König Maximilian zu gelangen, um sich seine Unterstützung in der Rivalität mit Kastilien zu sichern." Maximilian wollte seinerseits, wie Hermann Wiesflecker (1986: 449) betont, „(...) seine portugiesischen Verwandten gegen den damals noch sehr mißgünstigen König Ferdinand [von Aragón (J. P.)] unterstützen, um als künftiger christlicher Kaiser an den Entdeckungen Anteil zu haben." Tatsächlich ist es ganz offensichtlich, dass in dieser Phase ein gegenseitiges Interesse Portugals und Habsburgs an einem gemeinsamen Bündnis heranreifte. Im Dezember 1493 sandte D. João II Diogo Fernandes Correia zu Verhandlungen an den Hof Maximilians.[14] Eben jener Diogo Fernandes war bereits einige Wochen zuvor auf Vermittlung des Römischen Königs in Augsburg an die Fugger und Gossembrot herangetreten, um diese zur Mitfinanzierung einer überseeischen Expedition zu bewegen.[15] Ziel der geplanten Unternehmung war das im Münzer-Brief zitierte Cathay.

---

13 Behaim hatte einige Jahre zuvor Joana de Macedo, die Tochter des Gouverneurs der Inseln Pico und Faial, Josse van Hurtere, geheiratet. Daher wurden hier die Azoren wohl als Ausgangspunkt der Unternehmung genannt.

14 Haus-, Hof- und Staatsarchiv Wien, Maximiliana 2 (alt 1b), fol. 289 [siehe dort auch: Maximiliana, alter Zettelkatalog "Portugal" (18.12.1493)]; Böhmer (1990: 360).

15 Die Augsburger Handelshäuser waren hingegen lediglich zu einer Investition von 100 Gulden bereit. Vgl. Tiroler Landesarchiv, Innsbruck [TLA], Älteres Kopialbuch 1493,

Während die Verhandlungen des portugiesischen Gesandten mit den beiden renommierten Augsburger Handelshäusern eher enttäuschend verliefen, waren seine diplomatischen Gespräche am habsburgischen Hof von Erfolg gekrönt.[16] Am 23. Juni 1494 ratifizierte Maximilian in Köln die sogenannten *Capitolos de Pazes*, in denen sich der Kaiser und dessen Sohn Philipp der Schöne einerseits und D. João II andererseits immerwährende Freundschaft und gegenseitige Hilfe im Kriegsfall zusicherten.[17] Durch diese Allianz verschaffte sich der portugiesische Monarch ein diplomatisches Mittel, um den im Rahmen seiner Expansionspolitik ausgeübten Druck auf die Katholischen Könige verstärken zu können.[18] Auf der anderen Seite erhielt Maximilian durch den Bündnisvertrag mit Portugal die Möglichkeit, in die Politik auf der Iberischen Halbinsel einzugreifen, was sowohl den außenpolitischen als auch den dynastischen Interessen Habsburgs entgegenkam, nicht zuletzt wegen der Thronfolgefrage in Portugal.

In der Zwischenzeit war Martin Behaim aus Portugal in die Niederlande gereist. Ob er dabei in diplomatischer Mission unterwegs war und mit den Verhandlungen, die zu dem portugiesisch-habsburgischen Freundschaftsvertrag führten, in irgendeiner Verbindung stand, lässt sich nicht nachweisen. Sicher ist hingegen, dass er im März 1494 in Antwerpen in Zuckerhandelsgeschäfte seines Schwiegervaters, Josse van Hurtere, verwickelt war.[19] Daraufhin kehrte er zu diesem nach Lissabon zurück mit der Absicht, sich in Kürze auf den Azoren niederzulassen.[20]

Horst Pietschmann (2007: 384) hat zuletzt die besondere Rolle der Niederlande innerhalb der Wirtschafts- und Handelsinteressen Maximilians hervorgehoben, "(…) da sie eng mit dem ihm neu zugefallenen, wirtschaftlich potenten Raum verknüpft waren" und dabei versucht, eine Beziehung zu Martin Behaim herzustellen. Dieser könnte als Schwiegersohn eines in Portugal engagierten Flamen

---

fol. 5. Ich danke Herrn Dr. Christoph Haidacher für die freundliche Übermittlung der Quellenangabe.

16 Anfang Juni 1494 wurde Maximilian informiert, dass eine Gesandtschaft aus Portugal am kaiserlichen Hof angekommen sei. Sh. TLA, Maximiliana I/38 (2. Teil), fol. 156; Böhmer (1990: 379).
17 Biblioteca da Ajuda, Lissabon, 51-VI-38, fol. 114–117v.
18 D. João II ratifizierte den im Juni 1494 abgeschlossenen Teilungsvertrag von Tordesillas nicht sofort, sondern erst Anfang September 1494 kurz vor Ablauf der legalen Frist.
19 StadtAN, E11/II. FA Behaim, Nr. 569,4 (Brief Martin Behaims an Michel Behaim, "Brabant", 11.3.1494).
20 Dies geht aus einem Zusatz des zuvor angesprochenen Schreibens hervor, den Behaim nach seiner Ankunft in Lissabon zwischen Ostern und Pfingsten 1494 verfasste.

eine für Maximilian interessante Vermittlerfunktion eingenommen haben (*ibidem*). Leider verlieren sich nach 1494 bis kurz vor seinem Tod im Jahre 1507 fast sämtliche Spuren des berühmten Nürnbergers.

Im August 1494 brach Hieronymus Münzer zu einer ausgedehnten Reise auf, die ihn auf die Iberische Halbinsel führte, wobei er unter anderem als Gesandter Maximilians auftrat.[21] Nach einigen Stationen in Aragón und Kastilien bereiste Münzer mit seinen oberdeutschen Begleitern im November 1494 den portugiesischen Königshof in Évora, wo sie mit D. João zusammentrafen. Münzer hielt seine Eindrücke in einem Reisebericht fest, welcher Anfang des 16. Jahrhunderts erschien. In seinem "Itinerarium"[22] berichtete der Nürnberger Humanist davon, dass er sich mit D. João II ausgiebig über Fragen der Kosmographie unterhalten habe. Die portugiesischen Entdeckungsfahrten kamen dabei ebenso zur Sprache wie der Handel in Übersee. In seinem Reisebericht hob Münzer die großen handelspolitischen Fähigkeiten seines Gastgebers hervor sowie den für die portugiesische Krone äußerst gewinnträchtigen Warenhandel in Westafrika. Er betonte dabei, dass einige der europäischen Tauschwaren aus Nürnberg kämen. Martin Behaim wird hingegen in Münzers Bericht mit keinem Wort erwähnt. Dies ist insofern äußerst seltsam, als Behaim in dem an D. João gerichteten Brief des Humanisten aus dem Vorjahr noch eine zentrale Rolle hinsichtlich der geplanten Westfahrt nach Cathay gespielt hatte. Daraus darf man wohl folgern, dass eine solche überseeische Unternehmung in den Plänen der portugiesischen Regierung aktuell keine Rolle mehr gespielt haben dürfte, was angesichts der Bestimmungen des Teilungsvertrags von Tordesillas auch nicht sonderlich verwundern kann.

Die verbleibende Zeit in Portugal verbrachten die Reisenden aus Oberdeutschland in Lissabon, bevor sie sich Anfang Dezember 1494 auf den Heimweg begaben. Münzer untergliederte seine Reiseaufzeichnungen zwischen dem Besuch der königlichen Residenz in Évora und dem Beginn ihrer Rückreise in fünf Kapitel. Während sich die ersten drei Kapitel auf seine Beobachtungen *in loco* beziehen, fußen die beiden letzten ("*De Terra Portugaliae*", "*De Africa maritima*

---

21 Zu den vielfältigen Funktionen Münzers auf dieser Reise sh. Herbers (2000) und Hurtienne (2010: 66–69).

22 "*Itinerarium suie Peregrinatio Exellentissimi viri, artium as utriusaz medicine doctoris, Hieroni monetarii de Feltkirchen, Civis Nuremburgensis*". BSB, Clm (*Codex latinus monacensis*) 431, fol. 96–274v. – Nach René Hurtienne (2009: 255) ist der Reisebericht Münzers nach 1501 und vor 1507 entstanden. Für Albrecht Classen (2003: 318) handelt es hierbei um den aus deutscher Sicht "(...) für das gesamte Mittelalter und auch die Frühneuzeit wohl wichtigsten Reisebericht über die Iberische Halbinsel". Vgl. dazu auch Herbers (2005).

occidentali") lediglich auf Informationen, welche Münzer über Portugal und die portugiesischen Besitzungen in Afrika gesammelt hat. In allen Kapiteln fanden ökonomische und handelspolitische Fragen, vor allem im Zusammenhang mit den überseeischen Expeditionen der Portugiesen, die besondere Beachtung des Autors. So erwähnte Münzer unter anderem die Besichtigung der *Casa da Mina*[23], die beträchtlichen Goldlieferungen aus Innerafrika sowie die zahlreichen Negersklaven, die er in Lissabon sah.

Münzers Biograph E. P. Goldschmidt (1938: 66 f.) hatte bereits in den 30er-Jahren des vorigen Jahrhunderts bemerkt, dass wirtschaftliche Interessen im Mittelpunkt der Reise des Nürnberger Humanisten gestanden haben dürften. Allein die Tatsache, dass die Begleiter Münzers, die bereits genannten Kaspar Fischer und Nikolaus Wolkenstein sowie der Augsburger Anton Herwart allesamt Kaufleute waren, unterstreichen diese Annahme. Durch ihre Reiseerfahrungen erhielten die großen Nürnberger und Augsburger Handelshäuser eine wertvolle Informationsgrundlage für den künftigen Handel auf der Iberischen Halbinsel.

Außerdem lieferte Münzer Kaiser Maximilian, mit dem er im Frühjahr 1495 auf dem Reichstag in Worms zusammentraf, wichtige Hinweise bezüglich der portugiesischen Expansionspolitik (Rötzer, 2005: 226).[24] Portugal mit seinen überseeischen Territorien spielte eine grundlegende Rolle in der Weltreichskonzeption des Kaisers. Maximilian, der nach Horst Pietschmann (2011: 29) "(…) im Vergleich zu Karl V. der modernere und weitblickendere Herrscher war", strebte dabei eine dynastisch begründete ideelle Oberherrschaft über das portugiesische Kolonialreich an (Metzig, 2013: 38). Auch in seinen späteren Kreuzzugsplänen zählte er seinen Vetter D. Manuel I zu seinen wichtigsten Verbündeten. Mithilfe des Kaisers begannen die großen oberdeutschen Handelshäuser seit dem Beginn des 16. Jahrhunderts sich in umfassendem Maße in den portugiesischen Überseehandel einzuschalten. 1503 ließen sich die Augsburger Welser in Lissabon nieder. In den folgenden Jahren gründeten auch die Fugger und Höchstetter sowie die Nürnberger Imhoff und Hirschvogel Faktoreien am Tejo (Pohle, 2000: 97–134).

---

23 Die *Casa da Mina* war der Hauptumschlagplatz des Überseehandels in Portugal. Nach der Entdeckung des Seewegs nach Indien durch Vasco da Gama wurde sie 1503 in *Casa da Índia* umbenannt.

24 Während der Kaiser in der Anfangsphase seiner Herrschaft viele seiner Informationen über Portugal über den Nürnberger Humanistenkreis um Hieronymus Münzer bezog, tat er dies im 16. Jahrhundert vor allem über den Augsburger Gelehrten Konrad Peutinger (1465–1547) sowie über die am Gymnasium Vosagense in St. Dié tätige Humanistengruppe um Martin Waldseemüller († 1520) und Matthias Ringmann (1482–1511).

Ausgestattet mit großzügigen Privilegien, die sie von der portugiesischen Krone erworben hatten, investierten diverse oberdeutsche Firmen in die Indienfahrten der Jahre 1505 und 1506, womit die deutsch-portugiesischen Wirtschaftsbeziehungen einen vorläufigen Höhepunkt erreichten. Maßgeblich eingeleitet wurde diese Entwicklung am Ende des 15. Jahrhunderts, und zwar durch Nürnberger Kaufleute und Gelehrte, wie Klaus Herbers (2000: 182 f.) betont:

> Zusammenfassend ergibt sich eine deutliche Intensivierung der Kontakte im 15. Jahrhundert. Der Bericht Münzers scheint vor diesem Hintergrund einen gewissen Höhepunkt der Beziehungen zu dokumentieren. Das Netz, das bisher durch Pilger, Händler, Handwerker, Gesandte, Gelehrte und Ordensmitglieder geknüpft worden war, wurde dichter, vor allem wohl, weil die sogenannte „Europäische Expansion" in ihre Hochphase eingetreten war.
> (...) Der Atlantik gewann inzwischen eine neue Dimension, und Nürnberger, besonders Händler und Kosmographen, griffen dies auf.

Unter jenen Nürnbergern zeichneten sich, wie wir gesehen haben, in erster Linie Martin Behaim und Hieronymus Münzer aus, die durch die Übermittlung ihrer persönlichen Erfahrungen, die sie in Portugal machten, und die Verarbeitung ihrer Kenntnisse über die portugiesischen Entdeckungen nicht nur den Kaufmanns- und Gelehrtenkreis ihrer Heimatstadt, sondern auch Maximilian I. nachhaltig inspirierten.

Bibliographie

Bernecker, Walther L. 2000. „Nürnberg und die überseeische Expansion im 16. Jahrhundert". In: Neuhaus, Helmut (Hrsg.). *Nürnberg. Eine europäische Stadt in Mittelalter und Neuzeit*. Nürnberg: Selbstverlag des Vereins für Geschichte der Stadt Nürnberg, 185–218.

Biedermann, Johann Gottfried (Hrsg.). 1748. *Geschlechtsregister des Hochadelichen Patriciats zu Nürnberg*. Bayreuth: Dietzel.

Böhmer, J. F. (Hrsg.). 1990. *Regesta Imperii XIV: Ausgewählte Regesten des Kaiserreiches unter Maximilian I. 1493–1519*. Bd. 1, Wien: Böhlau.

Bräunlein, Peter J. 1992. *Martin Behaim: Legende und Wirklichkeit eines berühmten Nürnbergers*. Bamberg: BVB/ Bayerische Verl.-Anstalt.

Classen, Albrecht. 2003. „Die Iberische Halbinsel aus der Sicht eines humanistischen Nürnberger Gelehrten Hieronymus Münzer: Itinerarium Hispanicum (1494–1495)". *Mitteilungen des Instituts für Österreichische Geschichtsforschung* 111, 317–340.

Ehrhardt, Marion. 1989. *A Alemanha e os Descobrimentos Portugueses*. Lisboa: Texto Editora.

Focus Behaim-Globus. Ausstellungskatalog des Germanischen Nationalmuseums Nürnberg. 1992. 2 Teile, Nürnberg: Verlag des Germanischen Nationalmuseums.

Garcia, José Manuel. 2012. *O Mundo dos Descobrimentos Portugueses.* Vol. 2, Vila do Conde: QuidNovi.

Görz, Günther. 2007. „Altes Wissen und neue Technik. Zum Behaim-Globus und seiner digitalen Erschließung". *Norica* 3, 78–87.

Goldschmidt, E. P. 1938. *Hieronymus Münzer und seine Bibliothek.* London: Warburg Institute.

Grauert, Hermann. 1908. „Die Entdeckung eines Verstorbenen zur Geschichte der großen Länderentdeckungen. Ein Nachtrag zu Dr. Richard Staubers Monographie über die Schedelsche Bibliothek". *Historisches Jahrbuch der Görres-Gesellschaft* 29, 304–333.

Herbers, Klaus. 2000. „ ‚Murcia ist so groß wie Nürnberg' – Nürnberg und Nürnberger auf der Iberischen Halbinsel: Eindrücke und Wechselbeziehungen". In: Neuhaus, Helmut (Hrsg.). *Nürnberg. Eine europäische Stadt in Mittelalter und Neuzeit.* Nürnberg: Selbstverlag des Vereins für Geschichte der Stadt Nürnberg, 151–183.

Herbers, Klaus. 2005. „Die »ganze« Hispania: der Nürnberger Hieronymus Münzer unterwegs – seine Ziele und Wahrnehmungen auf der Iberischen Halbinsel (1494–1495)". In: Babel, Rainer (Hrsg.). *Grand Tour. Adeliges Reisen und europäische Kultur vom 14. bis zum 18. Jahrhundert.* Ostfildern: Jan Thorbecke, 293–308.

Hurtienne, René. 2009. „Ein Gelehrter und sein Text. Zur Gesamtedition des Reiseberichts von Dr. Hieronymus Münzer, 1494/95 (Clm 431)". In: Neuhaus, Helmut (Hrsg.). *Erlanger Editionen. Grundlagenforschung durch Quelleneditionen: Berichte und Studien* (Erlanger Studien zur Geschichte 8). Erlangen/Jena: Palm&Enke, 255–272.

Hurtienne, René. 2010. „Arzt auf Reisen. Medizinische Nachrichten im Reisebericht des doctoris utriusque medicinae Hieronymus Münzer († 1508) aus Nürnberg". In: Fuchs, Franz (Hrsg.). *Medizin, Jurisprudenz und Humanismus in Nürnberg um 1500. Akten der gemeinsam mit dem Verein für Geschichte der Stadt Nürnberg, dem Stadtarchiv Nürnberg und dem Bildungszentrum der Stadt Nürnberg am 10./11. November 2006 und 7./8. November 2008 in Nürnberg veranstalteten Symposien* (Pirckheimer Jahrbuch für Renaissance- und Humanismusforschung 24). Wiesbaden: Harrassowitz, 47–69.

Jakob, Reinhard. 2007. „Wer war Martin Behaim? Auf den Spuren seines Lebens". *Norica* 3, 32–47.

Kellenbenz, Hermann. 1958. „Portugiesische Forschungen und Quellen zur Behaimfrage". *Mitteilungen des Vereins für Geschichte der Stadt Nürnberg* 48, 79–95.

Kellenbenz, Hermann. 1967. „Die Beziehungen Nürnbergs zur Iberischen Halbinsel, besonders im 15. und in der ersten Hälfte des 16. Jahrhunderts". *Beiträge zur Wirtschaftsgeschichte Nürnbergs* 1, 456–493.

Kellenbenz, Hermann. 1969. „Martin Behaim". *Fränkische Lebensbilder* 3, 69–84.

Knefelkamp, Ulrich. 1992. „Martin Behaims Wissen über die portugiesischen Entdeckungen". *Mare Liberum* 4, 87–95.

Knefelkamp, Ulrich. 2007. „Die Neuen Welten bei Martin Behaim und Martin Waldseemüller". In: Kraus, Michael und Ottomeyer, Hans (Hrsg.). *Novos Mundos – Neue Welten. Portugal und das Zeitalter der Entdeckungen*. Dresden: Sandstein, 73–88.

Krendl, Peter. 2002. „Kaiser Maximilian I. und Portugal". In: Scheidl, Ludwig (Hrsg.). *Relações entre Portugal e a Áustria. Testemunhos históricos e culturais*. Coimbra: Assírio&Alvim, 111–135.

Lopes, Marília dos Santos. 2012. "Ao serviço do Império: a nobilitação de estrangeiros na corte joanina e manuelina". *Pequena Nobreza nos Impérios Ibéricos de Antigo Regime*. Lisboa: IICT, 1–9.

Metzig, Gregor. 2013. „Maximilian I. (1486–1519), Portugal und die Expansion nach Übersee". *Jahrbuch für Europäische Überseegeschichte* 11 (2011), 9–43.

Nascimento, Aires A. (Ed.). 2002. *Diogo Gomes de Sintra: Descobrimento Primeiro da Guiné*. Lisboa: Edições Colibri.

Pietschmann, Horst. 2007. „Bemerkungen zur ‚Jubiläumshistoriographie' am Beispiel ‚500 Jahre Martin Waldseemüller und der Name Amerika' ". *Jahrbuch für Geschichte von Staat, Wirtschaft und Gesellschaft Lateinamerikas* 44, 367–389.

Pietschmann, Horst. 2011. „Deutsche und imperiale Interessen zwischen portugiesischer und spanischer Expansion im 15. Jahrhundert". In: Curvelo, Alexandra (Hrsg.). *Portugal und das Heilige Römische Reich (16.–18. Jahrhundert) – Portugal e o Sacro Império (séculos XVI–XVIII)* (Studien zur Geschichte und Kultur der iberischen und iberoamerikanischen Länder 15). Münster: Aschendorff, 15–30.

Pölnitz, Götz Freiherr von. 1959. „Martin Behaim". In: Rüdinger, Karl (Hrsg.). *Gemeinsames Erbe. Perspektiven europäischer Geschichte*. München: Bayerischer Schulbuch-Verlag, 129–141.

Pohle, Jürgen. 2000. *Deutschland und die überseeische Expansion Portugals im 15. und 16. Jahrhundert* (Historia profana et ecclesiastica 2). Münster: Lit Verlag.

Pohle, Jürgen. 2007. *Martin Behaim (Martinho da Boémia): Factos, Lendas e Controvérsias* (cadernos do cieg 26). Coimbra: Cieg/MinervaCoimbra.

Pohle, Jürgen. 2012. "Martin Behaim (Martinho da Boémia) e os Açores". *Boletim do Núcleo Cultural da Horta* 21, 189–201.

Quiles, Daniel López-Cañete. 1995. "El Globo de Martin Behaim y las Memorias de Diogo Gomes". *Mare Liberum* 10, 553–564.

Ravenstein, Ernest George. 1908. *Martin Behaim, his life and his globe*. London: Philip & son.

Rötzer, Hans Gerd. 2005. „Kolumbus kam ihm zuvor. Hieronymus Münzer und der Seeweg westwärts". *Montfort. Vierteljahresschrift für Geschichte und Gegenwart Vorarlbergs* 57/3, 223–227.

Rücker, Elisabeth. 1980. „Nürnberger Frühhumanisten und ihre Beschäftigung mit Geographie. Zur Frage einer Mitarbeit von Hieronymus Münzer und Conrad Celtis am Text der Schedelschen Weltchronik". In: Schmitz, Rudolf/Krafft, Fritz (Hrsg.). *Humanismus und Naturwissenschaften* (Beiträge zur Humanismusforschung 6). Boppard: Boldt, 181–192.

Schedel, Hartmann. 2004. *Weltchronik 1493*. Augsburg: Weltbild [¹1493].

Stauber, Richard. 1908. *Die Schedelsche Bibliothek* (Studien und Darstellungen aus dem Gebiete der Geschichte 6). Freiburg: Herder.

Timann, Ursula. 2007. „Die Handwerker des Behaim-Globus". *Norica* 3, 59–64.

Werner, Theodor Gustav. 1965. „Nürnbergs Erzeugung und Ausfuhr wissenschaftlicher Geräte im Zeitalter der Entdeckungen. Das Martin-Behaim-Problem in wirtschaftsgeschichtlicher Betrachtung". *Mitteilungen des Vereins für Geschichte der Stadt Nürnberg* 53, 69–149.

Wiesflecker, Hermann. 1986. *Kaiser Maximilian I. Das Reich, Österreich und Europa an der Wende zur Neuzeit*. Bd. 5, München: Verlag für Geschichte und Politik.

Willers, Johannes. 1980. „Der Erdglobus des Martin Behaim im Germanischen Nationalmuseum". In: Schmitz, Rudolf/Krafft, Fritz (Hrsg.). *Humanismus und Naturwissenschaften* (Beiträge zur Humanismusforschung 6). Boppard: Boldt, 193–206.

Willers, Johannes. 1992. „Leben und Werk des Martin Behaim". In: Ders. (Hrsg.). *Focus Behaim-Globus*. Teil 1. Nürnberg: Verlag des Germanischen Nationalmuseums, 173–188.

Wuttke, Dieter. 2007. *German Humanist Perspectives on the History of Discovery, 1493–1534* (cadernos do cieg 27). Coimbra: Cieg/MinervaCoimbra.

Marília dos Santos Lopes

Importing Knowledge: Portugal and the Scientific Culture in Fifteenth and Sixteenth Century's Germany

Abstract: The paper works out the intensive relations between travelers, merchants and scholars from Portugal and the German Roman Empire from the end of the fifteenth century until the first half of the sixteenth century. Exploring some outstanding examples, it gives evidence how the German knowledge community was longing for the many new insights which Portuguese travelers had brought from all over the world, concerning its borders and outlines as well as the people they met and described. This transfer of knowledge was made possible not only through direct contact and participation, but also through the translation of travelogues and other texts concerning these New Worlds. They stood as a starting point for the remapping and reconceptualization of geography, the leading discipline and the most prominent field of knowledge at that time.

While tracing the knowledge networks in Early Modern Europe, historian Peter Burke (2000) decisively emphasizes the importance of trade and seaports in the transmission of knowledge, referring in particular to well-established business networks. In this context, Burke draws special attention to the importance of Lisbon as one of these sites for the production and dissemination of knowledge:

> The phrase 'importing knowledge' is intended as a reminder of the importance of trade and more particularly of ports in the spread of information [...]. The importance of Lisbon in the history of knowledge, especially in the fifteenth and sixteenth centuries, derived from its position as the capital of the Portuguese seaborne empire (Burke, 2000: 77).

Given its location and central role in the process of the expansion of European geographic knowledge and in the opening to a whole new world overseas, Lisbon proves to be an important station, both in the production and in the dissemination and propagation of information concerning the new outlines of the world, its peoples and nature, contributing to an increasing global database. It comes as no surprise that those agents interested in collecting this amazing and precious information seek to import new knowledge and to develop adequate instruments of knowledge transfer, such as texts, images or other information media.

The information newly gathered was quickly introduced, for example, in Portuguese cartography, as witnessed by the famous maps held by the Reinel family (ca. 1485), in which already in the second half of the fifteenth century the outlines of the Northern and Western African coast were duly adapted to the progress made in the time of Henry the Navigator when the Portuguese vessels arrived in Sierra Leone.

In the early sixteenth century the geographic representation of the world suffered a profound change witnessed in many maps circulating through Europe (Carlton, 2015). One example is the map acquired in Lisbon by an agent of the Duke of Ferrara, Ercole d'Este, the representative of a powerful line of merchants of Renaissance Italy: Alberto Cantino. The Italian merchant managed to gain the favors of one (or more than one) master of cartography, who provided him with a copy of the map displayed in the House of India, which included the latest Portuguese geographic discoveries. The correspondence between the Duke and his agent reveals not only the amount of 12 ducats of gold that he had paid for it, but it also indicates the date of November 19, 1502 when he passed through Rome with his precious purchase, so that he must have left Lisbon a few weeks before.

This Portuguese map by an anonymous cartographer would become known under Cantino's name and it is a document of particular importance for the study of the first stage of the Portuguese maritime expansion, insofar as it reflects the Portuguese geographic knowledge, in particular one of its most decisive moments, shortly after the voyages of Vasco da Gama and Pedro Álvares Cabral. Thus we can state, following Jerry Brotton, that this is the most comprehensive "up-to-date visualization of the Portuguese empires" (Brotton, 1997: 23). The map represents the political power over the world, but it also maps the domain and scope of knowledge available at the time. Works like this attest that seeing and observing mean knowing, giving rise to curiosity and interest in learning and obtaining these documents, as the example of Cantino's acquisition well illustrates. We further have to consider that these maps not only provide a new geographic image of the world, as they also disclose through their captions and symbols a wide variety of data about nature, institutions or commodities.

As Brotton claims, "[…] Possession of decoratively elaborate and aesthetically magnificent maps as *Cantino Planisphere* empowered their owners in making a series of claims to both worldly and other-worldly authority" (Brotton, 1997: 23). Therefore, these maps are widely admired, sold at the price of the most precious goods, such as pepper, and esteemed as wonders, as the scholar and collector Richard Hayklut writes in his book on the *The principal navigations, voyages and discoveries of the English nation* (1589) (see Brotton, 1997: 25).

But the admiration for the new observations was not limited to maps and cartography. One of the most vivid and outstanding examples of how information quickly turned into real and concrete knowledge in Portuguese daily life can be witnessed in the paintings of Vasco Fernandes, commonly known as Grão Vasco. In his famous "Adoration of the Magi" (1501–1506), Vasco Fernandes represents for the first time the attributes and characteristics of the Brazilian Indian: the feather headdress and adornments are displayed in a striking composition that reconciles different iconographic languages. Without wishing to enter the long debate about whether Vasco Fernandes' knowledge of new people derived from direct reception of sources concerning the first voyages to Brazil or from other references (Ribeiro, 1996), the fact remains that his painting witnesses the prompt use of these data and information in an innovative iconographic program, representing a new biblical characterization and cultural understanding of a scene like the Adoration of the Magi.

These noteworthy examples illustrate how information was acquired and how it quickly found technical and visual expression in popular works, admired by the contemporary public. You have to "import knowledge" to participate in this effective and innovative cultural change. And the merchants and literati of the Holy Roman Empire were no exception in this desire to know the new reports, and to use different strategies to acquire such documentary sources.

The examples presented so far already reveal some of the characteristics of the data revolution which would promote a new visibility of the world, representing different areas of scientific and technical knowledge, as in mapping and painting, building on threefold ways of dealing with the news: by collecting and editing, by translating and compiling as well as by classifying and mediating. These three important ways and strategies in dealing with, in understanding and in disseminating the new data would really give "new worlds to the world" – and consequently change the image of the world itself.

1. Collecting and editing

The historiography of travel literature has assiduously highlighted the importance that seeing and viewing take on travel accounts; seeing and viewing are the most crucial and fundamental features of a new time, in which knowledge is based on the visual as authentic experience and empirical observation (Barreto, 1989). Only the visual observation allows for the notion of a possibly new knowledge, as expressed by the navigator Diogo Gomes, in his famous statement: "And I say with truth that I have seen much of the world, but I have never seen anything similar" (Garcia, 1983: 29), annotated by Martin Behaim in his report on Gomes' first voyage to Guinea (ca. 1456).

This aspect is evident right from the early voyages along the African coast, but this attitude will be present also in other texts on other parts of the world. The rapporteurs of the news will also grow in their task and mission and thus will not limit their reports on what they themselves have seen, but will try to enlarge their vision with further information available at and about a certain place, so as to ensure better recording and registering, as stated by Duarte Barbosa († 1521): "(…) beyond what I have personally seen, I have always strived for asking the Moors, Christians and Gentiles for the uses and customs they were practicing, using this information to combine with one another to have a more accurate news of them, which has always been my main purpose, like it should be for all who write on similar issues" (Barbosa, 1989: 13).

In his book on Indian things, probably written between 1511 and 1516, Barbosa who claimed to be the best Portuguese interpreter, as he stated in a letter to King Manuel I, traces an accurate and comprehensive overview of the oriental world. Duarte Barbosa is one of the first witnesses of Portuguese presence in Malabar, because he arrived in India as early as 1500 in the company of his uncle Gonçalo Gil Barbosa and soon learned the Malabar language. Thanks to the direct contact with the land and its people, his work is the result of careful, curious and rigorous observation, as Barbosa writes himself. It is by seeing "to have exact news" that he collected his information. Later his report therefore served as an exercise book for geographers and ethnographers interested in knowing more about these new regions. However, this was only possible because his writing, resulting from careful observation, witnessed and revealed rigor and precision.

In a first survey of new realities many authors undertake and understand their works as a primary inventory, trying to witness and to represent what their eyes had been able to see. In many cases, as in the example of Duarte Barbosa, they try to avoid any evaluation or judgment, because it is the concrete and factual information that matters and therefore should be disclosed and reported in works that were able to accurately and precisely transmit the thing itself.

Furthermore, this attitude in viewing and describing which arose in the Modern Age takes into account what a possible reader could or should expect from such a publication. It is not an exercise in self-reflection, as being, first and foremost, a personal account; on the contrary, the authors never forget – and this aspect is very important – that their main task is to make this experience, as well as the acquired knowledge, available to those who cannot travel themselves; they want to share the experience and certainly the documentary fundus.

There is widespread interest in seeking information and even in collecting plants, stones or objects, as evidenced by the flourishing appearance of cabinets of art and wonder. The cabinet as an expression of a contemporary "habit of curiosity" (Benedict, 2001: 2) includes both a desire to possess and to see with one's own eyes. Seeing is thus considered to be the most valuable sense allowing for the appropriation of knowledge. Through the eyes, knowledge becomes visible; knowledge as a matter of acquisition, control and mastery.

This is the reason for the impressive number of collections and for the many efforts to edit reports and compendia of travelogues, given the very urgency of the news revealed by them. Thus, the German Hieronymus Münzer († 1508) came to Portugal already in 1494 in search of information concerning the maritime endeavor. In Lisbon he was received by King John II, who Münzer considered a well-educated and very sagacious man ("um homem instruidíssimo e em tudo muito sagaz"; Vasconcelos, 1931: 13). Münzer traces the portrait of an intelligent and curious king, while describing his sympathy and interest in foreign news, which he seeks to understand and confirm.[1] The sovereign, cosmopolitan connoisseur and patron, as expected of a perfect noble man, knighted one of Münzer's companions, Antonio Herwart of Augsburg, in a gesture of recognition, which would not be the only one (Lopes, 2012a). In his *Itinerary* to the Iberian Peninsula, Münzer draws a thorough picture of Portugal at the end of the fifteenth century, paying special attention to the presence of new and exotic people and things, like the assiduous appearance of black Africans at work or in the streets, as well as goods like sugar, or animals, as shown by the snake skin seen in the Cathedral of Évora. These elements, along with the Portuguese texts, confirm the importance of observation and of witnessing the news with one's own eyes. In addition to this report on Portugal and Spain, the doctor and humanist wrote a book about the sea journey along the West African coast under the heading *Inventione Africae* which resembles in a certain sense the report of Martin Behaim following information given by Diogo Gomes.

In Münzer's texts readers are informed about the Iberian maritime endeavor and how Lisbon is pervaded by goods and people from other nations. Such information will be reproduced in 1493 by the humanist Hartmann Schedel (1440–1514) in his important *Cronicae mundi* published in Nuremberg. In this sum of knowledge and, more precisely, in a chapter dedicated to Portugal, reference is already made to the voyage by Diogo Cão along the West African coast,

---

1 "É muito afável e amigo de indagar muitas cousas. Àqueles que o procuram e se gabam de empresas guerreiras, de navegação ou quaisquer outras, ouve-os atentamente, manda apresentar as provas ou demonstrações, e se os acha verídicos e valentes, não os deixa sem recompensa" (Vasconcelos, 1931: 13 f.). All translations from Portuguese or German on our own responsibility.

or to sugar from Madeira, among other things, showing how the information collected in Portugal was promptly imported into a groundbreaking encyclopedic work by one of the most influential literati of his time. In fact, Münzer himself actively participated in the production of the German version of this important sum of knowledge.

In the sixteenth century and following Hartmann Schedel, the imperial counsellor, political agent and lawyer, Konrad Peutinger (1465–1547), also a collector and a geography enthusiast, carried out geographic collections on the terrestrial orb, revealing the import of news about the Portuguese voyages as an important contribution to his work (Wuttke, 2007; Lopes, 2012b). In his *Sermons convivales* (1506) he already makes reference to the caravels in the Atlantic and the arrival in Calicut. He made every effort to convince Emperor Maximilian I to support the participation of German merchants in the maritime and trading company in eastern lands (Vogel, 1991). Married to Margarete Welser (1481–1552), from one of the most influential German trading houses, Konrad Peutinger realized the opportunity that the discovery of the Cape route could mean. This is most visible in the letters and reports found among his papers, especially the texts concerning Vasco da Gama's first voyage to India (1497), the expedition by Pedro Álvares Cabral (1501), the voyage of Amerigo Vespucci (1501), Vasco da Gama's second trip to India and also concerning the fleet under the command of Francisco de Almeida. And his curiosity was not restricted to reading these texts, since Peutinger translated into German the report on Vasco da Gama's second voyage (Lopes, 1999).

One might ask how Peutinger could have had access to this information. The answer leads us to Valentim Fernandes, one of the most important printers of the time, who lived in Lisbon (Anselmo, 1983; Anselmo 1984; Anselmo, 1991: 105–110, Marques, 1995; Dias, 1995).

Valentim Fernandes was the editor, translator and author of one of the first European editions of the book by Marco Polo, published in 1502, in the introduction to which one can read an enthusiastic hymn to the opening of the world made possible by the recent navigations. Valentim Fernandes originated from Moravia, and had met Konrad Peutinger before moving to Portugal, a contact that would be kept for life, as evidenced by the so-called "Manuscript Valentim Fernandes", a written collection of the Portuguese voyages that was part of Konrad Peutinger's collection in Augsburg. This manuscript is indeed a pioneering work in which Valentim Fernandes proposed himself to "rewrite the world" (Lestringant, 1993), a task that at the dawn of the sixteenth century is clearly early and unique, giving rise to one of the first and most outstanding anthologies of novelty (Radulet, 1991: 17–35).

By directly or indirectly collecting data on new worlds, many publishers and literary men now held the most recent and most current knowledge that had ever been available.

2. Translating and compiling

The justification for translating the Portuguese literature of travel is, in general, the great on-site experience that these reports provide to those who could only see the world through reading. Translators and editors praise the observation *ad vivum* and disseminate throughout Europe the news and the reports of the travelers who had been able to see new things with their own eyes (Lopes, 2002).

The same cultural dynamics which led the travelers to describe the world, invited others to map and create categories for organizing and classifying knowledge, whether in words, diagrams, sketches or works of art, thus expressing the same mental willingness to learn and to reflect, a feature of modern Europe. At different locations in Europe we find artists, authors, translators, editors, printers and writers meeting around printing shops interconnected by frequent and active networking, which allows for the identification of sites and emerging groups in the production and management of knowledge (Burke, 2000; Chartier, 2014).

Translations were made of different kinds of documents. They may be small texts, such as the epistles of King Manuel published in different editions (Matos, 1991) and also in various cities, which already shows the interest of the Holy Roman Empire in the news coming from Portugal (Lopes, 2012b); or they may be longer travelogues, which describe new territories and realities in a more detailed manner.

Edited in 1507 by the humanist Francazano Montalboddo, the collection of *Paesi novamente retrovati* ran across Europe in many different editions (Lopes, 2016: 2–8). A year later, for example, the collection came to light in Latin, with the title, *Itinerarivm Portugallensivm* (Matos, 1992) and in that same year of 1508, in German: *Newe unbekanthe landte und ein newe weldte in kurzer zeythe erfunden* (Ruchamer, 1508). Jobst Ruchamer, a physician from Nuremberg translated Montalboddo's anthology into German, witnessing his amazement concerning the stunning and unusual news from distant lands, where one could find colored people with habits and customs so different from the known. As he writes in the prologue, such astonishing things were the reason that led him to translate the book. The news would cause such admiration that he felt compelled to make them known in his country. The discovery of new areas of the world, inhabited after all, caused in his view an act of surprise similar to a *miraculum* (Lopes, 1990;

Lopes 2016: 8). It is worth to read the prologue in which the translator justifies his interest in writing:

> After one of my best friends had given me this little book (written in Italian), so that I could translate it into German in intent and purpose of its publication, as happened then, I read some of it and found wonderful things so far never seen that contradict in various parts of the writings on the nature of the ancient masters and scholars, for example, when writing that under many circles of the sky would not live any people. However, travel and navigations, which were made by the recommendation and order of the Royal Highnesses of Portugal and Spain, show a different reality, according to the contents of this little book, because they found in the same places unseen beautiful and graceful islands, populated with naked and black people with strange habits and uses, also birds and strange and wonderful animals, delicious trees, spices, many precious stones, pearls and gold, which are highly appreciated among us, but there among them are common. When I discovered these things in this little book, I intended to please my good friend already mentioned and decided to translate it into German; and, therefore, it is that I finally translated it slowly in my spare time whenever possible. Thus a lot of people can meet and discover the great wonderful miracles of Almighty God, who created and adorned the world with different human species, lands, islands and strange creatures (as mentioned above), which were totally unknown to Christianity and to our nation. It is also almost miraculous that Christians have made these voyages and distant, dangerous, unknown and wonderful navigations. These, according to the order of this little book to be called the New World, are splendidly presented below.[2]

---

2 "Depois que um dos meus melhores amigos me deu este livrinho (escrito em língua italiana), a fim de que eu o traduzisse para a língua alemã na intenção e propósito de o publicar, como então aconteceu, li uma parte dele e descobri coisas maravilhosas e até agora nunca vistas, que contradizem em várias partes os escritos sobre a natureza dos mestres antigos e muito eruditos, por exemplo, quando escrevem que não habitariam (na terra) pessoas abaixo de muitos dos círculos do céu. Isto mostram, todavia, as viagens ou navegações de forma diferente, as quais foram feitas pela recomendação e ordem das Altezas Reais de Portugal e da Espanha segundo o conteúdo deste livrinho, pois eles descobriram nos mesmos locais, ilhas maravilhosas, belas e divertidas com gente nua e negra, com maneiras e usos estranhos e nunca vistos, também aves e animais estranhos e maravilhosos, árvores deliciosas, especiarias, muitas pedras preciosas, pérolas e ouro, que são muito apreciados entre nós e lá entre eles são comuns. Quando eu descobri estas coisas neste livrinho, tive então a intenção de agradar ao meu bom amigo já referido e de o traduzir para a língua alemã; e, por isso, é que eu o traduzi até ao fim lentamente nos meus tempos livres, quando os tenho. Para que muita gente possa conhecer e descobrir os grandes milagres maravilhosos de Deus Omnipotente, que criou e ornou o mundo com diversas espécies humanas, terras, ilhas e criaturas estranhas (como foi acima referido), que eram totalmente desconhecidas para a cristandade e para a nossa nação. É também quase milagroso que os cristãos tenham feito

Given their factual and informative content, translations and their reading turn out to be "very practical and useful", as can be read in the prologue of the German version of the work on Prester John by the Franciscan author Francisco Álvares (Álvares, 1566). Many other works will be translated, with similar reasons as in the case of the German translator of Fernão Lopes Castanheda († 1559):

> [...] They [the kings of Castile and Portugal] found a lot of the other world we call the Antipodes and even opened the navigations and sea routes, which were previously considered impossible. In these new lands which they discovered they found several species of people never seen so far with strange customs and wonderful plants and animals; many of the soldiers they sent [...] have endeavored to describe and publish these trips and also translate it into other languages what the Spanish and Portuguese described in their languages, as these different trips are not few and are required reading for those who appreciate wonderful stories.³

And still in the words of the same translator:

> I can imagine that each applied reader, after having seen and recognized the five navigations that make up this book, will undoubtedly have a huge desire to also know the other [...] even if it often may seem unusual: Although being unusual things to be reported and never seen, we should not reject them as not being true, because there are many possible and true things which seem unbelievable to those who just hear about them since they are not ordinary things. This is however not their fault, but the fault of the people who consider impossible what did not come to their ears or eyes before. Many things were presented as possible and true that were previously regarded as fables and legends.⁴

---

    estas viagens ou navegações longínquas, perigosas, desconhecidas e maravilhosas. Estas, segundo a ordem deste livrinho que será denominado o NOVO MUNDO, serão esplendidamente apresentadas em seguida" (Lopes, 1990: 270 f.).

3   "[...] eles [os reis de Castela e Portugal] descobriram uma grande parte do outro mundo que denominamos Antípodas e ainda abriram as navegações e caminhos marítimos, que até então eram considerados impossíveis. Eles viram nessas terras novas que descobriram várias espécies de gentes até então nunca vistas, costumes estranhos, plantas e animais maravilhosos; os soldados por eles enviados tiveram inesperadamente muita sorte e muita desgraça, muitos deles esforçaram-se por descrever e publicar estas viagens e também por traduzir noutras línguas aquilo que os Espanhóis e Portugueses descreveram nas suas línguas, pois estas diferentes viagens não são poucas e são necessárias de ler para aqueles que apreciam histórias maravilhosas" (Lopes, 1990: 272).

4   "Posso imaginar que cada leitor aplicado, depois de ter visto e reconhecido as cinco navegações que compõem este livro, terá sem dúvida um enorme desejo de conhecer também as outras [...] mesmo que muitas vezes possam parecer insólitas: Embora nos sejam relatadas coisas invulgares e nunca vistas, não as devemos rejeitar de repente como não verdadeiras, pois são possíveis e verdadeiras muitas coisas que para aque-

These examples may help understand the crucial role that the editing and printing business played in communication and culture, a role of mediation that, as Roger Chartier (see Chartier, 2014) notes, highlights the very conceptualization of the written text. Translators, like authors, are key mediators and agents in the invention and building of knowledge that will be visualized and conceptualized. The interest in this collection of information will continue as one of the main reasons for the publication of Portuguese works across borders – as witnessed by the German publication of the work by Duarte Lopes and Filippo Pigafetta (1533–1604) on the Congo in 1597 (Lopes/Pigafetta, 1597).

3. Classifying and mediating

Recognizing the world and mapping the many areas that were recently revealed had to be understood as the result of perseverance, courage and confidence of a new sort of men who thus gave "new worlds to the world". The notion of a new and open world and the astonishing communication between the different geographic areas, shown on maps and in geographic works, would draw special attention to travel and travel literature which found widespread interest due to the vivid description of all kind of encounters abroad (Lopes, 2002: 135–170).

Martin Waldseemüller and his *Carta Marina Navigatoria* (Waldseemüller, 1516) is one of the first examples of a new representation of the world, a new visibility which urged dissemination throughout Europe. Significantly, the world map shows next to the Cape of Good Hope the King of Portugal, the *Cristianissimi Emanuelis regis Portogalie victoria*, as can be read in the caption (Waldseemüller, 1516). Sitting on a dolphin, the king of Portugal hoists the national flag representing the victorious maritime epic.

As a landmark in contemporary cartography, the map was reedited by other geographers, as a guide to a new formula of the world, as Laurentius Frisius showed in his *Vuslegung der Mercarthen, oder Cartha Marina* (Frisius, 1527) greatly popular through Europe between 1525 and 1530. In alphabetic order the work presents more than 200 locations mentioned in Waldseemüller's map with a brief note on land and people, as for example on the island of Madeira, or Meli in Africa. In the

---

les que as ouvem lhe parecem como inacreditáveis, porque não são coisas vulgares. Isso, não é todavia culpa sua, mas das pessoas que consideram impossível, o que não chegou antes aos seus ouvidos ou que não viram. Especialmente foram muitas coisas apresentadas como possíveis e verdadeiras que antes eram consideradas como fábulas ou lendas" (Lopes, 1990: 275).

latter case, the text refers again to formerly mentioned material, such as Alvise de Cadamosto's report on the early voyages to the West African coast, crucial for the description of African locations, compiled from Montalboddo's collection and its German translation by Jobst Ruchamer. In fact, at the beginning a map entitled *Tabvla prima navigationis Aloisii Cadamvsti* appears, where one can recognize the island of Madeira, as well as the West African coast, translating Cadamosto's verbal description into a cartographical representation which should become the most important tool in the reconfiguration of the worldview in Renaissance.

Given the surprise and the astonishment caused by the travelogues, many geographers initially considered these new reports and data like a different world, or, as some called it, an "extra-Ptolemy world", a formula used, for example, by Johannes Schöner (1477–1547) in his *Opusculum geographicvm*, published in Nuremberg (Schöner, 1533). Thus, it is a world beyond the knowledge of the Alexandrian geographer "In extra regionibus Ptolomaeum", or the world "invented" by the Portuguese, "Extremum Ptolemy cognitum Prassum Promontorium, verum Nostra Aetate tota haec portio to Portugalensibus invents est" (Schöner, 1533: E).

These new observations needed to be woven into a dialogue with the inherited and current knowledge, particularly with the knowledge supported by classical authorities. As I have already shown elsewhere (Lopes, 2012b and Lopes, 2016), most of the intellectual debate in the Renaissance had to deal with the challenge of considering both the value and importance of experience and the liability of a knowledge based on classical authority.

The authors of the sixteenth century faced, in fact, a difficult problem: how to insert and frame data of sea voyages into their historical context? The visible increase in the geographic space, inhabited after all, imposed the search for explanatory reasons for the origin of this phenomenon.

The observations and experiences of the travelers are not immediately included in the prevailing geographic system; Ptolemy, Pliny, Pomponius Mela would continue to dominate the order of knowledge. The authors of classical antiquity, the foundation of knowledge both with respect to the formation of the world and to the particular knowledge concerning each of its regions, build a first and indubitable level of knowledge.

Only the resilience of inherited knowledge can explain why the newly discovered lands were considered as an "extra-Ptolemaic world", such as defined by the German cosmographies in the first decades of the sixteenth century. The discovery of regions hitherto unknown means for these authors more than the experience of something additional; it means facing a completely different view for which there was no place in the Ptolemaic world map.

In 1534, Sebastian Franck published his *Weltbuch, Spiegel vnd bildniß des gantzen Erdbodens*, a work (Franck, 1534) which, as the title indicates, was intended to present the world in its entirety. Interestingly, it is in the chapter devoted to America that we find Alvise Cadamosto's report, the writings on the journey of Pedro Álvares Cabral, the letter of King Manuel and the texts by Amerigo Vespucci and Christopher Columbus. The publication of these writings is hardly surprising, since they had already been edited in the aforementioned anthology translated by Ruchamer (1508); unique, however, is its introduction in a chapter dedicated to the Americas. The proper chapter on Africa itself does not mention Cadamosto's report at all; instead his observations along the African coast as well as other texts concerning the Far East will be included in this chapter entitled America. This confirms our hypothesis that new information on sea travel made up an own and unique block of observations, which is not yet integrated into the traditional vision of the world. This is the deep meaning of a New World as an "extra-Ptolemaic world", as people in the sixteenth century understood it.

While designing a new geographic representation of the world, scholars develop a vivid admiration for the diversity of cultures which have come to light with the navigations, as expressed by the cosmographer Sebastian Münster: "And because they are used to live after the manner of their land, they live as well as we do, following the manner of our land" (Münster, 1545: Mcccxx).

In 1545, in the second edition of Münster's *Cosmographia* a chapter appears called "New Africa" (Münster, 1545). As the author tells us, the insufficient and fragmented knowledge concerning this continent until then, when the voyages to Calicut allowed for a better understanding of its real dimension and extension, made it necessary to include a chapter on "New Africa". One could only map, conceive and characterize its regional variety by means of the material recently made available. "Importing knowledge" is critical for the construction and formation of a cosmography that claimed to offer a comprehensive understanding of the whole world. Hence the dissemination of works, such as the highly appreciated travelogues, was of utmost importance. During the sixteenth century scholars undertook huge efforts to meet this need by importing into the mother tongue the most important examples of travel literature, able to respond to the thirst for knowledge about land and people.

As mentioned before, among the texts published in Germany there are several Portuguese authors, reporting on the different geographic regions, corresponding with their detailed and accurate information to the interest in knowing more and better facts concerning these distant realms and cultures. With regard to Africa, one may recall again the aforementioned book on the *Lands of Prester John* by

Francisco Álvares (Álvares, 1974), published in German in 1566. Álvares' book became one of the most important sources on Ethiopia, frequently quoted and referred to by Sebastian Münster. Further works were used by Münster, such as those by the Portuguese humanist Damião de Góis (1502–1574), especially *Fides, Religio, Moresque aethiopum* (Góis, 1540) or the aforementioned report by Duarte Lopes and Filippo Pigafetta, printed in the late sixteenth century, which was of capital importance for knowledge concerning the realm of Congo and its surroundings (Lopes/Pigafetta, 1597).

The publication of these writings, including their translations, made an important contribution to the importation of knowledge. In the late sixteenth century, German readers thus had a huge amount of geographic and cultural data at their disposal, as the Portuguese reports became an irreplaceable and invaluable source for German geographic and cosmographical works.

Interested in knowing "the flickering change of all human things", the German geographer will map and mediate the legacy of classical antiquity to the present. Beside Ptolemy, we can see in the pages of his cosmography names such as those of contemporary Portuguese travelers and authors in the desire to trace a new prose of the world. Münsters cosmography was a bestseller for nearly one hundred years between 1544 and 1628, and it was one of the mediators of classical and contemporary knowledge in a permanent effort of epistemological updating. In the last edition of 1628, Africa, modeled in the new light of knowledge on site, is now presented in its diversity, respecting its projection to the South hitherto unconsidered. On the basis of the information brought about by the travelogues, the description seeks a more complete and detailed picture. Thanks to a gradual process of knowledge making and assimilation, the *Cosmographia* could formulate a new state of arts. Without questioning the authority of ancient writings, such as those by Pliny, Ptolemy and Pomponius Mela, the *Cosmographia* constitutes a new sum of knowledge. In fact, describing the world in its entirety by taking into account the most recent information was the primary interest of cosmographers, an ambitious project to bring together the most comprehensive amount of data and information, in building a "true" description of the world, an encyclopedia of knowledge.

It is the humanist curiosity that launches the hard and thorny task of compiling a work of such pretension. Cosmographies should bring together all the elements for describing and defining human nature in its ongoing transformation. And that would only be possible with the thorough and dogged support of historical science; only this could inform in detail and concision about the manifestations and transformations of the cultural routes. Based on the methods of translation, compilation and mediation, cosmography should rebuild and represent "the true course of historical events", as one can read in a letter written by Sebastian Münster: "I gathered the new with the old, the old with the new, to guarantee new value to the old, reputation to the new, splendor to the used, light to darkness, grace to the despised, certainty, if possible, to the doubtful".[5] With such an attitude Sebastian Münster finds the ideal way to characterize the world in its historical-cultural route.

These examples witness Portugal's contribution to a new visibility of the world, in which experience, collecting data, publishing, translating insights, compiling and mediating all kinds of knowledge work together to build a new geographic and visual image of the world. In the various stages of this construction, knowledge brought about by Portuguese travel literature played a decisive role, as the Portuguese texts were often the first documents to witness and describe realities hitherto unknown, prompting the desire to collect and edit this new knowledge, forcing the design of a new global map and mediating between inherited and newly acquired knowledge.

Bibliography

Álvares, Francisco. 1566. *Warhafftiger Bericht von den Landen auch Geistlichem vnd Weltlichem Regiment des Mechtigen Koenigs in Ethiopien den wir Priester Johan nennen wie solches durch die Kron Portugal mit besondern vleis erkuendiget worden Beschrieben durch Herrn Franciscum Aluares so derhalben sechs Jar lang an gedachts Priester Johans Hoffe verharren muessen. Aus der Portugallischen vnd Jtalianischen Sprach in das Deutsche gebracht vnd zuuorn nie im Druck ausgangen*. Eisleben: Joachim Heller.

Álvares, Francisco. 1974. *A verdadeira Informação das Terras do Preste João. Introdução e notas de Neves Águas* (A aventura portuguesa 6). Lisboa: Agência-Geral do Ultramar.

Anselmo, Artur. 1983. *Les Origines de L'Imprimerie au Portugal*. Paris: Fundação Calouste Gulbenkian.

Anselmo, Artur. 1984. *L'Activité Typographique de Valentim Fernandes au Portugal (1495–1518)*. Paris: Fundação Calouste Gulbenkian.

---

5 „Neues habe ich mit Altem verbunden und Altes mit Neuem, um so dem Alten neuen Wert, dem Neuen Ansehen, dem Abgenutzten Glanz, dem dunklen Licht, dem Verschmähten Anmut, dem Zweifelhaften, soweit es möglich war, Gewißheit zu verleihen". Sebastian Münster to Gustav, King of Sweden, Basel January 1550 (Burmeister, 1964: 159).

Anselmo, Artur. 1991. *História da Edição em Portugal*. III vols. Lisboa: Lello & Irmão.

Barbosa, Duarte. 1989. *Livro das Coisas da Índia*. Lisboa: Publicações Alfa.

Barreto, Luís Filipe. 1989. "As viagens marítimas e a nova visão do mundo e da natureza". In: Albuquerque, Luís de (dir.). *Portugal no Mundo*. Vol. II. Lisboa: Alfa, 406–413.

Benedict, Barbara M. 2001. *Curiosity: a cultural history of early modern inquiry*. Chicago: University of Chicago Press.

Brotton, Jerry. 1997. *Trading Territories: Mapping the Early Modern World*. London: Reaktion Books Ldta.

Burke, Peter. 2000. *A social history of knowledge: from Gutenberg to Diderot*. Cambridge, UK: Polity Press.

Burmeister, Karl Heinz (ed.). 1964. *Briefe Sebastian Münsters: Lateinisch und Deutsch*. Frankfurt: Insel.

Carlton, Genevieve. 2015. *Worldly consumers: the demand for maps in Renaissance Italy*. Chicago: The University of Chicago Press.

Chartier, Roger. 2014. *The author's hand and the printer's mind: Transformations of the written word in Early Modern Europe*. Cambridge. UK: Polity Press.

Dias, João José Alves. 1995. "Os Primeiros Impressores Alemães em Portugal". In: Dias, João José Alves (ed.). *No Quinto Centenário da Vita Christi. Os Primeiros Impressores Alemães em Portugal*. Lisboa: BNL, 15–27.

Franck, Sebastian. 1534. *Weltbuch: Spiegel vnd bildniß des gantzen Erdbodens* […]. Tübingen: Morhart.

Frisius, Laurentius. 1527. *Uslegung der Mercarthen oder Cartha Marina. Darin man sehen mag wo einer in der wellt sey und wo ein yetlich Landt, Wasser und Stadt gelegen ist* […], Strassburg: Johannes Grieninger.

Garcia, José Manuel (ed.). 1983. *Viagens dos Descobrimentos*. Lisboa: Presença.

Góis, Damião de. 1540. *Fides, Religio, Moresqve Aethiopvm Imperio Preciosi Ioannis (quem vulgo Presbyterum Ioannem vocant) degentium* […]. Lovaina: Rutger Resch.

Lestringant, Frank. 1993. *Écrire le Monde à la Renaissance. Quinze Études sur Rabelais, Postel, Bodin et la Littérature Géographique*. Caen: Paradigme.

Lopes, Duarte/Pigafetta, Filippo. 1597. *REGNVM CONGO hoc est Wahrhaffte vnd Eigentliche Beschreibung des Königreichs Congo in Africa* […]. Frankfurt am Main: de Bry.

Lopes, Marília dos Santos. 1990. "Portugal. Uma fonte de novos dados. A recepção dos conhecimentos portugueses sobre África nos discursos alemães dos séculos XVI e XVII". *Mare Liberum* 1, 205–308.

Lopes, Marília dos Santos. 1998. *Coisas maravilhosas e até agora nunca vistas, Para uma Iconografia dos Descobrimentos*. Lisboa: Quetzal.

Lopes, Marília dos Santos. 1999. "O impacto da viagem de Vasco da Gama na Alemanha". In: Garcia, José Manuel (Coord.). *A Viagem de Vasco da Gama à Índia 1497-1499*. Lisboa: Academia da Marinha, 604-608.

Lopes, Marília dos Santos. 2002. *Da descoberta ao Saber. Os conhecimentos sobre África na Europa dos séculos XVI e XVII*. Viseu: passagem.

Lopes, Marília Santos. 2012a. "Ao serviço do Império: a nobilitação de estrangeiros na corte joanina e manuelina". In: Rodrigues, Miguel Jasmins (coord.). *Pequena Nobreza nos Impérios Ibéricos de Antigo Regime*. Lisboa: IICT, 1-9.

Lopes, Marília dos Santos. 2012b. "From Discovery to Knowledge: Portuguese Maritime Navigation and German Humanism". In: Berbara, Maria/Enenkel, Karl A. E. (ed.). *Portuguese Humanism and the Republic of Letters* (Intersections: interdisciplinary studies in early modern culture 21). Leiden: Brill, 425-446.

Lopes, Marília dos Santos. 2016. *Writing new worlds: the cultural dynamics of curiosity in early modern Europe*. Newcastle upon Tyne: Cambridge Scholars Publishing.

Marques, António Henrique de Oliveira. 1995. "Alemães e Impressores Alemães no Portugal de Finais do Século XV". In: Dias, João José Alves (ed.). *No Quinto Centenário da Vita Christi. Os Primeiros Impressores Alemães em Portugal*. Lisboa: BNL, 11-14.

Matos, Luis de. 1991. *L'Expansion Portugaise dans la Littérature Latine de la Renaissance*. Lisboa: Fundação Calouste Gulbenkian.

Matos, Luís de (ed.). 1992. [Fracanzano da Montalboddo]: *Itinerarivm Portugallensivm*. Lisboa: Fundação Calouste Gulbenkian.

Münster, Sebastian. 1545. *Cosmographia. Beschreibung aller Lender durch Sebastianum Munsterum in welcher begriffen. Aller völcker, Herrschafften, Stetten, vnd namhafftiger flecken, herkommen: Sitten, gebreüch, ordnung, glauben, secten, vn[d] hantierung, durch die gantze welt, vnd fürnemlich Teütscher nation. Was auch besunders in iedem landt gefunden, vnnd darin bescheen sey. Alles mit figuren vnd schönen landt tafeln erklert, vn[d] für augen gestelt*. Basel: Heinrich Petri.

Peutinger, Konrad. 1506. *Sermones convivales Co[n]radi Peutingeri: de mirandês Germanie antiquitatibus*. Argentinae: Johann Prüß der Ältere.

Radulet, Carmen. 1991. *Os Descobrimentos Portugueses e a Itália*. Lisboa: Vega.

Ribeiro, Maria Aparecida. 1996. "Penas de índio: a representação do 'Brasileiro' na arte portuguesa". *Máthesis* 5, 293-323.

Ruchamer, Jobst. 1508. *Newe vnbekanthe landte und ein newe weldte in kurz verganger zeythe erfunden*. Nürnberg: Stuchs.

Schöner, Johannes. 1533. *Opusculum Geographicum ex diversorum libris ac cartis summa cura & diligentia collectum, accomodatum ad recenter elaboratum ab eodem globum descriptionis terrenae.* Nürnberg: Johann Petreius.

Vasconcelos, Basílio de (ed.). 1931. [Münzer, Jerónimo]: *Itinerário.* Coimbra: Imprensa da Universidade.

Vogel, Klaus Anselm 1991. „Neue Horizonte der Kosmographie. Die kosmographischen Bücherlisten Hartmann Schedels (um 1498) und Konrad Peutingers (1523)". *Anzeiger des Germanischen Naionalmuseums* 67, 77–85.

Waldseemüller, Martin. 1516. *Carta Marina Navigatoria* Portvgallen *navigationes, atqve tocius cogniti orbis terre marisqve formam natvram sitvs et terminos nostris temporibvs recognitos et ab antiqvorum traditione differentes, eciam qvor vetvsti non meminervnt avtores, hec generaliter indicat.* Saint-Dié. [world map, online: https://www.loc.gov/item/2016586433/].

Wuttke, Dieter. 2007. *German Humanist Perspectives on the History of Discovery, 1493–1534.* Foreword by Marília dos Santos Lopes (cadernos do cieg 27). Coimbra: Centro Interuniversitário de Estudos Germanísticos.

Torsten dos Santos Arnold

Hermann Kellenbenz and the German-Portuguese Economic Relationships during the Sixteenth Century

Abstract: Die deutsch-portugiesischen Wirtschaftsbeziehungen im Verlauf des 16. Jahrhunderts sind eng mit der Geschichte der überseeischen Expansion Portugals verbunden. Seit der Eroberung von Ceuta im Jahr 1415 befand sich das Land auf einem kontinuierlichen Kurs zur Erweiterung seines politischen und wirtschaftlichen Einflussbereiches und wurde somit zum „Ersten Europäischen Seefahrtsimperium" (nach Malyn Newitt, 2010: 67–81), das von Brasilien über Westafrika und dem Indischen Ozean bis nach Osttimor und Macau reichte.
Der vorliegende Artikel soll die deutsch-portugiesischen Wirtschaftsbeziehungen in der Zeit des Aufstiegs des portugiesischen Imperiums behandeln: in etwa die Zeit von 1450 bis 1550. Dabei wird insbesondere auf die Studien des deutschen Wirtschaftshistorikers Hermann Kellenbenz (1913–1990) verwiesen, die noch heute eine wichtige Grundlage für dieses Gebiet der trans-nationalen Geschichte Europas darstellen.
Lagen die Ursprünge des Handels der Deutschen, inklusive der Hanse, mit Portugal in der Suche nach groben Meersalz für die einheimische Konservierung von Lebensmitteln, intensivierten sich diese Handelsbeziehungen seit dem 15. Jahrhundert zu einer gegenseitigen „Abhängigkeit" in der Lieferung von Massenwaren: Getreide und Holz für den Schiffbau gingen in Ost-West Richtung, Meersalz sowie nationale agrarische Produkte in West-Ost Richtung. Mit der kontinuierlichen Expansion Portugals erfuhren die deutsch-portugiesischen Beziehungen einen Aufschwung. Zentraleuropäische Halbfertigprodukte – vor allem Kupfer- und Messingwaren sowie Leinenstoffe – wurden auf den portugiesisch-afrikanischen Märkten vertrieben; im Gegenzug importierte man westafrikanische Gewürze und Zucker aus dem atlantischen Raum. An dem überseeischen Handel mit Gewürzen und Edelsteinen aus dem südostasiatischen Raum waren insbesondere Kaufleute aus dem deutschsprachigen Raum beteiligt. Und gerade hier war Antwerpen ein wichtiger Handelsplatz zwischen dem zentraleuropäischen Hinterland und Portugal mit seinen Überseegebieten.
Auch heute noch bilden die Forschungen und Publikationen von Hermann Kellenbenz, insbesondere aufgrund der Benutzung von internationalen Quellenmaterialien und dem Augenmerk auf die Aktionen historischer Figuren wie den Fuggern, Herwart und Hirschvogel eine wichtige prosopographische Grundlage für die weitere Erforschung der deutsch-portugiesischen Beziehungen des 16. Jahrhunderts.

In this article, we will examine the German-Portuguese economic relations during the sixteenth century in a rather wider scope. Portugal began its overseas discoveries and expansion with the conquest of Ceuta (1415) and, subsequently, after having discovered and expanded its sphere of political and economic influence in the Atlantic Basin, the Indian Ocean and as far as East Timor and Macau, became the "First European Maritime Empire", an expression used by Malyn Newitt. But what were the implications of the newly discovered maritime trade routes to India, the Spice Islands or Brazil, and how did the Portuguese organize their overseas trade? Who, in the European hinterlands participated and benefitted from these new opportunities?

This overview of German-Portuguese economic relations will therefore focus on the period from the mid-fifteenth century until the mid-sixteenth century: the period of the rise and consolidation of the Portuguese maritime empire and the profound changes in European trade patterns that it provoked.

In this perspective, the studies of Hermann Kellenbenz, a German economic historian, are still valuable sources for the research of the German-Portuguese economic relations of the early Renaissance period.[1]

I. Hermann Kellenbenz (1913–1990), the historian

Hermann Kellenbenz was a German economic historian and professor of social and economic history at the universities of Cologne and Nuremberg-Erlangen. Born in Süssen (Wurttemberg, Germany) in 1913, Kellenbenz graduated from Kiel University (Schleswig-Holstein, Germany) in 1940. Between 1939 and 1945, Kellenbenz was employed as a research assistant by the *Reichsinstitut für die Geschichte des Neuen Deutschlands* (Department for the Research of the Jewish Question of the Institute for the History of the New Germany) and studied the historical development of the Hamburg Jewish community during the sixteenth and seventeenth centuries.[2] From 1941 to 1943, he fought as a soldier at the eastern front (Kiev, Ukraine) and was sent back to Germany as a war-wounded. After World War II, Kellenbenz got his "Habilitation" with his dissertation on the Hamburg Sephardic community at the turn of the sixteenth towards the seventeenth century named "Sephardim of the Lower Elbe River" ("Sephardim an der unteren Elbe") at Würzburg University, scientifically based on his prior research. Still in Würzburg, Kellenbenz became the German supervisor of A. H. de Oliveira Marques' PhD thesis on the relations between Portugal and the Hanseatic League during the Late Middle Ages. At the same time, he developed professional and personal friendships with Portuguese historians such as A. H. de Oliveira Marques

---

1 Kellenbenz, 1970: 19–27.
2 Granda, 2015: 36–52; Heiber, 1966: 456 f.

(1933–2007), Virgínia Rau (1907–1973), Joel Serrão (1919–2008) and Avelino Teixeira de Mota (1920–1982).

During the decade of the 1960s, he was head of the department of economic and social history at Cologne University, as well as director of the Rhineland-Westphalian Economic Archives (Rheinisch-Westphälisches Wirtschaftsarchiv, Köln) and co-founder and co-editor (with Richard Konnetzke) of the peer-reviewed journal *Jahrbuch von Staat, Wirtschaft und Gesellschaft Lateinamerikas*.

During this period, Kellenbenz also organized the "Cologne Colloquia of International Social and Economic History" ("Kölner Kolloquien zur Internationalen Sozial- und Wirtschaftsgeschichte", cf. Kellenbenz, 1970 b), which were an important contribution not only for the research of a shared European history but also for the intensification of international research and collaboration post World War II.

During the following decade, Kellenbenz returned to Nuremberg and became Professor of social and economic history at the University of Nuremberg-Erlangen; since 1970, he also became director of the Fugger Archives in Dillingen, Germany.

In 1990, Hermann Kellenbenz died at the age of 77. In 2013, the National Library of Portugal held a temporary exposition commemorating his research and historiographical contributions regarding the German-Portuguese relations.[3]

Throughout his academic life, Kellenbenz investigated a variety of topics of Late Medieval and Early Modern international European social and economic history. When we look at the quite extensive list of his historiographic production, the German-Portuguese and German-Spanish social and economic relations represent a significant part of his publications.

Especially in his earlier works, Kellenbenz wrote history by applying the methods of the German traditional historicism and following the ideas of Leopold von Ranke (1795–1886): to write history the way it happened by solely using documents. At the same time, Kellenbenz was open to new approaches in research, concepts and theories such as the geo-economic research of Frederic C. Lane (1900–1984) and the *Géohistoire* by Fernand Braudel (1902–1985) of the Annales School. Therefore it seems like that Kellenbenz applied concepts such as *Géohistoire* or prosopography when it was more suitable for his objects of study.

Kellenbenz mainly studied the history of individuals and merchants with an emphasis on merchants originating from German-speaking areas. Yet, he was also open to more structural approaches, illustrated in his publications on certain commodities, such as copper, iron, etc.[4]

---

3   For further information on the life of Hermann Kellenbenz: Arnold and Granda, 2015.
4   Arnold and Granda, 2015: 1 f.

When international research of trans-national social and economic relations, and simultaneous use of material from several European archives was still in its beginnings, Kellenbenz already did this, and provided massive information on the correlations between political circumstances and long-distance international trade, e.g. by investigating Royal Iberian privileges granted to merchants such as the Ehinger, Fuggers, Herwart and Welser.

Hermann Kellenbenz himself said: "There are several publications concerning the [Portuguese] privileges granted to German merchants who lived and operated in Lisbon since the Late Middle Ages but we know relatively little about the persons, the individuals that enjoyed these privileges."[5] Therefore, his approach of a combined personal history of essentially southern Germans who, in one way or the other, were involved in the German-Portuguese or Indo-European trade, are of a microcosmic yet, at the same time, global perspective that we would call Glocalization today.

II. The German-Portuguese Economic Relations

After having observed the life and works of Kellenbenz in a more global view, we will now turn to the German-Portuguese economic relations. Although he himself had not spent much of his research on the early economic relations between Portugal and the Hanseatic League throughout the Late Middle Ages, his knowledge of the archival source material and publications still made him the German supervisor of the PhD thesis of A. H. de Oliveira Marques (Portuguese) titled "Hansa e Portugal na Idade Média" (The Hanseatic League and Portugal during the Middle Ages, 1959), still one of the most considered works on this subject.

Since the Late Middle Ages, Portugal's external trade was based on royal privileges granted to foreign merchants or merchant groups. German-Portuguese economic relations had existed for many decades before the sixteenth century. Since the late fourteenth century, individual ships or fleets of kogs and urcas set sail from Baltic seaports such as Danzig, Lübeck and Reval towards the Iberian Peninsula and Lisbon in particular. These voyages, also known as *Westfahrten* or voyages to the West were initiated by the merchants of the Hanseatic League who were in search of an additional supply of sea salt purchased at the Baie de Bourgneuf (France).

Hanseatic-Portuguese trade, during the Late Middle Ages was based on the exchange of bulky and mass quantities of merchandise, most prominently the export of grain and timber in an east-west direction and the export of the Setúbal

---

5   Free translation of: Kellenbenz, 1964: 171.

sea salt as well as agricultural produce such as fruits and wine in a west-east direction. It should be noted that these commodities were basic needs produce of a relatively availability and agricultural surplus on the one side and in demand on the other side.[6]

Timber and masts were in particular demand for the Portuguese shipbuilding industry. At the end of the fifteenth century, when Bartolomeu Dias returned from his successful voyage circumnavigating the Cape of Good Hope (1488), the Portuguese crown had to recognize that the Lateen sail rigged caravels, that had been in use until then, were insufficient and inappropriate for navigating in the Indian Ocean, and that a new ship type, the Nau, had to be developed. Therefore, large timber supplies had to be secured for the following years as the Naus were of bigger proportions and equipped with longer masts.

As stated in a letter dated March 9th, 1494, King John II of Portugal (1455–1495) granted a privilege to all merchants who could supply Portugal with logs and timber of at least 10 fathoms or cubits length during the following ten years;[7] and it is here where the records of the discharge letters of the Portuguese factory at Antwerp (*Real Feitoria Portuguesa de Antuérpia*), the *Cartas de Quitação*, clearly indicate the purchase of at least ten masts of the desired length by the agents of the Portuguese factory in Antwerp between 1495 and 1498.[8] These masts, most likely to have originated from the Baltic seaports, were transported to the Lisbon shipyard *Ribeira das Naus*, shortly before the departure of Vasco da Gama's fleet to India and Pedro Álvares Cabral's fleet to Brazil.

During the course of the following decades and under the reign of King Manuel I (1469–1521) and King John III (1502–1557), timber trade between Portugal and the *Easterling* merchants of the Baltic Sea regions continued as royal Portuguese privileges granted by King Manuel I and King John III confirm the exemption of several trade customs,[9] along with the several discharge letters and receipts of the Portuguese Antwerp factory that prove the purchase of "Riga", "Norwegian" or "Prussian" logs at the same time.

The German-Portuguese economic relations have entered a new phase of development since the last decades of the fifteenth century. Many of the international merchants and merchant families had established their branches and trade posts

---

6   For further information on the relations between Portugal and the Hanseatic League cf. Dürrer, 1953.
7   Stein, 1916: 462 f. and 507 f. – The only known original document of the time period is known to still exist today in the Gdansk States Archives, Poland: APG 300, D_17C_2.
8   Costa, 1997: 326 f.
9   Silveira, 1958: 19–25, 33–39, esp. 19 and 39.

in Antwerp which, during the following decades, was to become one of the most important market places in western and central Europe. It was also here that the Portuguese had established their trade post, the *Real Feitoria Portuguesa*.[10]

Closely accompanying with the process of Portuguese expansion in the Atlantic Basin, central European commodities such as brass and copper wares (esp. basins and manilhas), glass wares and linen textiles were exported via Portugal to satisfy the demands of the western African markets. According to Kellenbenz, the majority of the brass and copper basins and manilhas were actually produced by Aachen, Cologne or Nuremberg entrepreneurs such as the Wolf-Rechterghem-Schetz syndicate; Rechterghen was an Aachen merchant and Schetz, of Flemish origin, who based his trade activities in Antwerp.[11] In return, bullion and spices from the western African shores were imported into the Portuguese motherland and further distributed by the *Casa da Guiné e da Mina*, based in Lisbon, and founded in 1482.[12]

Sugar and plants for dye manufacturing from the Atlantic islands such as Madeira and Cape Verde were exported to satisfy the demands of the central European markets; especially during the first decades of the sixteenth century, the Nuremberg Welser company was one of the major trading partners of the Portuguese Atlantic sugar commerce and most of the color plants were exported to the Netherlands and Nuremberg.[13]

After the return from the voyage of discovery of the maritime trade route to India, the first oriental goods transported on the Portuguese Naus arrived at the Antwerp market and were sold to a German merchant, Nikolaus van Rechtergem from Aachen, the same who delivered brass and copper merchandise to the Portuguese Antwerp trade post and the Portuguese Atlantic trade.[14]

In contrast to the traditional maritime-terrestrial trade route linking Europe with the Orient via the Mediterranean Sea and the Levant, this new momentum in history, the direct maritime trade route between Europe and Asia, permitted the shipping of larger quantities of merchandise and a higher profit margin. Therefore, and as Kellenbenz pointed out, the trans-national trade patterns of the late fifteenth century shifted from the Republic of Venice and the Mediterranean changed towards Antwerp, and Lisbon, into the Atlantic at the beginning of the

---

10   Kellenbenz, 1970 a: 20.
11   Kellenbenz, 1977: 335–337.
12   For the importance of central European metal ware exports to West Africa: Strieder, 1927; Małowist, 2012: 339–369 and 371–393.
13   Kellenbenz, 1970 a: 11.
14   Godinho, ²1994: Vol. IV, 98. Cf. also Pohl, 1977.

sixteenth century.[15] Kellenbenz dedicated much of his time researching the historic events and actions of individuals during this specific time period in which several important changes took place as we will see later on.

As early as 1503, the Upper German Welser-Vöhlin company sent their agent Simon Seitz (or Simao Seyes) to Portugal to negotiate the terms of an exclusive privilege for the Welser-Vöhlin-company. This privilege, granted by King Manuel I actually states that the Welser-Vöhlin received special terms such as the payment of lesser taxes and customs, but the privilege was granted to all German merchants who were interested in participating in the Indo-Portuguese spice trade.

> [...] E mais porque o ditto Simão Seyes não somete procurou estas immunidades, e priuilegios, pera ditta sua companhia, mas tambem pera qualquer outra companhia d'Alemaes que em a nossa cidade de Lisboa quiser assentar casa, p ra a ditta essa nossa negoceaçaõ de trato, nos apraz por estes letras conceder a qualquer outra companhia der mercadores Alemaes estes nosos priuilegios, & immunidades que aqui saõ conteudos, & tambem singualarmente a qualquer Alemaõ mercador que pera sy so aqui quiser tratar comtanto que a faculdade do seu trato possa valer [...].[16]

These privileges served the purpose of attracting foreign investment in the Portuguese maritime enterprises and King Manuel I tried to diversify his potential sources by granting these not only to German but also to Italian merchants. In 1505, the Portuguese king granted the right of direct participation of German and Italian merchants, mainly Upper German merchants such as the Ehinger, Fugger, Gossenbrot, Hirschvogel and Welser, in the annual India bound fleets; this strategy was soon after revised and changed, due to the high profit margins that the private merchants made. German merchants now could only either deliver monetary deposits at the Lisbon India House, the *Casa da Índia*, or could purchase spices at the Antwerp market.[17]

On the other hand, Southeast Asian markets were in great demand of copper and silver supplies, precious metals that the Portuguese themselves had to import due to insufficient natural ore deposits in their kingdom and overseas territories.

At this time, the Fugger family had gained much of the mining and exploitation rights of central Europe's copper and silver ore deposits including Neusohl (Hungary) and Schwaz (Tyrol). It is documented that over the course of the first half of the sixteenth century at least half of the annual copper production of the Fugger-Thurzo company of Neusohl, Hungary, was exported towards Antwerp,

---

15 Kellenbenz, 1970 a: 19 f.
16 Silveira, 1958: 19 f.
17 Kellenbenz, 1989: 620.

where the Portuguese were one of the, if not the most important buyer of the semi-hemispherical or rectangular shaped copper ingots destined for the India trade.[18]

Until the mid-sixteenth century, Portugal was one of the major importers and distributors of oriental spices in Europe but, on the other hand, depended also on a stable demand in order to sell their overseas produce, too. Given the fact that Portugal was lacking natural copper and silver ore deposits and was dependent on foreign supplies for the annual shipment to India and Southeast Asia, copper-pepper trade contracts, due to the two major commodities that were transitioned between the Portuguese and central European merchants and the Fuggers in particular, were made.[19]

Again, Kellenbenz contributed significantly to the knowledge on the Indo-European and German-Portuguese copper trade as he himself organized the Cologne Colloquium (Kellenbenz, 1970 b) that dealt with the European copper production and trade during the period of 1500 and 1650. In times of the Cold War, the Cologne Coloquia were a meeting point of eastern and western European researchers and scholars. It should be clearly noted that this initiative of creating bonds and networks for the benefit of sciences, managed to overcome ideological borders and political systems.

Yet, there is still another aspect of the German-Portuguese economic relations that deserves our attention and it is here where the reader can clearly see one of the major relevance of Hermann Kellenbenz' glocalized research focus: the actions of southern German individuals in the trade with exotic merchandise from Portuguese India.

In times when the Indo-Portuguese and German-Portuguese economic relations were mostly directed towards the copper-pepper trade, at least two Upper German merchant families dedicated much of their commercial attention to the trade of luxury commodities. During the first three decades of the sixteenth century, the Hirschvogel from Nuremberg and the Herwart from Augsburg traded with diamonds, pearls and precious stones along with exotic animals. When comparing the history and the actions of some of the individuals, marriages between members of the Hirschvogel, Imhoff and Behaim families, created the basis for a wealthy and expanding Nuremberg merchant families' network since the mid-fifteenth century.[20]

---

18 Vlachovic, 1977: 154 f.
19 Kellenbenz, 1961: 7.
20 Schaper, 1970: 176 f.

The Herwarts of Augsburg, on the contrary, who had established direct relations with the Portuguese crown since the last decade of the fifteenth century, actually were not part of the group of Upper German merchant families that participated in the process of the outfitting and financing of the India bound fleet of 1505.[21]

However, the presence of German merchants themselves or their respective factors and agents in Lisbon was not permanent. As Kellenbenz noted, it was due to several waves of the Black Death during the early sixteenth century that they sometimes left the Portuguese capital for an unspecified time period. Nevertheless, trading opportunities with precious stones and possible profit margins were quite attractive to the Hirschvogel who actually maintained their factory open during the pest year of 1518. Around this time, the Hirschvogel were represented by Jörg Pock; while Lazarus Nuremberger (1499–1564), Pock's antecessor, was send to India. Joachim Pruner, who afterwards became an independent merchant in the precious stones business, together with Jobst Tetzel, served the Hirschvogel family at the same time. It is no wonder that both families, the Herwarts and Hirschvogel, actually operated in a joint venture during the years around the 1520s, a strategy that allowed both parties to obtain a higher profit margin on transactions with clients such as the Welser.[22]

Albeit the fact that the First Cologne Colloquium, organized by Kellenbenz, focused on the foreign merchants who traded in market places on the Iberian Peninsula in general, the articles of Christa Schaper and Kellenbenz himself are one of the richer sources regarding the merchant and family networks of the southern Germans, namely Behaim, Ehinger and Hirschvogel.

III. Conclusions

In conclusion, the Portuguese voyages of discoveries and overseas expansion during the sixteenth century are closely linked with the German-Portuguese economic relations during that same period. Since the mid-fifteenth century, Portugal regularly traded with merchants of the Hanseatic League and imported timber and logs for their shipbuilding industry and exported its Setúbal sea salt.

At the turn of the fifteenth towards the sixteenth century, the Portuguese Antwerp trade post or factory, the *Real Feitoria Portuguesa*, served the purpose of purchasing pre-standardized mass quantities of barter goods for the Luso-African and Indo-Portuguese trade. Portugal itself depended on foreign supplies of metal

---

21 Kellenbenz, 1967: 471; Kellenbenz, 1990: 619.
22 Kellenbenz, 1967: 472; Kellenbenz, 1990: 617; Schaper, 1970: 184 f.

objects such as brass and copper basins and manilhas due to its insufficient natural copper and silver ore deposits. In return, large quantities of commodities, such as sugar and spices were sold at the Antwerp market.

Finally, the German-Portuguese economic relations during the sixteenth century (1450-1550) accompanied and, partially actively shaped, a process of shifting trade patterns away from the Mediterranean Sea into the Atlantic Basin. In this sense, merchants and merchant families were economic and cultural intermediaries; market places such as Lisbon and Antwerp were *entrepôts* between the Northern and Southern hemispheres.

Hermann Kellenbenz was one of the most proficient scholars working in this field, highlighting the importance of the links between the emerging Iberian empires and the commercial and industrial cities in the southern provinces of the Holy Roman Empire. Kellenbenz had recognized some of the crucial structures of early glocalization, long before the term "glocalization" was coined. His publications are still an important basis for the research of the German-Portuguese economic relations during the sixteenth century. As a state of the research at that time and based on international archival research, his articles still provide an important inside view into this shared trans-national past.

Hermann Kellenbenz was a Professor of Economic and Social History who, during his early academic career collaborated with the NS Institute for the History of the New Germany. After World War II, he became a cosmopolitan historian, internationally recognized and appreciated. His organization of the Cologne Colloquia, a platform for the collaboration of international historians linking the East and the West during the Cold War, must be highlighted. Some of his research results were only to be revised and reanalyzed by the PhD theses of Jürgen Pohle (2000) and Jorun Poettering (2013).

Bibliography

Arnold dos Santos, Torsten/Granda, Jeanette. 2015. "Kellenbenz, Hermann". *Dicionário de Historiadores Portugueses: Da Academia Real das Ciências ao Final do Antigo Regime.* http://dichp.bnportugal.pt/historiadores/historiadores_kellenbenz.htm.

Costa, Leonor Freire. 1997. *Naus e Galeões na Ribeira das Naus – A construção naval no século XVI para a Rota do Cabo.* Cascais: Patrimonia Historica.

Dürrer, Ingrid. 1953. *As Relações Económicas entre Portugal e a Liga Hanseática desde os últimos Anos do Século XIV até 1640.* Coimbra: [s.n.].

Godinho, Vitorino Magalhães. ²1994. *Os Descobrimentos e a Economia Mundial.* Lisbon: Presença.

Granda, Jeanette. 2015. *Hermann Kellenbenz (1913–1990). Individualismus, Internationalismus, Universalismus. Ein (Wirtschafts-)Historiker im 20. Jahrhundert.* Dissertation (Ph.D.) Friedrich-Schiller University. Jena [unpublished working paper].

Heiber, Helmut. 1966. *Walter Frank und sein Reichsinstitut für Geschichte des neuen Deutschlands* (Quellen und Darstellung zur Zeitgeschichte 13). Stuttgart: Deutsche Verlagsanstalt.

Kellenbenz, Hermann. 1954. *Unternehmerkräfte im Hamburger Portugal- und Spanienhandel. 1590–1625.* Hamburg: Verlag der Hamburgischen Bücherei.

Kellenbenz, Hermann. 1957. „Der Brasilienhandel der Hamburger Portugiesen zu Ende des 16. und in der ersten Hälfte des 17. Jahrhunderts". *III Colloquio Internacional de Estudos Luso-Brasileiros.* Lisboa: [s.l.], 277–296.

Kellenbenz, Hermann. 1961. "Os mercadores alemães de Lisboa por volta de 1530". sep. *Revista Portuguesa de História IX*, 5–20.

Kellenbenz, Hermann. 1964. "A estadia de dois 'Ulrich Ehinger', mercadores alemães em Lisboa nos princípios de séc. XVI". *Actas do Congresso Histórico de Portugal Medievo, promovido pela Câmara Municipal de Braga.* Braga: Bracara Augusta XVI–XVII, 171–176.

Kellenbenz, Hermann. 1967. „Die Beziehungen Nürnbergs zur Iberischen Halbinsel, besonders im 15. und in der ersten Hälfte des 16. Jahrhunderts". *Beiträge zur Wirtschaftsgeschichte Nürnbergs.* Nürnberg: [s. n.], 465–493;

Kellenbenz, Hermann. 1970 a. „Wirtschaftsgeschichtliche Aspekte der überseeischen Expansion Portugals". *Scripta Mercaturae 2*, 1–39.

Kellenbenz, Hermann (Ed.), 1970 b. *Fremde Kaufleute auf der Iberischen Halbinsel* (Kölner Kolloquien zur Internationalen Sozial- und Wirtschaftsgeschichte 1). Köln/Wien: Böhlau Verlag.

Kellenbenz, Hermann, 1977. „Europäisches Kupfer, Ende 15. bis Mitte 17. Jahrhundert. Ergebnisse eines Kolloquiums". In: Kellenbenz, Hermann (ed.). *Schwerpunkte der Kupferproduktion und des Kupferhandels in Europa, 1500–1650: Kölner Kolloquien zur Internationalen Wirtschafts- und Sozialgeschichte 3.* Köln/Wien: Böhlau Verlag, 335–337.

Kellenbenz, Hermann. 1989. "The Portuguese Discoveries and the Italian and German Initiatives in the Indian Trade in the first two Decades of the 16th Century". In: *Bartolomeu Dias e a sua Época, Economia e Comércio Marítimo.* Porto: Comissão Nacional para as Comemorações dos Descobrimentos Portugueses, 609–623.

Kellenbenz, Hermann. 1990. "The Herwarts of Augsburg and their Indian Trade during the first half of the Sixteenth Century". In: Mathew, K. S. (ed.). *Studies in Maritime History.* Pondicherry: Pondicherry University Press, 69–83.

Małowist, Marian. 2012. "The Foundations of European Expansion in Africa in the 16th Century. Europe, Maghreb and Western Sudan". In: Batou, Jean/Szlajfer, Henryk (eds.). *Western Europe, Eastern Europe and World Development, 13th–18th Centuries. Collection of Essays of Marian Małowist*. Chicago: Haymarket Books, 339–369.

Małowist, Marian. 2012. "Portuguese Expansion in Africa and European Economy at the Turn of the 15th Century". In: Batou, Jean/Szlajfer, Henryk (eds.). *Western Europe, Eastern Europe and World Development, 13th–18th Centuries. Collection of Essays of Marian Małowist*. Chicago: Haymarket Books, 371–393.

Marques, A. H. de Oliveira, 1959. *Hansa e Portugal na Idade Média*. Lisboa: [s. n.].

Newitt, Malyn, 2010. *Portugal in European and World History*. London: Reaktion Books.

Poettering, Jorun. 2013. *Handel, Nation und Religion: Kaufleute zwischen Hamburg und Portugal im 17. Jahrhundert*. Göttingen: Vandenhoeck und Ruprecht.

Pohl, Hans. 1977. *Die Portugiesen in Antwerpen (1568–1648): Zur Geschichte einer Minderheit* (Vierteljahresschrift für Sozial- und Wirtschaftsgeschichte Beiheft 63). Wiesbaden: Steiner.

Pohle, Jürgen. 2000. *Deutschland und die überseeische Expansion Portugals im 15. und 16. Jahrhundert* (Historia profana et ecclesiastica 2). Münster: Lit Verlag.

Schaper, Christa. 1970. „Die Hirschvogel von Nürnberg und ihre Faktoren in Lissabon und Sevilla". In: Kellenbenz, Hermann (ed.). *Fremde Kaufleute auf der Iberischen Halbinsel:* (Kölner Kolloquien zur Internationalen Sozial- und Wirtschaftsgeschichte 1). Köln/Wien: Böhlau Verlag, 176–196.

Silveira, Luís. 1958. *Privilégios concedidos a Alemães*. Lisbon: Instituto Superior das Bibliotecas e Arquivos.

Stein, Walther. 1916. *Hansisches Urkundenbuch. Elfter Band*. Munich/Leipzig: Verlag von Duncker und Humblot.

Strieder, Jakob. 1927. „Deutscher Metallwarenexport nach Westafrika im 16. Jahrhundert". In: *Historische Aufsätze: Aloys Schulte zum 70. Geburtstag*. Düsseldorf: Schwann, 179–189.

Vlachovic, Josef. 1977. „Die Kupfererzeugung und der Kupferhandel in der Slowakei vom Ende des 15. bis zur Mitte des 17. Jahrhunderts". In: Kellenbenz, Hermann (ed.). *Schwerpunkte der Kupferproduktion und des Kupferhandels in Europa, 1500–1650* (Kölner Kolloquien zur Internationalen Wirtschafts- und Sozialgeschichte 3). Köln et al.: Böhlau Verlag, 148–171.

Yvonne Hendrich

«De insulis et peregrinatione lusitanorum» – Valentim Fernandes als Vermittler von Informationen zwischen Portugal und Oberdeutschland zu Beginn des 16. Jahrhunderts

Abstract: This article deals with the Lisbon-based typographer, Valentim Fernandes of Moravian-German descent, who, at the turn of the 16th century, not only printed religious and secular books but also worked as a notary and interpreter for German merchants residing in Portugal's capital. Fernandes contributed significantly to the dissemination of information about the *"Novos Mundos"* (New World) through his correspondence with merchants and humanistic scholars from Nuremberg and Augsburg, Upper Germany's trade centers and printing capitals of the time.
Fueled by his own interest in Portugal's maritime expansion efforts in Africa and Asia, Fernandes collected reports from sailors and navigators during the first decade of the 16th century. In his manuscript, known as the *Códice Valentim Fernandes* (accessible through the Bavarian National Library in Munich), Fernandes gives accounts of cultural contacts between the European colonizers and indigenous peoples, while simultaneously, revealing his own fascination with the flora and fauna of remote and uncharted territories.
This article analyzes Fernandes' written correspondence with his merchant friends in Nuremberg and the humanistic scholar Konrad Peutinger in Augsburg, by which the following research questions emerge: What was the role of the bilingual Fernandes in the mediation of knowledge between Portugal and Upper Germany? Who participated in the process of disseminating information and news spreading? What were the cultural and commercial interests of his correspondence partners in Germany?
With regard to the artistic perception of the exotic imagery, Fernandes' letters include an astonishing piece of information: In 1515, an Indian sovereign presented Portugal's King D. Manuel I. with an Indian rhinoceros. Fernandes wrote about this gift in his correspondence with Nuremberg upon which none other than Albrecht Dürer himself immortalized the rhinoceros in one of his famous xylographs.

In einer Zeit, als die überseeische Expansion das Bewusstsein Europas zu verändern und den Horizont zu erweitern begann, zählt der aus Mähren stammende Buchdrucker Valentim Fernandes[1] zweifellos zu den bemerkenswertesten und

---

1 Des Weiteren existieren Namensversionen mit toponymischen Attributen wie Valentim Fernandes Alemão oder Valentim Fernandes de Morávia. Vgl. zu den von Valentim

vielseitigsten Persönlichkeiten im deutsch-portugiesischen Kontext von Kaufleuten, Gelehrten und Künstlern zu Beginn des 16. Jahrhunderts, wenngleich – nicht zuletzt aufgrund der lückenhaften Quellenlage – in der deutschsprachigen Forschung weitaus weniger wahrgenommen als in der portugiesischen. Neben seinem Wirken als Typograf und Verleger war Fernandes als Dolmetscher, Makler (*corretor*) und Notar (*tabelião*) der in Lissabon ansässigen und Handel treibenden oberdeutschen Kaufleute tätig (Hendrich, 2007: 179–181). Über seine Bemühungen hinaus, eine umfangreiche portugiesischsprachige Quellensammlung zur Geschichte der portugiesischen Expansion anzulegen, kam Valentim Fernandes eine wichtige Rolle als Vermittler von Informationen von Portugal nach Oberdeutschland zu. Mittels brieflicher Korrespondenzen ließ er Kaufleuten und Gelehrten in Augsburg und Nürnberg aktuelle Informationen über Ereignisse aus der portugiesischen Hauptstadt sowie aus Übersee, zu denen er in seiner Funktion als Notar Zugang besaß oder denen er als Augenzeuge sogar selbst hatte beiwohnen können, zukommen.

> [...] era Valentim Fernandes, que, na sombra, movendo influências na Corte de Lisboa, cumpria a preceito o seu papel de mediador, numa alteridade nova para os Portugueses e nova para os Alemães (Anselmo, 1987: 31).

Erhalten und dokumentiert sind einige Briefwechsel zwischen Valentim Fernandes und Konrad Peutinger (1465–1547) – Jurist, Politiker und Berater Kaiser Maximilians I. – in Augsburg sowie dem Nürnberger Kaufmann Stephan Gabler. Darüber hinaus stand er, wie aus verschiedenen Schreiben ersichtlich, in Nürnberg mit dem an Hartmann Schedels Weltchronik beteiligten Nürnberger Arzt Hieronymus Münzer († 1508) sowie Willibald Pirckheimer (1465–1530), seines Zeichens Humanist und Freund Albrecht Dürers (1471–1528), in Kontakt. Die Tatsache, dass Fernandes mit Peutinger und Pirckheimer auf Latein korrespondierte, lässt Rückschlüsse auf seinen fundierten Bildungsstand im Sinne des klassischen Kanons zu, der über dem Durchschnitt der meisten seiner Druckerkollegen gelegen haben dürfte[2]:

> Ignoramos onde fez os seus estudos, que não só foram muito além do simples saber ler e escrever, como até, sem exagero, podemos classificar de universitários (Roque, 1979: 364).

Fernandes gebrauchten Namensvarianten in Kolophonen und Unterschriften Hendrich, 2007: 35 f.
2 Nur wenige Drucker hatten eine Universität besucht und einen akademischen Grad erworben. In der Regel haben die Drucker der Inkunabelzeit sowie des frühen 16. Jahrhunderts wohl über eine gewisse Bildung sowie ausreichende Lateinkenntnisse verfügt. Vgl. Geldner, 1970: 17 f.

Im vorliegenden Beitrag soll anhand von Fernandes' Korrespondenzen mit seinen Briefpartnern in Oberdeutschland der Rezeptions- und Verarbeitungsprozess der Informationen aus den "Neuen Welten" aufgezeigt werden: Wer war in welcher Weise an der Wissenszirkulation beteiligt und von welchen Interessen geleitet? Für die am Überseehandel beteiligten oberdeutschen Handelshäuser standen aus kommerziellem Interesse selbstverständlich die Informationen über die in den betreffenden Ländern und Regionen vertriebenen Handelsgüter sowie die aus den Expansionsbestrebungen resultierenden macht- und territorialpolitischen Veränderungen im Vordergrund:

> Eine Berichterstattung [...], wie sie der mittelalterliche Kaufmann seinen auf Reisen befindlichen oder in europäischen Städten stationierten Mitarbeitern auferlegte – sofern er sich nicht selbst orientierte – wurde noch notwendiger, als sich im Zeitalter der Entdeckungen der europäische Handel auf fremde Erdteile auszudehnen begann. Der Kaufmann in Europa, der am kolonialen Handel teilnahm oder teilnehmen wollte, musste bemüht sein, sich Nachrichten über Verlauf und Erfolge überseeischer Expeditionen zu verschaffen und Informationen einzuholen über die wirtschaftlichen Verhältnisse und die Art der Bevölkerung entdeckter Gebiete, über die Kolonisation und Handelstätigkeit dort sowie über die Auseinandersetzungen und Kämpfe um die koloniale Vorherrschaft. Vor allem wollten natürlich die unmittelbar an der Expedition beteiligten Unternehmer eingehende Berichte haben (Werner, 1975: 14).

In diesem Zusammenhang versorgte auch Valentim Fernandes von Lissabon aus seine in Handelsaktivitäten involvierten Briefpartner in Oberdeutschland mit Informationen geostrategischer, wirtschaftlicher und ethnografischer Art, wie folgende Exzerpte verdeutlichen. Die erste Textstelle ist einem Schreiben an Konrad Peutinger in Augsburg vom 16. August 1506 entnommen:

> Classis 30 navium anno praesenti in Indiam est profecta ad domandum regem de Calicud et ad construendum tria fortissima castra: primum in Zoffala [Sofala], terra Aethiopiae 400 leucas ultra caput Bonae Sperancae, ubi auri copia [...] (Fernandes *apud* König, 1923: 57).

Die folgende Passage stammt aus einem Brief an Stephan Gabler in Nürnberg (26. Juni 1510):

> [...] das er mit seynen schifffñ mit fortnā kam lauffñ in die insul Seylon [*am Rand: da allayn wechst die canella] / vnd daß ist die insul taprobana da die altñ so vil ab habñ geschribñ /[...] (Germanisches Nationalmuseum Nürnberg [zukünftig zitiert als: GMNM], HistA, RSt Nürnberg XI, 1d, fol. 1 v.; Brásio, 1960: 343).

Des Weiteren ist in jenen Nachrichten aus den aus europäischer Sicht als "Neue Welten" empfundenen überseeischen Territorien eine neugierige Faszination gegenüber den "coisas maravilhosas e até agora nunca vistas" (*O livro de Marco*

Paulo, 1502, Biblioteca Nacional de Portugal, Res. 431 V, fol. aj v.) zu spüren, die sich nicht zuletzt materiell im possessiven Akt des Sammelns sowie im haptischen "Begreifen" exotischer Tiere und Artefakte widerspiegelt:

> Die Empfindung der Neugierde hat [...] ein Gerüst von moralischen, ästhetischen und emotionalen Elementen für die frühmoderne Wissenschaft errichtet, es sorgte für die Auswahl der Gegenstände, der Inhalte und der Haltung: fremdartige Gegenstände – oder landläufige, welche verfremdet wurden –, welche in angespanntester Aufmerksamkeit durch Menschen untersucht wurden, die untereinander häufig lediglich durch ihren Geschmack an solchen Gegenständen und durch die Kultivierung dieser Haltung verbunden waren (Daston, 1994: 55).

Deutlich wird dies an folgendem Auszug aus einem Schreiben vom 7. April 1507, worin Valentim Fernandes' Korrespondenzpartner Konrad Peutinger gegenüber dem Humanisten Sebastian Brant († 1521) in Straßburg seine Freude über die von den oberdeutschen Handelsagenten aus Indien mitgebrachten Papageien und zahlreichen Exotika zum Ausdruck bringt:

> Vellem, ut aliquando videres papageios meos humane loquentes (psitacos non appello, cum alium quem Plinius describit colorem habeant) multaque alia ex India a nostris factoribus quae mihi transmissa sunt: ligna, arcus, pela, conchae et alia. Iterum vale. Si longior fui, veniam dabis. [...] (Peutinger *apud* König, 1923: 77 f.).

In den vorangegangenen Zitaten lassen sich einige Referenzen konstatieren, welche die Wahrnehmung der "Neuen Welten" seitens der humanistischen Gelehrten zu Beginn des 16. Jahrhunderts als konstruktiven Prozess kultureller Aneignung vor dem Hintergrund der Rückbesinnung auf den Kenntnisstand der griechisch-römischen Antike charakterisieren, so wie es Fernandes im oben zitieren Brief an Gabler formuliert, "die altñ so vil ab habñ geschribñ" (*op. cit.*, fol. 1 v.). Der frühneuzeitliche Diskurs ist gekennzeichnet durch christliche Norm-, Werte- und Moralvorstellungen sowie durch das tradierte kosmografisch-ethnografische und historiografische Wissen antiker Autoren, wie etwa Ptolemäus, Plinius dem Älteren, Strabo oder Herodot:

> Europas [...] Diskurs ist im 16. Jahrhundert nach wie vor ein doppelter, nämlich der antike und der christliche. Kreative Aneignung der Neuen Welt durch Europa heißt entweder Einfügung in das antike und das christliche Weltbild oder seltener Konfrontationen mit ihnen. In der Regel wird das Angebot an Einpassungsmöglichkeiten mit für uns bisweilen grotesk anmutender Findigkeit ausgeschöpft (Reinhard, 1993: 12).

In jener frühen Phase, in der die europäischen Gelehrten mit Nachrichten aus den "Neuen Welten" konfrontiert wurden, koexistierten, kollidierten und verschmolzen folglich die überkommenen, zu jenem Zeitpunkt *de facto* längst obsolet gewordenen geografischen Vorstellungen und fantastisch-fabulösen Bilder aus den mittelalterlichen Reiseberichten eines Marco Polo oder Jean de Mandeville mit den aktuellen Informationen aus Übersee, zumal bereits

> ein Kontinuum von Auffassungen, Kenntnissen und Konzeptionen [existierte], von denen aus die Berichte von neuen Entdeckungen beurteilt und die dabei ihrerseits erweitert, reflektiert und angepasst wurden (Vogel, 1995: 12).

Die kulturelle Transferleistung sowohl der kognitiven als auch materiellen Verarbeitung der neuen Informationen sollte von einem langwierigen Transitions- und Lernprozess begleitet werden. Dies wird vor allem daran deutlich, dass tradierte Konzeptionen und aktuelle Erkenntnisse sich zunächst gegenseitig überlagerten. Zwar hatte man im Zuge des Humanismus bereits die Vorstellung überwunden, das menschliche Streben nach Erweiterung des Erfahrungswissens und die daraus resultierende Grenzüberschreitung als verwerflich zu erachten (Wuttke, 1992: 9–25). Doch trotz des infolge der Entdeckungsfahrten sukzessiv wachsenden empirischen Erfahrungswissens, das den überkommenen Vorstellungen widersprach und sie somit in Frage stellte, dauerte die Rezeption der antiken und mittelalterlichen Autoren im 16. Jahrhundert nach wie vor an. 1502 druckte Valentim Fernandes eine portugiesische Übersetzung des Marco Polo mit einem Vorwort zu den portugiesischen Unternehmungen in Übersee.[3] Zwischen 1508 und 1525 erschienen mehrere lateinische Übersetzungen der *Geographia* des Ptolemäus:

> Wie sehr man antike und zeitgenössische Kenntnisse gleichwertig behandelte, kann man an der Illustrationspraxis erkennen. [...] Die Fabelwesen, selbstverständlicher Bestandteil der antiken Heidenwelt bevölkern unbezweifelt die unbekannten Regionen neuer Welten der Kosmographie. Es ist nicht Naivität, nicht nur drucktechnische Ökonomie, die hierin zum Ausdruck kommt. Vielmehr zeigt sich gerade an diesem Beispiel, wie man bemüht war, alte und neue Erkenntnisse miteinander zu verbinden (Lopes und Hanenberg, 1994: 160).

Zugleich begann man aber zunehmend damit, die Inkompatibilität des überlieferten kosmografischen Wissens der Antike mit den jüngsten, durch portugiesische und spanische Reisen gewonnenen Erkenntnissen zu problematisieren, wie

---

3 "O que cousa tam maravilhosa que ho vosso muy nobre porto de Lyxboa he ja feyto porto da India ho qual nom soo sobrepoja todollos portos da nossa Europa, mas ainda os de Affrica e Asya. [...] As quaaes adições tirey de huũ liuro de latim em lingoaem portugues. [...] porque os simprizes e nom letrados melhor sejam informados das repartições daquellas prouincias do vosso titulo real [...]" (*O livro de Marco Paulo*, 1502, Biblioteca Nacional de Portugal, Res. 431 V, fls. aii, aj v./ aij).

der Nürnberger Gelehrte Jobst Ruchamer 1508 im Vorwort zu *Newe unbekandte landte*[4] formulierte:

> [...] so wunderbarliche und bysshere unerhörte Dinge, welche auch an etlichen Orten den geschrifften der alten Natürlichen Mayster und hochgelerten wyderwertigc sein in deme, das ist, so sie geschryben haben unther etlichen kraissen des hymels (auff de erdtriche) kain wonungc der menschen zu sein. Welches dyse raysse ader schyeffarthe so gethan ist worden auss geschicke ader bevelhe der allerdurchleuchtigsten küngen zu Porthugal und Hispania [...] (Ruchamer *apud* Sadji, 1983: 35).

Der Buchdruck trug, nicht zuletzt in Hinblick auf die Verbreitung von Übersetzungen, zweifellos zur rascheren Zirkulation von Informationen innerhalb Europas bei. Allerdings darf nicht außer Acht gelassen werden, dass nur ein kleines elitäres Netzwerk von Gelehrten, Druckern, Verlegern und Künstlern, das neben den Kaufleuten Zugang zu Informationen aus Übersee besaß, diese rezipierte, diskutierte und somit an der Wissensverbreitung, wenngleich innerhalb eines limitierten Adressatenkreises, beteiligt war:

> Bis zum Beginn des 16. Jahrhunderts waren die Nachrichten, die aus den überseeischen Gebieten Portugals nach Deutschland strömten, nur wenigen Reichsangehörigen vorbehalten. Zu diesem erlesenen Kreis der Informierten zählten einige deutsche Humanisten und Politiker und natürlich diejenigen Kaufleute, die damals mit der portugiesischen Krone in Handelsverbindungen standen (Pohle, 2000: 230).

Diese reziprok wirksame Beteiligung von Kaufleuten und Gelehrten in Hinblick auf Rezeption, Kompilation und Reproduktion von Wissen kommt im folgenden Exzerpt eines Schreibens des Druckers Johann Grüninger in Straßburg an seinen Kollegen Hans Koberger in Nürnberg (25. Juli 1524) zum Ausdruck, wobei in Bezug auf die Verbreitung nicht zuletzt auch verlegerische Gründe eine Rolle spielten:

> [...] hoff als vff den ptholemeus vnd vff daß Carthamarina büch würt ein Cronic der welt, ob ir hulffen darzü durch birckheimer [Willibald Pirckheimer] vnd andere hystorici vnd kaufflüt me zu erfarn auch von kauffluten wie ich üch etlich figuren zu besichtigen geschickt, mein es würd ein kürtzwylig buch werden (Grüninger *apud* Hase, 1885: CXXXI).

Die Intention, Nachrichten über die jüngst befahrenen Gebiete zu sammeln, zu edieren und dem lesekundigen Publikum zugänglich zu machen, dürfte wohl

---

4   1508 erschien in Nürnberg unter dem Titel "*Newe unbekanthe landte un ein newe weldte in kurtz verganger zeythe erfunden*" die von Jobst Ruchamer angefertigte deutschsprachige Übersetzung von Fracanzano da Montalboddos "*Paesi novamente retrovati*" (1507), die u. a. die Reiseberichte des Cadamosto, von Kolumbus, Vespucci sowie von Vasco da Gama enthält (Sadji, 1983).

auch hinter Valentim Fernandes' Motivation gestanden haben bezüglich der Handschriftensammlung, dem sogenannten *Códice* oder *Manuscrito Valentim Fernandes*, die sich heute im Besitz der Bayerischen Staatsbibliothek in München befindet (*Codex hispanicus 27*)[5]:

> E porque tenho esprito muytas cousas destas ilhas e sua gente e de seus custumes y ydolatrias amtes que fossem conquistadas pellos christãos por ysso quero ho aqui poer por nom perder meu trabalho e nom menos os leentes folgarem de ouujr (Códice Valentim Fernandes, 1997: 123).

Das von Valentim Fernandes im Zeitraum von ca. 1505 bis 1510 zusammengetragene und redigierte Manuskriptkonvolut enthält insgesamt 33 Texte, die sich thematisch-geografisch in einen, weitaus umfangreicheren, afrikanischen sowie einen asiatischen Part untergliedern lassen (Andrade, 1972: 533; Hendrich, 2007: 197–213) und einen Zeitraum von fast einhundert Jahren abdecken. Der afrikanische Teil beinhaltet neben der Zusammenfassung der Chronik von Gomes Eanes de Zurara († 1474) über die Eroberung Ceutas mehrere Reiseberichte portugiesischer Seefahrer entlang der afrikanischen Küste, darunter, als einziger Text der Sammlung auf Latein, der Bericht von Diogo Gomes (1420–1502) und Martin Behaim (1459–1507). Der asiatische Teil umfasst eine Beschreibung der Malediven sowie das Bordbuch Hans Mayrs, der 1505/1506 die unter oberdeutscher Beteiligung ausgerüstete Flotte von D. Francisco de Almeida nach Indien begleitet hatte.[6] Darüber hinaus finden sich im *Códice* einige von Valentim Fernandes selbst angefertigte Skizzen geografischer Umrisse und eines Pelikans. Es ist trotz fehlender dokumentarischer Belege davon auszugehen, dass Fernandes beabsichtigte, die Manuskriptsammlung in den Druck zu bringen, dieses Vorhaben jedoch aus verschiedenen Gründen nicht mehr realisieren konnte (Hendrich, 2007: 201 f.). Wie die Handschriften in Konrad Peutingers Hände gelangten, ist ebenfalls

---

5   1807 wurde die Handschriftensammlung, die sich zuvor im Besitz Konrad Peutingers und danach im Jesuitenkolleg in Augsburg befunden hatte, von der Bayerischen Staatsbibliothek in München erworben. Eine 1848 angefertigte Abschrift des Manuskriptes ist in der *Biblioteca Nacional* in Lissabon vorhanden (*Manuscritos Iluminados* [IL], 154). Eine quellenkritische Bearbeitung erfolgte 1940 durch António Baião unter dem Titel *O manuscrito "Valentim Fernandes"*. Im Jahre 1997 veröffentlichte die *Academia Portuguesa da História* unter dem Titel *Códice Valentim Fernandes* eine von José Pereira da Costa paläografisch überarbeitete Ausgabe.
6   Dieselbe Fahrt wurde auch von dem Welser-Handelsagenten Balthasar Sprenger begleitet, dessen Aufzeichnungen 1509 in gedruckter Version, illustriert mit Holzschnitten Hans Burgkmairs d. Ä., erschien. (Erhard und Ramminger, 1998: 9–38; vgl. auch Horst, 2009).

ungeklärt, aber es ist anzunehmen, dass sie nach Valentim Fernandes' Tod als Konvolut und nicht – wie vielfach angenommen – als einzelne Zeitungsbriefe zu seinen Lebzeiten nach Augsburg kamen. Peutingers Besitzvermerk, der auf dem Einband zu erkennen ist, greift den ersten Satz des Gomes-Behaim-Berichts auf: "De insulis et peregrinat[ione] lusitanorum. Liber Chuonradi Peutinger Augustani V 1 Doctoris [...]" (Códice Valentim Fernandes, 1997: XXIV; Anselmo, 1987: 28 f.). Nachdem der portugiesische Humanist und Diplomat Damião de Góis (1502–1574) während seiner Reise durch Nord- und Mitteleuropa in Peutingers Haus in Augsburg besagte Manuskriptsammlung zu Gesicht bekommen hatte, bat er Hans Jakob Fugger (1516–1575) in einem Schreiben vom 11. April 1542 darum, ihm unbedingt eine Kopie davon zu beschaffen:

> Caeterum quod ad librum Lusitanicum attinet, uidi quae scribis pro quibus tibi gratias ago maximas. Librum enim egomet apud Peutingerum uidi truncatim legi, qui re uera ei cum linguam nostram ignoret, minime usui est. Quare te iterum atque iterum oro ut uel librum ipsum mihi impetres, uel eiusdem exemplar unum; quod si feceris, historiae nostrae quam in manibus de rebus Indicis habemus, magnum adferes adiumentum, et nobis facies rem pro qua tibi perpetuo deuincti erimus (Góis *apud* Torres, 1982: 188).

In seiner Argumentation führt Damião de Góis an, dass die Sammlung für die portugiesische Historiografie von großer Bedeutung, für Peutinger aufgrund mangelnder portugiesischer Sprachkenntnisse allerdings nur von geringem Nutzen sei. Zwar kam Fugger der Bitte nach, wenn auch nur mit mäßigem Erfolg. Nach anfänglichem Zögern erlaubte er ihm zwar Einsicht in das Handschriftenkonvolut, untersagte ihm jedoch die Anfertigung einer Kopie (Hendrich, 2007: 216 f.). Góis' Bemühungen um die Handschriftensammlung unterstreichen die Relevanz und somit auch Fernandes' privilegierte Position bezüglich der Weitergabe von Informationen. Aufgrund seiner notariellen Funktion (*tabelião*) – nicht nur für die oberdeutschen Kaufleute, sondern auch für die portugiesische Krone – besaß Valentim Fernandes Zugang zu Nachrichten über die Vorgänge in Übersee. Wie er selbst im *Códice* vermerkt, sind die Berichte über die dort enthaltenen Reisen aus der ersten Dekade des 16. Jahrhunderts gemäß den Aussagen der jeweiligen Seefahrer aufgezeichnet: "escripta per mym Valentym Fernandez [....] de palavra de [...]" (Códice Valentim Fernandes, 1997: 19). Ähnlich formuliert es Fernandes auch in einem Schreiben vom 26. Juni 1510 an den befreundeten Kaufmann Stephan Gabler in Nürnberg, dass, nachdem die zurückgekehrten Kapitäne dem portugiesischen König Bericht erstattet hatten, er ebenfalls an den Auskünften teilhaben könne: "[...] sayn die pylotñ noch alle bay dem konig vnd habñ zu schaffñ dornoch werñ sie mir auch zu tayl" (GNMN, HistA, RSt Nürnberg XI, 1d, fol. 3). Exemplarisch sollen an folgender Textstelle aus besagtem Brief an

Stefan Gabler Konstellationen und Interessenslage in Bezug auf die Informationsübermittlung aufgezeigt werden, vor deren Hintergrund Valentim Fernandes als eine Art Dreh- und Angelpunkt des sozialen und kommunikativen Netzwerkes zwischen Portugal und Oberdeutschland fungierte:

> Ao muyto honrrado Sñor Steffan Gabler mercador alemã Em Nurenberga. [...] Frūtlichñ gruß mit willigen dinst eyn vor lieber Sñor cōpadre stevā gabler. vnd gute frūd Ich hoff zu got yr sayn frisch vnd gesūt daß gleichñ bin ich auch vnd alle die mayn / Item vlrich ehinger hat mir eynen briff gewißñ darin yr mich grussñ lassñ das danck ich euch zu tausent mal / auch schraibñ yr dass ich euch newe mer võ India schreibñ sol / So waiß ich nichts besūders den das das der ehinger seynen herrñ schreybt / doch ob es sich[er] wer euch die selbñ nicht bald wurdñ so habt hie mayn sch[reiben] vor gut / wie wol ich hab euch dißes Jar .2. briff gesch[rieben] eynen durch den ehinger vnd den anderñ durch calixto mit niclas großñ briff / vnd ist mir recht so hab ich euch mit Yeronimo holtschuer auch gesch[rieben] / vnd als ich versta so habt ir sie nicht enphangñ das mich ser verdrossñ hab./ [...] Ist mayn frūtlich bit yr mir eyn astrolabio schickñ das gut ist vnd etlich brieff die mā d[r]außen druckt võ Lisbonen dingñ vnd mir newer / mer wollñ schr[eiben] võ ytalia ungerñ pehm vnd polñ etc. vnd mich etwa kūth machñ mit aynem gelertñ mā dem vol say in der astronomay und kosmographay saynd daß ich maynen guten doctor Jeronīo vorlorñ hab /[...] vnd so will ich euch die kost võ India schickñ biß yn malacka vnd die mayl mit den aylandñ [...] / vnd ob ayn buchel vorhanden wer daß hayst petrus de elyaco in cosmog[ra]phia schickñ es mir / [...] got der almechtig der spar euch lang gesūt vnd grust mir alle gute gesellñ vnd herrñ wo yr sayn die mich kennen / vnd ewer göt ist frisch vnd gesūt mit der mutter got say gedanckt./ Ewer cōpadre vnd williger diener Valentyn Fernandez [...] Ich bit euch frūtlich / ich hab vorstandñ daß ma den ptholomeo hat corregirt vnd allen new gedruckt daß yr etwa eynen gelertñ astronomo wollñ fragñ wie dem ist / vnd mir beschayd schraiben etc. (GNMN, HistA, RSt Nürnberg XI, 1d, fls. 1, 3/3 v.).

Deutlich wird Fernandes' zentrale Position u. a. an der namentlichen Erwähnung einiger Repräsentanten der in Lissabon stationierten Augsburger und Nürnberger Handelshäuser, mit welchen er offenbar vor Ort in Kontakt stand.[7] In diesem Zusammenhang sind auch die logistischen Schwierigkeiten erkennbar, welche den – aus unserer heutigen Sicht verhältnismäßig mühsamen – Nachrichtenaustausch über eine Distanz von über 2000 km behinderten und wertvolles Dokumentenmaterial verloren gingen ließen. Was den Inhalt des Schreibens an Stephan Gabler betrifft, sind es in erster Linie geo- und handelspolitische Interessen, die Fernandes mittels der kolportierten Informationen bedient. Er liefert einen detail-

---

[7] Genannt sind Ulrich Ehinger (Höchstetter, Augsburg), Calixtus Schüler (Imhoff, Nürnberg) sowie Hieronymus Holzschuher, Vertreter des gleichnamigen Nürnberger Handelshauses (Hendrich, 2007: 233 f.).

lierten chronologischen Überblick der zwischen 1506 und 1510 stattgefundenen Indienexpeditionen und die im Zuge der militärischen Vorstöße erzielten territorialen Erfolge der Portugiesen. In Bezug auf die aus Indien eingetroffenen Waren gibt Fernandes zum Teil bemerkenswert genaue Auskünfte über Mengen- und Gewichtsangaben sowie Ladekapazitäten der Schiffe:

> Item a d[ie] 24 Junho [1510] Sctī Johīs ist ayn schiff auß India komen vō 400 tonel vō der armada des maretschal [= Fernando Coutinho] mit pip[er] vnd etlich klayn spetzaray suder [d. h. insbesondere] canel [= Zimt] [...] (GNMN, HistA, RSt Nürnberg XI, 1d, fol. 2 v.).

Zudem weist er bei der geografischen Bestimmung der von den jeweiligen Flottenverbänden angelaufenen Häfen auf die dort vorzufindenden Handelsgüter hin. Der Brief schließt mit einigen persönlichen Anliegen seitens Valentim Fernandes, anhand derer die bereits erwähnte "Kontinuität der Veröffentlichung alter Texte, einem hartnäckigen Fortleben inzwischen veralteter Beschreibungen der Welt" (Vivanti, 1993: 283) zum Ausdruck kommt. Neben einem aus Nürnberger Produktion stammenden Astrolabium gibt Fernandes in Auftrag, ihm ein Exemplar des *Tractatus de Imagine Mundi* des Pierre d'Ailly sowie eine überarbeitete Fassung der *Geographia* des Ptolemäus zu beschaffen.[8] Mit "maynen guten doctor Jeronĩo" (*op. cit.*, fol. 3) ist zweifellos Hieronymus Münzer, Nürnberger Arzt und Diplomat, gemeint, der 1494/95 zusammen mit einer Gruppe von Kaufleuten eine Reise über die Iberische Halbinsel unternommen hatte und dem Fernandes – wie er im *Códice* schildert – während dessen Aufenthalt am Hofe von D. João II. Dolmetscherdienste erwiesen hatte: "doctor Jeronimo Monetario alemão cuja lingoa eu era" (Códice Valentim Fernandes, 1997: 174).[9] Zwar existieren keine Dokumente aus der Korrespondenz zwischen Fernandes und Münzer, doch ist davon auszugehen, dass beide miteinander in Kontakt standen. Wie aus dem Schreiben an Stephan Gabler hervorgeht, sorgte sich Fernandes nach Münzers Tod (1508) um einen Nachfolger, mit dem er den Austausch auf wissenschaftlicher Ebene fortführen könne. Gabler scheint dem Anliegen nachgekommen zu sein, denn der folgende, wahrscheinlich auf 1511 zu datierende lateinische Brief

---

8   Die erste Ausgabe dieser aus der Feder des französischen Kardinals Pierre d'Ailly (Petrus de Alliaco) stammenden kosmografischen Abhandlung (1410), in der die wichtigsten klassischen Autoren redigiert sind, wurde 1483 in Leuwen gedruckt. Sie gehörte zu jenen Werken, auf denen die um 1457 von dem florentinischen Gelehrten Paolo Toscanelli aufgestellte Theorie der Möglichkeit eines westlichen Seeweges nach Asien basierte. (Andrade, 1972: 752 f.).

9   Vgl. zu Münzers Itinerarium Vasconcelos, 1931; Herbers, 2000.

aus Willibald Pirckheimers Feder ist zwar laut Editionsvermerk an einen ihm persönlich unbekannten Empfänger adressiert, doch handelt es sich dabei ohne jeglichen Zweifel um Valentim Fernandes (Reicke, 1956: 126–129).[10] Gabler wird von Pirckheimer explizit als derjenige gemeinsame Freund, *"amicus communis"* benannt, der ihn über die Nachrichten aus Indien, die *"rebus Indicis"* in Kenntnis setzte. Ebenso wird der von Fernandes betrauerte Hieronymus Münzer, an dessen Stelle Pirckheimer treten sollte, als hochgeschätzter Freund betitelt:

> Quamvis, quis sim, penitus ignores, ego quoque te nunquam viderim, has tamen ad te litteras dare cogor. Steffanus enim Gabler, amicus communis, quoniam me rebus Indicis admodum delectari novit, tuas mihi legendas tradidit litteras, ex quibus haut parvam accepi delectationem, non solum, quod in his multa inerant rerum curiosus scrutator, iam ante intellexeram – sed quod occasionem mihi oblatam esse cernebam nanciscendi amici, qualem pridem desideraram. Accesit praeterea, quod Hyeronimo (sic) Monetario, viro mihi quondam amicissimo, te charissimum esse intellexi. Hortatus me insuper est Steffanus Gabler, imo rogavit, ut quoniam desiderio tuo in coemendis libris desideratis minus satisfacere posset, tibi de his scriberem. His omnibus permotus, parvi lubens homini amico meque his litteris amicum tibi dedo, quo deinde ex arbitrio tuo uti potes. Scias igitur, Petrum de Elyaco in cosmo[graphiam] hic venalem non reperiri. Dabo tamen operam, ut desiderio tuo aliquando satisfiat. Ptolomeum noviter impressum esse aiunt, qui nondum ad manus meas pervenit nec ego id admodum curo, quoniam Graeco utor, quem Florenciae transcribi feci. Scio etenim, vix pure illum ab quoquam transferri posse etiam Graecis litteris doctissimo, [ni]si mathematicis disciplinis non solum imbutus, sed egregie quoque sit instructus. Novi enim plerosque, qui hanc provinciam agressi sunt, quibus haut feliciter temerarius cessit conatus. Est tamen in urbe nostra quidam, qui illum a Jo. de M. [Johanne de Monte sc. regio] translatum et manu propria descriptum possidet, quem huc usque perpellere non potui, ut illum ad comunem utilitatem quandoque in lucem prodire sineret. Spero tamen aliquando me auspice latebras invidiossisimi evadet hominis (Pirckheimer *apud* Reicke, 1956: 126 f.).

Es könnte somit, wenngleich sich dies aufgrund der dürftigen Quellenlage nicht mehr rekonstruieren lässt, entweder Stephan Gabler oder Willibald Pirckheimer gewesen sein, über wen die Informationen über das indische Panzernashorn, das 1514 als Geschenk des Herrschers von Cambaia an D. Manuel I. nach Lissabon verschifft worden war, an Albrecht Dürer gelangten, der 1515 eine Federzeich-

---

10 Leider ist kein Datum ersichtlich, aber aufgrund des Papierbogens sowie derselben Tinte, die Pirckheimer auch in einem an einen seiner Neffen gerichteten Schreiben verwendete, lässt sich der Brief an Fernandes auf 1511 bzw. 1512 datieren. Für 1511 spricht ferner der Umstand, dass in jenem Jahr in Venedig eine neue Auflage der von Pirckheimer erwähnten *Geographia* des Ptolemäus erschien (Reicke, 1956: 128 f.; Hendrich, 2007: 240).

nung (Abb. 1) sowie den berühmten *"Rhinocerus"*-Holzschnitt des Dickhäuters fertigte (Abb. 2). Um sich in der bereits bei Plinius und Strabon geschilderten Feindschaft zwischen Elefant und Nashorn zu überzeugen, ließ D. Manuel I. einen Kampf zwischen beiden Tieren veranstalten.[11] Aller Wahrscheinlichkeit nach hatte Valentim Fernandes dem Spektakel beigewohnt und eine nicht mehr existente Skizze angefertigt, um seine Korrespondenzpartner in Oberdeutschland über das fremdartige Tier zu informieren. Sowohl Dürers Federzeichnung als auch der Holzschnitt sind jeweils mit deskriptiven Begleittexten der Charakteristika des Nashorns versehen, die aufgrund bestimmter Abweichungen Aufschluss über die mögliche Provenienz geben könnten (Hendrich, 2007: 264–270):

Abb. 1: Albrecht Dürers Federzeichnung "Rhinoceron", Nürnberg, 1515. The Trustees of the British Museum, London, Inv.-Nr. 5218-161; Hendrich: 2007, 266.

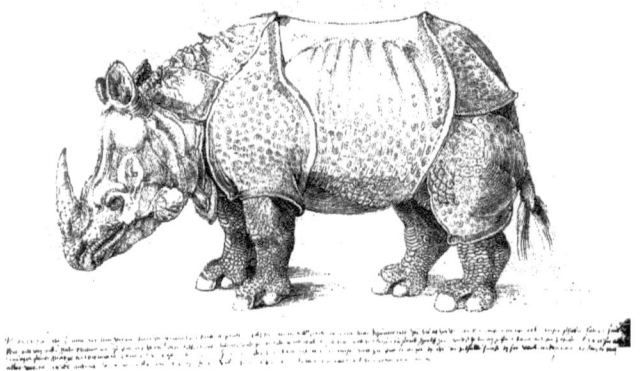

It im 153 [1513][12] jor adi may hat unserm küng von portigall gen lisboa procht ein solch lebendig tir aws India das nent man Rhynocerate. Das hab ich dir van wunders wegen

---

11 Auch Damião de Góis wurde Zeuge des Duells zwischen Elefant und Nashorn: "Elephantes quoq; quinq; vel fex tempore felicis Regis Emanuelis indeallati funt quorum dm puer ipfius prudentissimi Regis ab intimis cubiculis essem treis simul vidi, atque Rhinocerotem vnum qui omnes Regem ipsum equitantem praecedebant. Certamini etiam unius istorum Elephantum cum Rhinocerote interfui spectaculum sane admiratione dignum, in quo elephas succubuit. Quem ludum Rex ipse felicissimus Emanuel Vlyssipone praebuit anno (si bene memini) MDXV vel XVI" (Góis, 1542: 58 f.).
12 Vermutlich ist Albrecht Dürer bezüglich der Jahresangabe ein Lesefehler unterlaufen, denn sowohl bei der Zeichnung als auch beim Holzschnitt gibt er in der jeweiligen Beschreibung statt 1515 irrtümlicherweise 1513 als dasjenige Jahr an, in dem das Nashorn nach Lissabon gelangte.

müsen abkunterfet schickn hat ein farb wie ein krot und van dickn schaln überlegt fast fest und ist in dr gros als ein hellfant aber nydrer und ist des helfantz tott feint es hat forn awff der nasen ein starck scharbff horn und so dz tir an helfant kumt mit im zw fechten so hat es for albeg sein horn an den steinen scharbff geweszt und lawft dem helfant mit dem kopff zwischn dy fordern pein dan reist er den helfant awff wo er am dünsten hawt hat und erwürgt in also der helfant fürcht in ser übell den Rhynocerate dan er erwürgt in albeg wo er den helfant ankumt dan ist er woll gewapent und ser freidig und behent dz tir würt Rhinocero in greco et latino indico vero gomda [genannt] (Dürer apud, Costa, 1937: 19 f.).

Abb. 2: Albrecht Dürers Holzschnitt "Rhinocerus", Nürnberg, 1515. Sammlung-Otto-Schäfer-II, Schweinfurt, Inv.-Nr. D-273; Hendrich: 2007, 267.

Nach Christus gepurt. 1513 [sic].[13] Jar. Adi.j.May. Hat man dem großmechtigen Kunig von Portugall Emanuell gen Lysbona pracht auß Jndia/ ein sollich lebendig Thier. Das nennen sie Rhinocerus. Das ist hye mit aller seiner gestalt Abconderfet. Es hat ein farb wie ein gespreckelte Schildtkrot. Vnd ist võ dicken Schalen vberlegt fast fest. Vnd ist in der größ als der Helfandt Aber nydrertrechtiger von paynen/vnd fast werhafftig. Es hat ein scharff starck Horn vorn auff der nasen/ Das begyndt es albeg zu wetzen wo es bay staynen ist. Das dosig Tier ist des Helffandtz todt feynd. Der Helffant furcht es vbel/ dann wo es Jn ankumbt/ so laufft Jm das Thier mit dem kopff zwischen dye fordern payn/ vnd reyst den Helffandt vnden am pauch auff vñ erwürgt Jn/ des mag er sich nit erwern. Dann das Thier ist also gewapent/ das Jm der Helffant nichts kann thun. Sie sagen auch das der Rhynocerus Schnell/ Fraydig vnd Listig sey (GNMN, HistA, Inv.-Nr. H 5582, Kapsel 15a).

Beim Vergleich der beiden Textpassagen fällt auf, dass Albrecht Dürer bei der zur Federzeichnung gehörenden Beschreibung des Nashorns Valentim Fernandes' ursprünglichen Wortlaut übernommen zu haben scheint, worauf sowohl

---

13 Es müsste 1515 statt 1513 heißen. Vgl. Anm. 12.

die Verwendung des Possessivpronomens in der Ersten Person Plural als auch der deiktische Verweis, bei dem sich der Absender direkt an seinen Empfänger wendet, hindeuten: "[...] hat unserm küng von portigall gen lisboa procht ein solch lebendig tir aws India das nent man Rhynocerate. Das hab ich dir van wunders wegen müsen abkunterfet schickn [...]" (Dürer *apud* Costa, 1937: 19). Zudem ist neben dem griechischen Namen des Dickhäuters auch die aus dem Bengali bzw. Hindi (*"gomda, ganda"*) stammende Bezeichnung angeführt. Die dem Holzschnitt beigefügte Schilderung dagegen wurde, wohl von Dürer selbst, entsprechend modifiziert und angepasst: "[...] dem großmechtigen kunig von portugall [...] das ist hye mit aller seiner gestalt Abconderfet" (Dürer *apud*, Costa, 1937: 24 f.; Ehrhardt, 1989: 35). Trotz fehlender Quellen sind die Informationen über das indische Panzernashorn mit ziemlicher Sicherheit Valentim Fernandes zu verdanken, worauf die beiden folgenden Schreiben vermuten lassen. Dabei handelt es sich einerseits um ein nur noch in italienischer Abschrift erhaltenes, an Nürnberger Kaufleute adressiertes Schriftstück (*Lettera scripta da Valentino Moravia germano a li mercatanti di Nurimberg*) von Juni oder Juli 1515 (Gubernatis, 1875: 389–392)[14], das neben einer geografischen Darstellung vom Golf von Aden bis zum Chinesischen Meer sowie einer Nachricht über die Errichtung eines portugiesischen Forts bei Sofala eine Beschreibung des im Mai 1515 in Lissabon eingetroffenen indischen Nashorns enthält:

> Cariss. fratello. Nelli dì. 20. di questo mese di magio. 1515. giunse qui in Lisbona cità Nobilissima di tuta la Lusitania emporio al presente Excell. uno animale chiamato da greci Rhynoceros et dalli Indi Ganda mandato dal re potentissimo de India della Cità di Combaia a donare a questo serenissimo Emanuel Re di portogallo, il quale animale al tempo de Romani Pompeo Magno ne suoi guochi come dice Plinio fu mostrato nel circo con altri diversi animali [...] (Gubernatis, 1875: 389).

Zum anderen existiert ein lateinischer Brief mit dem Titel "*Descriptio Indiae*" aus dem Nachlass Peutingers, dem zu entnehmen ist, dass Valentim Fernandes auch seinen Korrespondenzpartner in Augsburg über den in Lissabon angekommenen Dickhäuter in Kenntnis gesetzt hat, da besagtes Schreiben nicht nur vergleichbare geografische Informationen über Südasien, sondern ebenfalls eine Deskription des Rhinozeros beinhaltet:

> Weitgehende Übereinstimmung besteht zwischen diesem merkwürdigen Schriftstück und einem von Valentim Moravus anscheinend an Nürnberger Kaufleute gerichteten

---

14 Die Handschrift befindet sich in der *Biblioteca Nazionale Centrale* in Florenz (Codice Strozziano 20, Ora C1 - XIII 80). Eine portugiesische Übersetzung findet sich bei Costa; 1937: 29–36.

Schreiben, das später nach Florenz gelangte und in italienischer Übersetzung vorliegt. Möglicherweise hat hier der Absender wie bei der Mitteilung des Inschriftenfundes zugleich Augsburg und Nürnberg mit seinen Neuigkeiten bedacht (Lutz, 1958: 56; Pohle, 2000: 226 f.).

An folgender Passage aus dem Brief an Konrad Peutinger wird abermals deutlich, wie innerhalb des Gelehrtendiskurses zu Beginn des 16. Jahrhunderts die Einpassung der "Neuen Welten" in das vom Kenntnisstand der antiken Autoren dominierten Weltbild von statten ging:

> [...] De Rinoceronte vide Plin. et Strabonem libro XVI ante finem, portatus est Ulixbonam de anno 1515 die XX mensis aii quem misit Campaye rex potentissimus regi Emanueli Portugallensi et appellatur animal hoc lingua Indica Ganga [...] (Peutinger *apud* Lutz, 1958: 56).

Aus europäischer Sicht erfolgte neben der kulturell-materiellen Aneignung nicht zuletzt auch eine ikonografische Konstruktion der "Neuen Welten" durch die künstlerische Transformation, wofür metonymisch der Dürer'sche *"Rhinocerus"*-Holzschnitt steht.[15] Der Rezeption der Exotika ist eine dichotome Relation von Eigenem und Fremden, von Identität und Alterität inhärent, bei der eigene Wertparadigmen als Bewertungskriterien angelegt wurden und die nachhaltig die stereotypisierten Vorstellungen in der europäischen kollektiven Wahrnehmung prägen sollten. Motiviert durch wirtschaftliche und kommerzielle Gründe, durch wissenschaftliches Interesse und Neugier, waren es Kaufleute, Gelehrte und Buchdrucker wie Valentim Fernandes, die durch ihre Beteiligung an Kulturtransfer und Wissenszirkulation über die Geschehnisse in Übersee zum Wandel tradierter Denkschemata und somit im wahrsten Sinne des Wortes zur Horizonterweiterung beitrugen.

Quellen- und Literaturverzeichnis

Andrade, António Alberto Banha de. 1972. *Mundos Novos do Mundo. Panorama da difusão, pela Europa, de notícias dos Descobrimentos Geográficos Portugueses*. 2 vols., Lisboa: Junta de Investigações do Ultramar.

Anselmo, António. 1987. "Valentim Fernandes ou a mediação na alteridade". *Revista da Biblioteca Nacional*, 2.ª série, 2/2, 27–32.

Barbas, Helena. 2000. "Monstros: o rinoceronte e o elefante: Da ficção dos bestiários à realidade testemunhal". In: Siepmann, Helmut (Hrsg.). *Portugal, Indien*

---

15 "Se o elefante era um portento, o rinoceronte ultrapassa-o em raridade e maravilha. A notícia do evento espalha-se pela Europa – de Lisboa chega uma das suas descrições à Alemanha atribuída à pena de Valentim Fernandes" (Barbas, 2000: 108).

und Deutschland: Akten der V Deutsch-Portugiesischen Arbeitsgespräche, Köln, 1998 (Portugal, Índia e Alemanha: Actas do V Encontro Luso-alemão, Colónia, 1998). Köln: Zentrum Portugiesischsprachige Welt, Universität zu Köln, Lisboa: Centro de Estudos Históricos, Universidade Nova de Lisboa, 103–122.

Bayerische Staatsbibliothek (BSB), München
- Cod. hispanicus 27.

Biblioteca Nacional de Portugal (BNP), Lisboa
- Manuscritos Iluminados [IL], 154.
- Reservados 431 V (*O livro de Marco Paulo*).

Biblioteca Nazionale Centrale, Firenze
- Codice Strozziano 20, Ora C1 – XIII 80.

Brásio, António. 1960. "Uma carta inédita de Valentim Fernandes". *Boletim da Biblioteca da Universidade de Coimbra* 24, 338–358.

Códice Valentim Fernandes. 1997. Oferecido pelo académico titular fundador Joaquim Bensaúde. Leitura paleográfica, notas e índice pelo académico de número José Pereira da Costa. Lisboa: Academia Portuguesa da História.

Costa, Abel Fontoura da. 1937. *Deambulações da Ganda de Modofar, rei de Cambaia de 1514 a 1516*. Lisboa: Divisão de Publicações e Biblioteca, Agência Geral das Colónias.

Daston, Lorraine. 1994. „Neugierde als Empfindung und Epistemologie in der frühmodernen Wissenschaft". In: Grote, Andreas (Hrsg.). *Macrocosmos in microcosmo. Die Welt in der Stube. Zur Geschichte des Sammelns 1450 bis 1800*. Opladen: Leske + Budrich, 35–39.

Ehrhardt, Marion. 1989. *A Alemanha e os Descobrimentos Portugueses*. Lisboa: Texto Editora.

Erhard, Andreas und Ramminger, Eva. 1998. *Die Meerfahrt: Balthasar Springers Reise zur Pfefferküste*. Mit einem Faksimile des Buches von 1509 (Die Merfart vñ erfarung nüwer Schiffung vnd Wege zu viln onerkanten Inseln vnd Künigreichen). Innsbruck: Haymon.

Geldner, Ferdinand. 1970. *Die deutschen Inkunabeldrucker. Ein Handbuch der deutschen Buchdrucker des 15. Jahrhunderts nach Druckorten*, Bd. 2: Die fremden Sprachgebiete, Stuttgart: Hiersemann.

Germanisches Nationalmuseum – Historisches Archiv (GNMN, HistA), Nürnberg
- RSt Nürnberg XI, 1d.
- Inv. Nr. H 5582, Kapsel 15a.

Góis, Damião de. 1542. *Hispania Damiani a Goes equitis Lusitani*. Lovanii: Excudebat Rutgerus Rescius.

Gubernatis, Angelo de. 1875. *Storia dei viaggiatori italiani nelle Indie Orientali*. Livorno: Vigo.

Hase, Oskar von. 1885. *Die Koberger. Eine Darstellung des buchhändlerischen Geschäftsbetriebes in der Zeit des Überganges vom Mittelalter zur Neuzeit*. 2., neubearb. Aufl., Leipzig: Breitkopf & Härtel.

Hendrich, Yvonne. 2007. *Valentim Fernandes – Ein deutscher Buchdrucker in Portugal um die Wende vom 15. zum 16. Jahrhundert und sein Umkreis* (Mainzer Studien zur Neueren Geschichte 21). Frankfurt am Main: Peter Lang.

Herbers, Klaus. 2000. „ 'Murcia ist so groß wie Nürnberg' – Nürnberg und Nürnberger auf der Iberischen Halbinsel: Eindrücke und Wechselbeziehungen". In: Neuhaus, Helmut (Hrsg.). *Nürnberg. Eine europäische Stadt in Mittelalter und Neuzeit* (Nürnberger Forschungen 29). Nürnberg: Verein für Geschichte der Stadt Nürnberg, 151–183.

Horst, Thomas. 2009. "The voyage of the Bavarian explorer Balthasar Sprenger to India (1505/1506) at the turning point between the Middle Ages and the Early Modern Times: his travelogue and the contemporary cartography as historical sources". In: Billion, Philipp u. a. (Hrsg). *Weltbilder im Mittelalter – Perceptions in the World of the Middle Ages*. Bonn: Bernstein-Verlag, 167–197.

König, Erich (Hrsg.). 1923. *Konrad Peutingers Briefwechsel* (Veröffentlichungen der Kommission für Erforschung der Geschichte der Reformation und Gegenreformation. Humanistenbriefe 1). München: Beck.

Lopes, Marília dos Santos und Hanenberg, Peter. 1994. „Das antike Erbe und die Welt der Entdeckungen". In: Füssel, Stephan u. a. (Hrsg.). *Artibvs. Kulturwissenschaft und deutsche Philologie des Mittelalters und der frühen Neuzeit. Festschrift für Dieter Wuttke zum 65. Geburtstag*. Wiesbaden: Harrassowitz, 151–166.

Lutz, Heinrich (Hrsg.). 1958. *Conrad Peutinger. Beiträge zu einer politischen Biographie* (Abh. zur Geschichte der Stadt Augsburg 9). Augsburg: Verlag Die Brigg.

O manuscrito "Valentim Fernandes". 1940. Oferecido à Academia por Joaquim Bensaúde. Leitura e rev. das provas de António Baião. Lisboa: Academia Portuguesa da História.

Pohle, Jürgen. 2000. *Deutschland und die überseeische Expansion Portugals im 15. und 16. Jahrhundert* (Historia profana et ecclesiastica 2). Münster: Lit.

Reicke, Emil (Hrsg.). 1956. *Willibald Pirckheimers Briefwechsel*. Bd. 2 (Veröffentlichungen der Kommission für Erforschung der Geschichte der Reformation und Gegenreformation. Humanistenbriefe 5). München: Beck.

Reinhard, Wolfgang. 1993. „Einführung". In: Ders./Prosperi, Adriano (Hrsg.). *Die Neue Welt im Bewußtsein der Italiener und Deutschen des 16. Jahrhunderts.* Berlin: Duncker & Humblot, 7–14.

Roque, Mário da Costa. 1979. *As pestes medievais europeias e o "Regimento proueytoso contra ha pestenença": Lisboa, Valentim Fernandes (1495–1496). Tentativa de interpretação à luz dos conhecimentos pestológicos actuais.* Paris: Fundação Calouste Gulbenkian.

Sadji, Uta (Hrsg.). 1983. *Entdeckungsreisen nach Indien und Amerika: d. Druck der dt. Übers. von 1508: Newe unbekanthe landte und ein newe weldte in kurtz verganger zeythe erfunden*, Bd. III–VI (Litterae: Göppinger Beiträge zur Textgeschichte 83). Göppingen: Kümmerle.

Torres, Amadeu (ed.). 1982. *Noese e crise na epistolografia latina Goisiana. Vol. 1: As cartas latinas de Damião de Góis.* Introdução, texto crítico e versão. Prefácio de José V. de Pina Martins. Paris: Fundação Calouste Gulbenkian, Centro Cultural Português.

Vasconcelos, Basílio de. 1931. *"Itinerário" do Dr. Jerónimo Münzer: Excertos.* Coimbra: Imprensa da Universidade.

Vivanti, Corrado. 1993. „Die Humanisten und die geographischen Entdeckungen". In: Reinhard, Wolfgang/Prosperi, Adriano (Hrsg.). *Die Neue Welt im Bewußtsein der Italiener und Deutschen des 16. Jahrhunderts.* Berlin: Duncker & Humblot, 273–290.

Vogel, Klaus A. 1995. „ 'America': Begriff, geographische Konzeption und frühe Entdeckungsgeschichte in der Perspektive der deutschen Humanisten". In: Kohut, Karl (Hrsg.). *Von der Weltkarte zum Kuriositätenkabinett: Amerika im deutschen Humanismus und Barock.* Frankfurt am Main: Vervuert, 11–43.

Werner, Theodor Gustav. 1975. „Das kaufmännische Nachrichtenwesen im späten Mittelalter und in der frühen Neuzeit und sein Einfluß auf die Entstehung der handschriftlichen Zeitung". *Scripta Mercaturae* 2, 3–52.

Wuttke, Dieter. 1992. „Humanismus in den deutschsprachigen Ländern und Entdeckungsgeschichte 1493–1534". In: Füssel, Stephan (Hrsg.). *Die Folgen der Entdeckungsreisen für Europa. Akten des interdisziplinären Symposions vom 12./13. April 1991 in Nürnberg* (Pirckheimer-Jahrbuch 7). Nürnberg: Carl, 9–52.

Gabriele Kaiser

Leonhard Thurneysser zum Thurn (1531–1596) und sein Nachlass in der Staatsbibliothek zu Berlin[*]

Abstract: Leonhard Thurneysser zum Thurn (1531–1596) was a real Renaissance-man who excelled in various fields like chemistry, metallurgy, botany, mathematics, astronomy and medicine. He was the personal doctor of the Brandenburg Elector alchemist, pharmacist, astrologer, writer and printer. He was born in Basel in 1531, worked as a goldsmith and mercenary as well as in a mine, travelled to Scotland, England and Portugal, lived in Tyrol, Münster and since 1570 in Brandenburg. There he had the most successful time of his life. Thurneysser first met Johann Georg, the Elector of Brandenburg (1571–1598) in Frankfurt/Oder where he succeeded in healing the Elector's wife. Therefore he was appointed to be the prince's personal doctor with a salary of 1352 Taler and invited to work at the court in Berlin. Thurneysser was even granted to use for his work a part of former Franciscan monastery, called on "Graues Kloster". For Thurneysser this monastery became his residence, his library, his printing shop (the first of Berlin!) and his laboratory – where he produced medicine which made him a rich man. He sold astrological calendars, horoscopes and talisman for the ordinary people and printed books with different alphabets, like Hebrew, Arabic, Syriac, Ethiopian and Armenian. Besides that, he ran the first scientific cabinet of the Brandenburg, a botanic garden and a little zoo with exotic animals. A journey to his hometown Basel in 1579 changed his life another time. He married his third wife and transferred his possessions to Basel. But only a year later, in 1580, Thurneysser returned to Berlin – disappointed and robbed by his wife to whom he lost all his properties. In 1584 his last journey began: he left Berlin, converted to Catholicism, lived in Rome and died in 1596. The Berlin State Library (Staatsbibliothek zu Berlin) holds the legacy of this adventurous and impressing man – more than 20 volumes comprising hundreds of letters, manuscripts and drafts. Among them is the description of Portugal and its nature, important for this workshop. Apart from his papers we own books from his personal library and mostly all the books printed by him.

---

[*] Der vorliegende Beitrag stellt eine geringfügig überarbeite Version des von der Autorin auf dem Workshop in Lissabon gehaltenen Vortrags zum Thurneysser-Nachlass in der Staatsbibliothek zu Berlin dar. Für weiterführende Hinweise sei jedoch auf Ihre zahlreichen Publikationen zu Thurneysser (insbesondere: Spitzer, 1996 mit zahlreichen Quellennachweisen) verwiesen.

„Thurneysser ist eyn WUNDERMAN, von Gott erwecktt groß Dingk zu thun, Sein Kunst und groß Geschicklichkeytt gehtt vber die Vernunftt gar weit."[1] Diese mehrseitige Lobeshymne von Simon Rotter (1524–1595), dem Bürgermeister der Stadt Brandenburg, welche Thurneysser, eitel wie er war, später auch gleich in seinen eigenen Büchern abdruckte, macht deutlich, dass der weitgereiste und geschäftstüchtige Leibarzt, Astrologe, Alchimist, Metallurge und Drucker Leonhard Thurneysser zum Thurn (1531–1596), dessen umfangreiche Biographie[2] im Rahmen dieses Beitrags hier nicht vorgestellt werden kann, eine höchst widersprüchliche und zugleich geheimnisvolle Person war. So vielfältig wie die Schreibweise seines Namens, so vielfältig sind die Geschichten über ihn.[3] In der Überlieferung wird er als ein Teufelsknecht, Scharlatan, Goldmacher und Wucherer bezeichnet. Für uns ist er eine jener grandiosen Gestalten der Spätrenaissance, deren Wirken unendlich viele Facetten zeigt und manche Überraschung bereithält. Im Folgenden soll diese außerordentliche historische Persönlichkeit in fünf Punkten vorgestellt werden.

Abbildung 1 und 2: „Thurneysser ist ein WUNDERMAN". Handschriftliches Manuskript aus dem Nachlass Thurneyssers (Staatsbibliothek zu Berlin, Ms. Bor. Fol. 685, Bl. 118 r), gedruckt in der „Confirmatio Concertationis" (Thurneysser, 1576: 2–4).

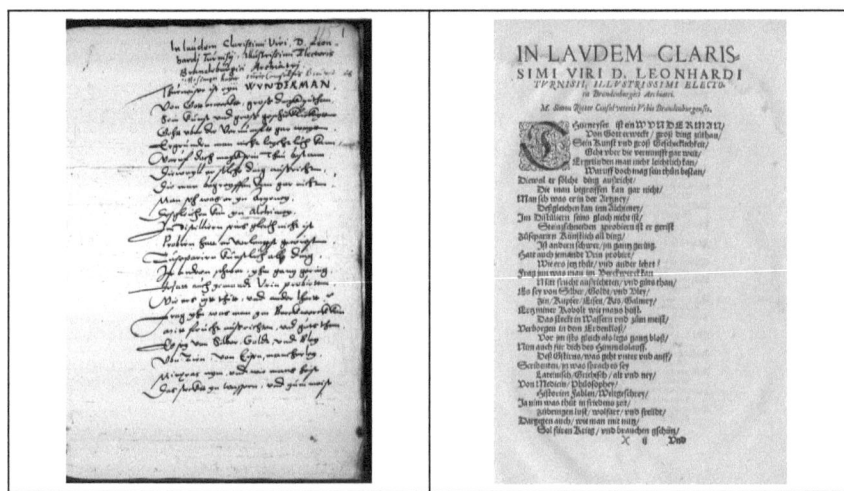

---

1 Staatsbibliothek zu Berlin, Ms. Bor. Fol. 685, Bl. 118 r–120 r.
2 Hierzu sei nachdrücklich die Lektüre von Schumacher, 2011 empfohlen.
3 Vgl. Eikermann/Kaiser, 2012 a.

1. Thurneysser – ein Basler Goldschmiedssohn zieht in die Welt

Thurneysser wurde 1531 in Basel geboren, sein Vater Jakob († 1560) war Goldschmied und auch Leonhard muss zunächst eine Goldschmied-Lehre beginnen. Er treibt sich in der Stadt herum, mit Wissensdurst, Entdeckerdrang und Neugierde. Als Famulus bei dem Arzt Johannes Huber (1507–1571) hilft er bei der Zubereitung von Medikamenten und sieht die berühmt gewordene erste Anatomie eines Leichnams von Andreas Vesal (1514–1564).

Er gerät an die falschen Freunde, ist wild und uneinsichtig und verlässt fluchtartig Basel, um sich vor Gläubigern zu retten. Seine Schulden wollte er mit einem vermeintlichen Goldbarren aus Blei bezahlen. Die Staatsbibliothek besitzt einen Brief von Jonas Jud zu Wyl, der am „denstag nach Johannis Im 1559 Jor" aufzählt, „was du mir auf deine Handschrift schuldig bist: 7 Taler, 41 Gulden, 51 ½ Kronen in Gold und 28 einfache Kronen"[4]. Das belastende Dokument verzeichnet eine lange Liste seiner Verbindlichkeiten.

Es folgen Lehr- und Wanderjahre in England, Frankreich und Deutschland, er kämpft als Söldner in der Schlacht bei Sievershausen 1553 und arbeitet als Goldschmied und Metallurge in verschiedenen Bergwerken.

Um 1555 könnten zwei junge Männer aus Basel Lissabon erreicht haben. Ich glaube, Adam Seidel war immer dabei. Der Freund aus Basler Kinderzeit ist Schreiber und Diener, ein Freund der Familie, auch sein Sohn Daniel wird später als Holzschneider für Thurneysser arbeiten. Ich sehe seine Schreiber-Hand in der handschriftlich überlieferten Naturbeschreibung Portugals, dem „Verzeichnis Vnnd beschreibung Etzlicher Kreutter, Stauden vnnd früchten, so Fürnemlich Inn Lusitania befunden, bey Vnns aber nicht viel oder gar wenig gesehenn worden."[5]

Adam Seidel war dabei, in Tirol, in Russland, in Straßburg oder wieder zu Hause in Basel. Er sorgte dafür, dass die Thurneysserschen Papiere mit dem Pferdewagen nach Münster in Westfalen gelangten, wo Thurneysser 1569/1570 kurzzeitig als Apothekerarzt für den Bischof Johannes III. von Hoya (1529–1574) arbeitete. Seidel begleitete Thurneysser nach Brandenburg, nach Frankfurt an der Oder und später nach Berlin. Er war der Bote in guten und schlechten Zeiten. Im Alter von 40 Jahren hat es Thurneysser schließlich geschafft: er tritt eine honorige Stelle an.

---

4 Staatsbibliothek zu Berlin, Ms. Bor. Fol. 686, Bl. 140 r–141 r.
5 Staatsbibliothek zu Berlin, Ms. Germ. Fol. 97.

2. Thurneysser in der Mark Brandenburg und in Berlin

Thurneysser reiste 1570 in die Mark Brandenburg und ließ in der weit bekannten Druckerei von Johann Eichorn (1524–1583) in der Universitätstadt Frankfurt an der Oder seine beiden Bücher *Praeoccupatio, ein Buch über Harnproben* (Thurneysser, 1571) und *Pison* (ein Buch über die Flüsse Europas, Thurneysser, 1572), drucken. Diese Bücher waren vom Feinsten, es wurden prächtige Foliobände, mit seinen Porträts und großen Holzschnitten ausgestattet.

Zwei der von Thurneysser im *Pison* beschriebenen Flüsse fließen durch Berlin: die Havel und die im Kapitel 79 beschriebene Spree. Das Wasser der Havel „helt faul wasser, / dauon etliche weiber die es trincken / gar böse / scharpffe und lügenhafftige zungen vnerkommen, …, und dis wasser Sprew ist grünferbig und lauter / es führet seinem Schlich Gold und eine schöne Glasur!"[6]. Diese Aussage sorgt für Aufregung. Als „Goldmacher" hat er sich zwar nie ausgegeben, doch verstand er es bestens, sich in Szene zu setzen. Der Kurfürst Johann Georg von Brandenburg (1525–1598) hörte von dem weitgereisten Besucher beim Drucker Eichorn und ließ sich dessen Werke zeigen. Der berühmte Zufall sorgt dafür, dass die stets kranke Gattin des Kurfürsten, Sabina von Brandenburg-Ansbach (1529–1575) nach einer Thurneysserschen Kur rasch gesundete.

Die Hoffnung auf Gold für die verschuldeten Staatsfinanzen bescherten Thurneysser den Posten seines Lebens: Er wurde Leibarzt des Kurfürsten von Brandenburg und erhielt für seinen Haushalt ein leerstehendes Kloster, das Graue Kloster in Berlin, zudem das Alchimistenlabor im Keller des Schlosses zur Benutzung und ein exorbitantes Gehalt von 1352 Talern.

Die Residenzstadt Berlin war damals ein „Dorf" mit knapp 10.000 Einwohnern. Eine faustische Gestalt wie Thurneysser belebte das provinzielle Berlin mehr als ihm lieb war: Der gelernte Goldschmid, der nie eine Universität besucht hatte, residierte im Grauen Kloster mit einem regelrechten Hofstaat von Angestellten und Dienern. Er lässt Medikamente und Tinkturen herstellen und verkaufen, obwohl er kein studierter Apotheker ist. Er richtet sich eine Druckerei, Setzerei und Schriftgießerei ein, um die eigenen Schriften zu verlegen und beschäftigt sich mit der Beobachtung der Sterne auf dem Dach seines Hauses.

Natürlich verfügt Thurneysser auch über eine Wunderkammer und einen Garten mit fremden Pflanzen; schwarze Doggen, einen Elch, Kutschen und Pferde. Sein Lebensstil ist glamouröser als jener des Kurfürsten. Erinnert sei insbesondere an die berühmt gewordenen seidenen Strümpfe, die Thurneysser jeden Tag trug, der Kurfürst aber nur sonntags.

---

6 Thurneysser, 1572: 363 (Havel) und 355 (Spree); vgl. Eikermann/Kaiser, 2012 b.

Thurneyssers Frau und Kinder werden nach Berlin geholt, dazu erscheint auch der unsägliche Bruder Alexander (1521–1601), ein Trinker und Betrüger. An die 200 Personen gehören zum Betrieb im Kloster, womit Thurneysser der erste Unternehmer in Berlin war.

Und es kam, was kommen musste: ein Verhältnis mit der Haushälterin, ein uneheliches Kind, Streit, Krankheit, Tod und die Pest 1576 in Berlin und sehr viele Neider.[7] Für den bisherigen und erst recht für seinen nun folgenden Lebenswandel wird er von der historischen Yellowpress abgestraft. Thurneysser wird als Goldmacher und Betrüger verschrien, und sorgt mit seinen cholerischen, derben Streitschriften selbst dafür, dass ein Vorurteil zum Urteil wird.

Hier setzt das Heimweh nach Basel ein, er nimmt Urlaub und trägt sich mit dem Gedanken der Rückkehr. Erste Wagen mit Gut und Geld, Kunstwerken, Büchern und seinen Handschriften werden nach Basel geschickt.

3. Thurneysser und seine Handschriften

Die Staatsbibliothek besitzt annähernd 10.000 Seiten seiner Hinterlassenschaft und dies ist nur ein kleiner Teil seines Erbes; es sind wohl die Materialien, welche in Berlin geblieben sind.

Diese Handschriften gehören zum Gründungsbestand der Königlichen Bibliothek (gegründet 1661), die im sogenannten Apothekerflügel des Berliner Schlosses untergebracht war. Die Briefbände tragen das Exlibris von Hofapotheker Michael Aschenbrenner (1549–1605), der bei Thurneysser in die Lehre gegangen war. Weitere Bände kamen durch Johann Carl Wilhelm Moehsen (1722–1795) in die Bibliothek; dieser war Leibarzt vom Preußen-König Friedrich II. („der Große", 1712–1786) und der erste Biograph Thurneyssers.

Es haben sich zwei Arten von Dokumenten zu Thurneysser erhalten:

a) Briefe an Thurneysser

Diese sind für die Jahre von 1564 bis 1583 überliefert (fast alle mit einem Namensregister versehen) und wurden jahrgangsweise gebunden. Diese Bände gehören zum Bestand der „Manuscripta Germanica"[8]. Die Familienbriefe wurden aus diesem Bestand später herausgelöst und in der Signaturengruppe „Manuscripta Borussica" extra verzeichnet[9].

---

7 Vgl. Spitzer, 1996.
8 Staatsbibliothek zu Berlin, Ms. Germ. Fol. 420–426.
9 Staatsbibliothek zu Berlin, Ms. Bor. Fol. 680–687 und 691.

Für Thurneysser arbeiteten mehrere Schreiber. Auch jetzt können wir sagen: Adam Seidel war immer dabei, er legte diese Register an, schrieb unzählige Briefe, Briefentwürfe und Manuskriptseiten. Pro Band sind das etwa 300–400 Seiten. Lücken bei den alten Blattzählungen in roter Schrift weisen darauf hin, dass viele Briefe entfernt wurden. Bereits Thurneysser beklagte sich, dass ihm Briefe von prominenten Personen wie Königen, Fürsten und Herzögen gestohlen wurden. In den erhaltenen Bänden finden sich Grafen, Geschäftsleute, Gelehrte, Händler und Künstler als Absender, aber auch alle Gewerke des Druckwesens. Glasmacher, Goldschmiede und Boten, Messeagenten, Papiermühlenbesitzer, Hofangestellte und Stadtprominenz, Ärzte und viele Kranke, Bergwerksangestellte, Vorarbeiter und Sammler, Kurfürstinnen und Kindermädchen sind hier vertreten (insgesamt rund 500 Personen!). Es ist kein Gelehrtenbriefwechsel zwischen Akademikern, es sind keine theologischen Disputationen, keine Ärztebriefe. Der Wert besteht in der Tatsache, dass diese Briefe ein breites Bild jener Zeit liefern: Alltägliches, Geschäfte, *Newe Zeitung*, Rezepte, Patientenbriefe mit Krankengeschichten, Politik und Neuigkeiten, Kunst und Geld, Haushaltsführung und Scheidungsgeschichten.

Abbildung 3: Brief des Hans Harttman Hyrus an Thurneysser, geschrieben in Lissabon am 1. Januar 1582 (Staatsbibliothek zu Berlin, Ms. Bor. fol. 685, Bl. 48 r).

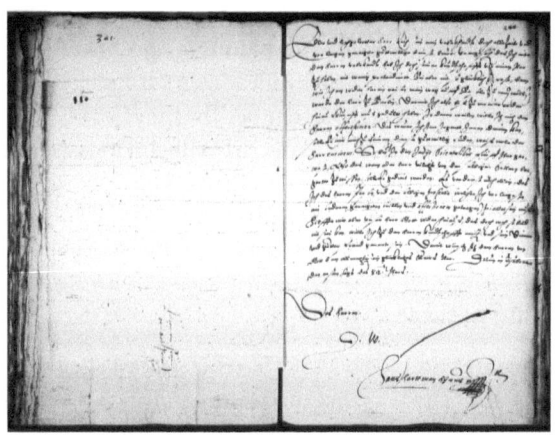

Auch hier hat der Zufall eine Überraschung parat: Im Band mit Briefen über die Haushaltung und die Ausgaben für Kinder und den Garten (zum Beispiel 1000 Rebstöcke für 3 Taler und das in Berlin!) fand ich einen Brief aus Lissabon vom 1. Januar 1582, verfasst von Hans Harttman Hyrus. Dieser schreibt, dass man sich zwar nicht kenne, aber dass er Thurneysser gern von seinen Reisen berichten will. Hyrus stammte aus Konstanz, war für Konrad Rot in Lissabon tätig und sagt, er hätte davor fünf Jahre in Indien gelebt.[10]

b) Manuskripte von Thurneysser

Zudem haben sich unzählige Handschriften von Thurneysser erhalten, wobei zu betonen ist, dass selten ganze Seiten eigenhändig geschrieben wurden; oftmals finden sich lediglich die Anmerkungen Thurneyssers in roter Tinte. Die Bandbreite dieser Manuskripte reicht von Alchemie über astrologische Berechnungen für Prognostiken, Arzneibücher mit Rezepten, Anwendungen und den Preisen für die Medizin, medizinische Kur-Beschreibungen, Harnprobenanalysen verschiedener Patienten aus den Jahren 1571 bis 1577 bis hin zu handschriftlichen Listen mit Pflanzennamen in fünf Sprachen. Auch ein Manuskript zur Kriegslehre und zur Einrichtung einer Bibliothek sowie die Naturbeschreibung Portugals und mehrere Seiten eines Druckmanuskripts mit Anmerkungen finden sich darunter.[11]

4. Thurneysser und seine Bücher

Abbildung 4: Thurneysser besiegt seine Feinde. Holzschnitt von Jost Amman [1531–1591], aus Thurneyssers Werk Kai Ekplerosis und Impletio oder Erfüllung der Verheissung [1580], vgl. Boerlin, 1976: 85, Anm. 252 und 199, Abb. 195.

---

10 Staatsbibliothek zu Berlin, Ms. Bor. Fol. 685, Bl. 48 und 49.
11 Vgl. Spitzer, 1996: 139–143: Verzeichnis der handschriftlichen Quellen.

Thurneysser ist der Autor von über 30 Büchern.[12] Als druckender Autor schafft er ein beachtliches Werk. Natürlich bedient er den öffentlichen Geschmack: Als erste Drucke erscheinen eine Apothekertaxe 1574, großformatige Plakate zur Hinrichtung des Juden Lippold am 28. Januar 1573 in Berlin, das Pestregiment als Plakat zum Anschlag in der Öffentlichkeit und zur Erscheinung des großen Kometen 1572, seine Kräuterbücher und das Onomasticum, ein Verzeichnis der Paracelsischen Begriffe in mehreren Sprachen (Thurneysser, 1583).

Als Auftrag des Kurfürsten wird das Breviarium der nun lutherischen Kirche in mehreren Bänden gedruckt, dazu Gelegenheitsschriften und die sehr erfolgreichen Kalender. Immer großformatig und prächtig sind die alchemistischen und botanischen Schriften. Es werden Boten mit Drucken zu Fürsten und gekrönten Häuptern in ganz Europa gesandt, manche Korrespondenz liest sich wie aus einem Lehrbuch für Direktmarketing. Wir haben 2012 insgesamt 80 Titel ermittelt, die Thurneysser als Autor nennen, dazu gehören auch Nachdrucke bis 1612, fast 20 Jahre nach seinem Tod (1591).

Ein wichtiges Dokument aber kommt nicht aus dem Nachlass, sondern aus der Autographensammlung von Ludwig Darmstaedter: Adam Seidel wird darin instruiert, wie er sich auf der Frankfurter Messe verhalten soll. Dieses Dokument führt den Schreiber Adam Seidel und den Autor Thurneysser zusammen, wir sehen beide Handschriften auf einem Brief.[13]

Abb. 5: Auch die Instruktionen für Adam Seidel von 1571 zeigen die eigenhändige Handschrift Thurneyssers (Staatsbibliothek zu Berlin, Sammlung Darmstaedter, G I, 1571 Leonhart Thurneysser).

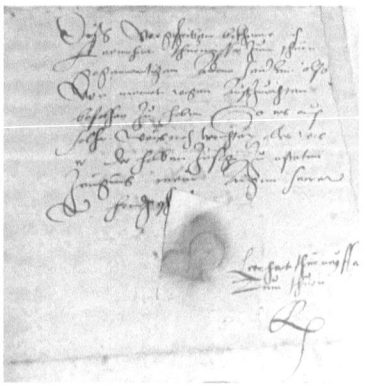

---

12 Eikermann/Kaiser, 2012 a.
13 Staatsbibliothek zu Berlin, Sammlung Darmstaedter, G I, 1560 Leonhart Thurneysser.

Zu den großen Verkaufserfolgen gehören als absolute Neuigkeiten sein Werk mit den anatomischen Klapptafeln (*Confirmatio Concertationis*, 1576), acht Planetentafeln mit drehbaren Scheiben[14] zu seinem Werk *Archidoxa* (1575) und die Typentafel zum zweiten Teil des *Onomasticums* (1583) mit 32 Alphabeten, in einer Größe von 90 x 50 cm.

Die Staatsbibliothek zu Berlin besitzt fast alle Druckwerke Thurneyssers. Einige stammen sogar aus seiner eigenen Bibliothek; in Leder gebunden, auf beiden Einbanddeckeln mit einem geprägten Supralibros, seinem Druckersignet von Jost Amman, mit der Umschrift „Festina Lente". Das Pestregiment von 1576, als Plakat für die Berliner Öffentlichkeit gedacht, hat nur in einem Exemplar die Wirren dieser Zeit überstanden: Es findet sich, nicht in der Staatsbibliothek Berlin, sondern als Beilage zum Briefwechsel mit dem Schweizer Gelehrten Theodor Zwinger der Ältere (1533–1588) in der Universitätsbibliothek Basel.[15]

5. Thurneyssers letzte Jahre und Perspektiven für die weitere Erschließung seines Nachlasses

Schließlich geht Thurneyssers hoher Anspruch, ohne akademisches Studium ein anerkannter und wohlhabender Gelehrter zu werden, nicht auf: Anfeindungen von allen Seiten beantwortet er mit üblen Schmähschriften. Auch die Hoffnung auf ein schönes Leben mit einer neuen wesentlich jüngeren Frau Marina Herbrott (um 1533–1610) erfüllt sich nicht. Nach kurzer Zeit verstößt er sie aus dem gemeinsamen Haushalt und schickt sie mit Adam Seidel 1582 zu ihren Eltern nach Basel zurück. Der Scheidungskrieg ruiniert ihn. Das neue Haus in Basel bewohnt er kaum, das Bürgerrecht wird ihm nicht zugesprochen. Als letztes Buch wird in Berlin im Grauen Kloster ohne Nennung des Druckernamens das *Ausschreiben*, eine ordinäre, wüste Abrechnung mit den Basler Bürgern und Gerichten, seiner letzten Frau Marina, seinem Bruder Alexander und der Welt überhaupt gedruckt; die Akten dazu liegen in Basel: es sind über 1.000 Seiten Elend.

Und wieder einmal sucht er das Weite, er geht nach Rom, konvertiert zum katholischen Glauben und gibt eine Reise- und Kriegsapotheke heraus.

---

14 In den „Digitalisierten Sammlungen" befindet sich das Thurneyssersche Astrolabium mit seinen Planetentafeln unter: http://digital.staatsbibliothek-berlin.de/werkansicht/?PPN=PPN645230227&DMDID=DMDLOG_0011.
15 Eikermann/Kaiser, 2012 b.

Thurneysser stirbt 1596 in Köln. Selbst dem Tod ringt er noch Bedeutung ab: nach seinem Wunsch wurde er neben Albertus Magnus begraben. Als Auftraggeber lässt er seine eigene Himmelfahrt in Form einer kreisförmigen Federzeichnung darstellen. Das Blatt in der Öffentlichen Kunstsammlung Basel zeigt in 50 Bildern verschiedene Lebensstationen.[16]

Was bleibt? Viele Namen, große Dramen, viele Bücher. Einer muß unbedingt noch genannt werden, nämlich Theophrastus Aureolus Bombastus von Hohenheim genannt Paracelsus (1493–1541)[17], das große Vorbild für Thurneysser. Auch er wohnte in Basel, im Haus am Kohlenberg. Das ist genau jenes Haus, welches Thurneysser bei seiner Rückkehr kaufte.

Es bleiben viele offene Fragen zu seinem Nachlass: Wo sind die Briefbände aus den Jahren Jahre 1565 bis 1578, die nach Basel gebracht wurden? Für rund 5.000 geschätzte Seiten muß es auch Antwortbriefe oder Vorgängerschreiben gegeben haben. Thurneyssers eigene Briefe sind selten, im Autographenhandel tauchen schon seit Jahren keine Briefe mehr auf, in der Datenbank Kalliope gibt es lediglich sechs Nachweise.

Welche Geld- und Münzgeschäfte wurden im europäischen Rahmen getätigt? Thurneysser verlieh Geld gegen Zinsen. Die Briefe seines Goldschmiedes Andreas Hindenberg berichten zudem von vielen Münzarbeiten.

Auch Rochus Quirinus Graf Lynar (1525–1596) gehört zu den Korrespondenten Thurneyssers; der gebürtige Florentiner und Freund der Medicis wurde Festungsbaumeister in Berlin, was sagt die Korrespondenz dazu?

In den nächten Jahren sollen alle Namen der Schreiber in den Briefbänden in der Datenbank Kalliope der Staatsbibliothek zu Berlin zugänglich gemacht werden. Damit werden Thurneyssers internationale Vernetzungen aufgezeigt werden und dies wird weitere umfangreiche Forschungen zum Thema Thurneysser ermöglichen.

Literatur- und Quellenliste

Boerlin, Paul H. 1976. *Leonhard Thurneysser als Auftraggeber. Kunst im Dienste der Selbstdarstellung zwischen Humanismus und Barock.* Basel/Stuttgart: Birkhäuser Verlag.

---

16 Boerlin, 1976: 161, Anm. 452 a.
17 Ein mögliches Datum des Geburtstages von Paracelsus, den 10. November 1483, verdanken wir Thurneyssers Angaben; s. Meier, 1993: 198 (er rechnet zehn Jahre dazu).

Eikermann, Diethelm/Kaiser, Gabriele. 2012 a. „Die Druckwerke von Leonhard Thurneysser zum Thurn (Basel 1531–Köln 1596)". *Gutenberg-Jahrbuch* 87, 171–198.

Eikermann, Diethelm/Kaiser, Gabriele. 2012 b. *Die Pest in Berlin 1576. Eine wiederentdeckte Pestschrift von Leonhart Thurneisser zum Thurn (1531–1596).* Rangsdorf: Basilisken-Presse.

Meier, Pirmin. 1993. *Paracelsus. Arzt und Prophet. Annäherungen an Theophrastus von Hohenheim.* Zürich: Ammann.

Schumacher, Yves. 2011. *Leonhard Thurneysser: Arzt – Abenteurer – Alchemist.* Zürich: Römerhof Verlag.

Spitzer, Gabriele. 1996. *... und die Spree führt Gold: Leonhard Thurneysser zum Thurn, Astrologe – Alchimist – Arzt und Drucker im Berlin des 16. Jahrhunderts* (Beiträge aus der Staatsbibliothek zu Berlin, Preußischer Kulturbesitz 3). Ausstellungskatalog. Wiesbaden: Dr. Ludwig Reichert Verlag.

Thurneysser zum Thurn, Leonhart. *Handschriftlicher Nachlass in der Staatsbibliothek zu Berlin*:

– Ms. Bor. Fol. 680–687 und 691 [Korrespondenz in 9 Bänden: Briefe an Thurneysser von fürstlichen Personen und Familienbriefe aus der Zeit von 1570 bis 1583].

– Ms. Germ. Fol. 97 [Naturbeschreibung Portugals].

– Ms. Germ. Fol. 176, Bl. 195 r–204 v [Thurneyssers Manuskript über das Bibliothekswesen].

– Ms. Germ. Fol. 420–426 [Korrespondenz in 11 Bänden: Briefe an Thurneysser aus der Zeit von 1564 bis 1583].

– Sammlung Darmstaedter, G I, 1560 Leonhart Thurneysser [Instruktionen für Adam Seidel].

Thurneysser zum Thurn, Leonhart. 1571. Προκαταληψις *Oder Praeoccupatio, Durch zwölf verscheidenlicher Tractaten, gemachter Harm [!] Proben, Durch Leonhart Thurneisser zum Thurn erfunden, vo[n] gemeinem nutz zu gutem an tag geben. Das 59. Buch.* Franckfurt an der Oder: Eichorn.

Thurneysser zum Thurn, Leonhart. 1572. *Pison. Das erst Theil. Von Kalten, Warmen Minerischen vnd Metallischen Wassern, sampt der vergleichunge der Plantarum vnd Erdgewechsen 10. Buecher: Durch Leonhart Thurneisser zum Thurn, mit grosser m[ue]he vnd Arbeit, gemeinem nutz zu gut an tag geben. Mit Röm. Kay. Freyheit auff 10. Jar. 1572.* Gedruckt zu Franckfurt an der Oder: durch Johan Eichhorn.

Thurneysser zum Thurn, Leonhart. 1576. Βεβαιωσις αγωνισμου *[Bebaiōsis agōniomu]. Das ist Confirmatio Concertationis, oder ein Bestettigung deß Jenigen so Streittig Håderig/ oder Zenckisch ist, wie dann auß vnuerstandt die*

Neuwe vnd vor vnerhörte erfindung der aller Nützlichestē vnd Menschlichem geschlecht der Notturftigesten kunst dess Harnnprobirens ein zeitlang gewest ist. [...]. Berlin: im Grauwen Closter.

Thurneysser zum Thurn, Leonhart. 1580. *Kai Ekplerosis und Impletio oder Erfüllung der Verheissung.* [Nürnberg]: [Heußler].

Thurneysser zum Thurn, Leonhart. 1583. *KAI 'EPMHNEIA Das ist ein ONOMASTICVM vnd INTERPRETATIO oder außführliche Erklerung, Leonharten Thurneyssers zum Thurn, Churfürstischs Brandenburgischs bestalten Leibs Medici. Vber Etliche frembde vn[d] vnbekante Nomina, Verba, Proverbia, Dicta, Caracter, Zeichen vnd sonst Reden. Deren nicht allein in des theuren Philosophi vnd Medici Aurelii Theophrasti Paracelsi von Hohenheim, Sondern auch in anderer Authorum Schrifften hin vnd wider weitleufftig gedacht, welche hie zusamen, nach dem Alphabet verzeichnet. Das Ander theil.* Berlin: Voltz, Nikolaus.

Thomas Horst

A Rediscovered Manuscript about Portuguese Plants and Animals: Preliminary Observations*

Abstract: Der Beitrag behandelt die umfangreiche Sammelhandschrift Ms. Germ. Fol. 97 der Staatsbibliothek zu Berlin-Preußischer Kulturbesitz. Diese befindet sich im Nachlass des am Hofe des Brandenburger Kurfürsten Johann Georg (1525–1598) wirkenden „Wunderdoktors" Leonhard Thurneysser zum Thurn (1531–1596), der nach eigener Angabe in den Jahren 1555/1556 in Lissabon im Hause des portugiesischen Humanisten Damião de Goís (1502–1574) verweilte und sich nicht nur auf den Gebieten der Alchemie, Magie, Medizin, Pharmazie, Anatomie, Astrologie, des Bibliothekswesens und der Chronologie, sondern vor allem in der Botanik hervorgetan hat.
In einem ersten Teil wird die Forschungsgeschichte dieses für die Flora und Fauna Portugals eindrucksvollen Dokumentes behandelt und die wiederholten Versuche seiner Transkription näher erläutert. Anschließend wird ein knapper Überblick über die gesamte Sammelhandschrift gegeben, um ein besseres Verständnis über die Hintergründe ihrer Erstellung und insbesondere die Frage der Datierung des gesamten Kodex zu ermöglichen. Dieser umfasst rund 459 Seiten und enthält neben dem portugiesischen Teil mehrere Exzerpte von zeitgenössischen Drucken, so etwa aus den Kräuterbüchern des Rembert Dodoens (1517–1585), aus dem „*Gifftjager*" von Wilhelm Klebitz (1576) oder dem „*Herbarium*" des Paracelsus in der Ausgabe des Michael Toxites (1570).
Im Zentrum der vorliegenden Betrachtungen, die lediglich erste Eindrücke vermitteln können, stehen jedoch Thurneysser's Ausführungen zu iberischen Pflanzen, Kräutern und Tieren sowie „*Miscellanea. Historica Geographica medica e*[t] *varie mixta*" (fol. 129 r–144 v),

---

* This work is the result of a collaboration between three researchers: Professor Dr. Bernardo Jerosch Herold (Centro de Química Estrutural, Instituto Superior Técnico, University of Lisbon), Professor Dr. Henrique Leitão, and Dr. Thomas Horst (both: Centro Interuniversitário de História das Ciências e da Tecnologia, University of Lisbon). It is presented here under a single authorship to be consistent with the fact that it was Thomas Horst who gave the first presentation about the topic. Its contents, however, are the result of the collaborative work. The text presented here incorporates several aspects that were not mentioned in his original oral presentation. Special acknowledgements go to Dr. Annemarie Jordan Gschwend (Centro de História d'Aquém e d'Além-Mar, Lisbon and Switzerland), who was so kind to correct the English of this text, and to our colleague Dr. Samuel Gessner, who has given a lot of useful information to our group.

worin in drastischen Worten nicht nur die Riten der norwegischen Seefahrer, sondern auch der Sklavenmarkt in Lissabon beschrieben wird.

Erst die geplante Gesamtedition des Portugal betreffenden Teils der Handschrift wird viele bislang ungeklärte Fragen klären können. Dabei wird anhand dieses eindrucksvollen Manuskriptes herausgestellt werden, dass der Autodidakt Thurneysser ein „Renaissance Craftsman" war, dessen Leben und Wirken es noch besser zu erforschen gilt.

The "Nachlass" of the Swiss goldsmith, alchemist, pharmacist, astrologer, printer and entrepreneur Leonhard Thurneysser zum Thurn (1531–1596)[1], who made a journey to Portugal in 1555/1556, and who later worked as a miracle doctor at the court of Elector John George of Brandenburg (1525–1598), but who was often seen as a charlatan[2], is only partially preserved in the manuscript room of the Staatsbibliothek zu Berlin – Preußischer Kulturbesitz (cf. Kaiser, 2016, in this volume). Among many other documents and treatises by Thurneysser, which deal with alchemy[3], botany[4], esotericism, occultism and magic[5], medicine (urine diagnostics of illnesses)[6], pharmacy[7], anatomy[8], as-

---

1   For the life and work of Thurneysser see especially the older studies of Moehsen, 1783/1976; Becker, 1838 and the researches of Macco, 1934; Wallich, 1934/1967: 311–319; Bugge, 1939; Harms, 1963; Boerlin, 1976: 11–30; Spitzer, 1996; Spitzer, 1997; Gantenbein, 2011; Schumacher, 2011 and Eikermann, 2012. Eikermann/Kaiser, 2012 a and 2012 b give a good overview about his biography and prints; more prosaic, but historically incorrect are the novels by Kulemeyer, 1942 and Peuckert, 1956. Thurneysser was also an important printer, cf. for example the entry by Gabriele Spitzer, in: Pehlivanian, 2006: 192 f.
2   Cf. Partington, ²1969: 152–155, here: 153: „Thurneysser was essentially a charlatan and a business man who [...] stood near his master Paracelsus in 'Geist und Betriebsamkeit'. He had no real knowledge of medicine but was an unprincipled quack, selling all kinds of nostrums at immense prices".
3   Cf. Hofmeier, 2007; Kahn, 2014; Maar, 2000; Mittler, 1986: 351–353 (E 22.7: „Thurneissers Archidoxa – medicoalchemische Heimlichkeiten in Form eines Traumberichts") – and more general Moran, 2004 a and Moran, 2013.
4   Cf. Baumann, 1998: 76, 85–89, 94, 117 f. (reference to fig. 22 and 30), 122 (fig. 58) and 125 (fig. 77); Bulang, 2013 and Zaunick/Wein, 1938.
5   Scholz-Williams, 2013: 3–5 showed that Thurneysser in his autobiography from 1584 predated the publication of the *Faust* chapbook by three years. Cf. also Thurneysser, 1591.
6   Cf. Bleker, 1976. See also Reber, 1906 and Quecke, 1950 (Thurneysser was often seen as quack). – For uroscopy in Early Modern Europe see Stolberg, 2010 and Stolberg, 2015: 11–14 ("Consultation by Letter"), 66–70 ("Paracelsian Uroscopy"), 73 f., 102 (footnote 121), 134–141 ("Attempting Modernization") and 165.
7   Morys, 1982; Schmitz, 1988 and Eikermann, 2008.
8   Boerlin/Münster, 1960.

trology[9], librarianship[10] and chronology[11], a recently rediscovered manuscript (Manuscripta Germanica Fol. 97) in Berlin stands out, because it contains the earliest known description in German of Iberian flora and fauna. This is why a research group in Portugal (Prof. Dr. Bernardo Jerosch Herold, Prof. Dr. Henrique Leitão and Dr. Thomas Horst)[12], based at the Centro Interuniversitário de História das Ciências e da Tecnologia (CIUHCT) at the University of Lisbon, has established a collaboration with the Berlin State Library[13] with the aim of transcribing and editing the text of this cimelium.

1. Research history of the manuscript

The knowledge about this unique manuscript itself, dealing for the first time with Portuguese plants and animals in detail, is not completely new: It was the German librarian Hermann Degering (1886–1942), who first gave a short inventory of the miscellany, published in his "Kurzes Verzeichnis der germanischen Handschriften der Preussischen Staatsbibliothek (Berlin), volume I: Die Handschriften in Folioformat" in Leipzig (Degering, 1925: 11; Nr. 97)[14]. It is most likely that because of this entry, an economic geographer of the University of Bonn, Professor Otto Quelle (1879–1959, cf. fig. 1), whose mentor was Ferdinand Freiherr von Richthofen (1833–1905)[15], later became interested into this special topic.

---

9   Cf. for example his print "Deß Menschen Circkel vnd Lauff" (Thurneysser zum Thurn, 1575) with astrological tables (designed by the Swiss artist Jost Amman [1531–1591], who was celebrated for his woodcuts), described by Mazzini, 1946 and Mittler, 1986, 353 f. (E 22.8); to Amman cf. also Andresen, 1973.
10 Cf. Spitzer, 1996: 29 (relating to Ms. Germ. Fol. 176).
11 Thurneysser printed several "almanachs" in his last years, cf. Juntke, 1978 und 1980 and the old study of Sudhoff, 1908.
12 A detailed description of the manuscript was given by Professor Bernardo Jerosch Herold in a presentation at the Academy of Sciences in Lisbon on April 7th 2016, cf. Herold/Horst/Leitão, 2016 a and b [unpublished working papers].
13 We would like to thank in particular Dr. Gabriele Kaiser and the director of the Departement of Manuscripts at the Staatsbibliothek zu Berlin – Preußischer Kulturbesitz, Prof. Dr. Eef Overgaauw, for their kind collaboration and support of our project.
14 He already has inventoried the manuscript in 1922 cf. Degering, 1922.
15 For Richthofen cf. Beck, 1982: 149–163 ["Ferdinand Freiherr v. Richthofen – vorbildlicher China-Forscher und anerkanntester Geograph seiner Zeit (1833–1905)"].

Fig. 1: Otto Quelle (1879–1959) already wanted to edit the manuscript. Photo withdrawn from Die Erde. Zeitschrift der Gesellschaft für Erdkunde zu Berlin 85, Bd. 6, Heft 3–4 [Festschrift für Otto Quelle zum 75. Geburtstag].

Quelle has studied geography at the Universities of Göttingen and Berlin, together with Walter Emmerich Behrmann (1882–1955)[16], who later tells us in a "Festschrift" (cf. Behrmann, 1954, here: 97), which was published on occasion of Quelle's 75th birthday in the journal of the Geographical Society of Berlin[17], that they both were students of the famous geographer and map historian Hermann Wagner (1840–1929)[18] and assistants of the German geomorphologist Albrecht Penck (1858–1945)[19]. One of their colleagues was the young Hermann Wilhelmy (1910–2003), who later published not only important overviews in the field of cartography (Wilhelmy, 1966 in seven editions!), but also has made a significant impact in the area of Latin America regional geography (Wilhelmy, 1980 and 1982), with a special focus on climatic geomorphology (Wilhelmy, 1975–1978) and especially morphogenic urban geography (Wilhelmy, 1952/1968; Wilhelmy/

---

16 For Behrmann and his geographical works see Schultze, 1957.
17 In the same year he was honored with the "Carl-Ritter-Medaille". In 1956 he got the Federal Cross of Merit, cf. Kilwa, 2004: 161 and Kellenbenz/Schneider, 1987: 44. An incomplete list of Quelle's scientific work can also be found therein (cf. Schindler, 1954), as well as articles by his colleagues Hermann Lautensach, Herbert Wilhelmy, Edwin Fels, Erika Freitag and others. For Quelle see also Brauer, 1968; Kalwa, 2004: 159–166 and the entries in the database http://kalliope.staatsbibliothek-berlin.de/de/eac?eac.id=116317817.
18 For the cartographic circle around Wagner cf. *Hermann-Wagner-Gedächtnisschrift*, 1930 and Böhm, 1974.
19 For Penck see Beck, 1982: 191–212 ["Albrecht Penck – Geograph, bahnbrechender Eiszeitforscher und Geomorphologe (1858–1945)"]; Schaefer, 1989, Pinwinkler 2011 and Schultz, ²2011.

Borsdorf, 1984/1985).[20] Another student of Penck was Hermann Friedrich Christian Lautensach (1886–1971), who has done significant research about the geography of the Iberian Peninsula with a special focus on Portugal (Lautensach, 1941; Lautensach, 1964)[21].

In 1918 Otto Quelle became "Extraordinarius" for economical geography at the "Rheinische Friedrich-Wilhelm-Universität" in Bonn, Germany (Brauer, 1968: 216). Two years later he was appointed full professor (Kalwa, 2004: 160). In this period he founded a private Ibero-American Research-Institute (Liehr, 1992: 644; Quelle, 1930) in 1923, which was integrated together with his own library (more than 10,000 books!) into his university in 1925 (Kalwa, 2004: 165). From October 1924 onwards, Quelle also was the editor of the "Ibero-Amerikanisches Archiv. Zeitschrift für Sozialwissenschaften und Geschichte", a multidisciplinary scientific journal, which appeared until 1944 in 18 volumes twice a year – and again in a new series from 1975 to 2000 with articles in the German, Spanish, Portuguese and English language.[22]

Already in the time of the Weimar Republic one can witness various efforts in Prussia to establish a central Institute for Ibero-American Research in Berlin. This idea was realized in 1930, when the new Ibero-American Institute (IAI)[23] was founded, independent but part of the Prussian Ministry of Education. The new institution moved into the representative wing of the former Royal Stable building ("Neuer Marstall") in the centre of Berlin (Bock, 1963: 330) and Quelle became a member of its research staff[24]. His library was incorporated therein, as well as 25,000 books about Mexico from the Geographical Institute in Marburg and a huge book collection (around 82,000 volumes) belonging to the Argentinean

---

20 For Wilhelmy, who was really fascinated by Alexander von Humboldt (cf. Wilhelmy, 1970), see Schröder, 1970; Borsdorf/Leser, 2003 and Kohlhepp, 2004.
21 Cf. especial his "Bibliografia geográfica de Portugal" (Lautensach, 1948/1982) and Daveau, 1987–1991. We thank Prof. Dr. João Carlos Garcia (Porto) for this information.
22 The journal was founded with the help of Otto Matteis (the Brazilian ambassador in Bonn) and the bookseller Antonio Lehmann (1871–1941), cf. Kalwa, 2004: 164 f.
23 For the history of this institute during the Nazi period cf. Kalwa, 2004: 169–188.
24 Quelle also worked for a short time from 1909 to 1911 on the editorial staff of the important German journal "A. Petermann's Mitteilungen aus Justus Perthes' Geographischer Anstalt" in Gotha (cf. Brauer, 1968: 215 and Kalwa, 2004: 159). In 1930 he was appointed as a professor of geography at the "Technische Hochschule" in Berlin (cf. Liehr, 1992: 645; Kalwa, 2004: 161). It is important to state here, that Quelle joined the NSDAP (National Socialist German Workers' Party), but was suspended from the National Socialistic party because of his former sympathy for Freemasons, cf. Gliech, 1990: 12.

scholar and lawyer Ernesto Ángel Quesada (1858–1934), which was acquired already by the Ministry of Education in 1927 (Quesada, 1930; Bock, 1963: 326; Carreras, 2004).[25]

Professor Otto Quelle was a scientific consultant at this institution and ten years later, when he was nominated full professor for "Geographie, Volks- und Landeskunde Spaniens und Ibero-Amerikas" at the Friedrich-Wilhelm-University in Berlin, he published a cultural-historical description of Iberian Culture while analyzing select tapestries in Vienna, which bore references to Portugal (cf. for example the coat of arms of the Emperor Charles V in tapestries in: Quelle, 1940: 5 f. and plate I–V; this book also appeared in Spanish and Portuguese). Moreover, his history of Iberoamerica (*Geschichte von Iberoamerika*: Quelle, 1942), which he published during World War II, is still informative about this topic.

Quelle's extensive researches on Iberian culture in Berlin[26] finally brought to his attention the manuscript by Leonhard Thurneysser zum Thurn, which we want to focus on: In 1944 he published a short article about this topic in the *Revista do Instituto de Cultura Alemã. Zeitschrift des deutschen Kulturinstituts in Lisbon* (Quelle, 1944 a). In the same year, Quelle also announced the scientific publication and complete edition of the whole Thurneysser manuscript in the *Zeitschrift für Politik* (Quelle, 1944 b).[27] This is why another book entitled *Oito Séculos de História Luso-Alemã* (printed by the Iberian-American Institute in Berlin as a "Festschrift"), also cites this information (Strasen/Gândara, 1944: 163 f.).

---

25 For the role of the Ibero-American Institute during the Nazi regime, when it was lead by General Wilhelm Faupel (1873–1945), see Liehr/Maihold/Vollmer, 2003.

26 In April 1939 the Staatsbibliothek held an exhibition in Berlin about *Portugal in Vergangenheit und Gegenwart* ("Portugal in Past and Present"); at the opening the German ambassador in Lisbon, Oswald Theodor Baron von Hoyningen-Huene (1885–1963), who appealed to the nationalist sentiments of António de Oliveira Salazar, was present together with many well-known Portuguese authors and journalists. Maybe the Thurneysser manuscript was shown there, even if not cited in the report by Richert, 1939 [for the role of the well-known teacher of Romance languages and literature Gertrud Richert (1885–1965), who also is important for Iberian art history, see Kalwa, 2004: 182–188]. – For the German cultural strategy in Portugal during Nazi period cf. Ninhos, 2012.

27 "Die Ergebnisse seiner [Thurneysser's] Forschungen – der ersten wissenschaftlichen auf portugiesischem Boden – sind niedergelegt in einem mehrere hundert Seiten umfassenden Folioband, dessen wissenschaftliche Veröffentlichung mir hoffentlich bald ermöglicht wird" (cf. Quelle, 1944: 116. This article was written in the National Socialistic spirit of the time).

But it seems so that Quelle could not pursue his plan of editing the manuscript, even if we know that he had access to it until 1944. In any case, the only bigger publications by him after World War II are a "Festschrift", which he edited for the 125th anniversary of the "Gesellschaft für Erdkunde zu Berlin" (Quelle, 1953 a) and a detailed, but little known study about Portuguese manuscript atlases of the seventeenth century preserved in the Austrian National Library (Quelle, 1953 b)[28]. Even if we do not know his motives, it seems so that he did not work any more on the manuscript; perhaps the Renaissance palaeography was too difficult to read for him.

We only hear again about the Thurneysser manuscript in 1960, when the Swiss diplomat Henry Béat de Fischer (1901–1984), who was ambassador in Lisbon from 1953 to 1959, published his "Dialogue luso-suisse: essai d'une histoire des relations entre la Suisse et le Portugal [du XIVe siècle à la Convention de Stockholm de 1960]". Therein, Fischer tells us that Thurneysser came to Lisbon in 1555/1556, at the age of 25, where he resided at the house of the famous Portuguese Humanist Damião de Góis (1502–1574) and wrote his text, "la première étude de science naturelle dans ce pays en general" (de Fischer, 1960: 150). Fischer also informs us that Thurneysser's description of Portugal was at that time in the university library of Tubingen; the manuscript only returned to Berlin in 1967 (Ziesche, 2002: VIII).[29]

There it was studied by the German Lusitanist at the University of Coimbra, Professor Albin Eduard Beau (1907–1969)[30], who had been aware of the manuscript since 1941 (Beau, 1941: 176 f.) and who later came in the possession of a microfilm copy. After Quelle's death he even wanted to translate the manuscript into Portuguese, together with his wife Dr. Ursula Beau, born Becken (1906–1987).[31] But this project also did not materialize and the whole topic fell into oblivion.

---

28 Previous cartographic studies to that topic were published in the journal "Ibero-Amerikanisches Archiv", cf. Quelle, 1933 and Quelle, 1939.

29 During World War II valuable manuscripts and incunabula (Manuscriptae Germanicae) were transferred from Berlin to the Beuron Archabbey in Baden-Württemberg. After the war, the cimelia in the French occupation zone were united in the library of Tubingen, used as deposit, before they finally came back to Berlin in 1967, cf. also Breslau, 1995 and Voigt, 1995 and more general: Koch, 2003: 30 f.

30 For Beau, who was General Secretary of the "Deutsches Kulturinstitut" in Lisbon since 1943, cf. Delille 1969. – For the scientific exchange between the "Third Reich" in Germany and the Salazar regime in Portugal cf. Hausmann, 2001: 334-352 ("Das deutsche Kulturinstitut in Lissabon") and Hausmann, ²2008: 522-538.

31 Cf. de Fischer, 1960: 150, footnote 4: "L'Institut de Haute Culture, à Lisbonne, en a acquis unc copie et a prié le Prof. A. Beau, à Coïmbre, de le traduire en portugais et de

It is the merit of Gabriele Spitzer (now: Kaiser), who wrote a dissertation about Thurneysser in 1987, to have revived the interest in Thurneysser. She alluded to the manuscript in her study "… und die Spree führt Gold" (Spitzer, 1996: 29 f. and 139) – as did recently also Yves Schumacher in his new biography of Thurneysser (Schumacher, 2012: cf. in particular 50–53: "Lusitanische und andere Streifzüge")[32], which fortunately came to the hands of Prof. Dr. Bernardo Jerosch Herold and initiated our project in Lisbon. So, nearly 70 years later, we are following Quelle's footsteps to make a full analysis, transcription and edition of the Portugal-related parts of the manuscript, which still remain a desideratum until today.

2. Description of the Miscellany Ms. Germ. Fol. 97

To understand this interesting miscellany it is important to have a closer look at it from a codicological point of view: The manuscript was written in the second half of the sixteenth century, as the palaeographical analysis of the handwriting suggests. We do not know when exactly it came to the "Kurfürstliche Bibliothek" of Prussia, but this surely happened already during the Renaissance. Altogether the manuscript contains 457 leaves (which are in total more than 900 folio pages)[33] and is divided into eight parts, dealing with various topics. The mutual connection of the different parts is the natural scientific content.

The major part (around 136 leaves)[34] incorporates four chapters dealing with the flora and fauna of Portugal (fol. 1 r–144 v and 317 r–353 v). Aside from this,

---

le commenter, ensemble avec Mme Beau, tant au point de vue de l'histoire de la civilisation portugaise qu'à celui des sciences naturelles". – In a personal communication Professor Bernardo Jerosch Herold told us that maybe the quality of the microfilm, which does not exist anymore, was the main problem, because Mrs. Ursula Beau told him on occasion that she was not able to work with microfilms.

32 Schumacher, 2011: 82 also writes about a second journey of Thurneysser through the Iberian Peninsula, which he probably did in 1561 (cf. also Spitzer, 1996: 17), as well as other travels which took him to England and Scotland, but also to Italy, Hungary, Greece, Palestine, Egypt, Asia Minor and North Africa (cf. Boerlin, 1970: 15). Thurneysser cultivated a nomadic lifestyle, but it seems also he tended to exaggerate. This is why it is uncertain if he has really visited all these places.

33 While inventorying the manuscript, a librarian made a mistake in numbering the pages, which he did with a pencil: After page "284" follows "185". We will not follow this erroneous numbering and continue in our edition with page 285.

34 Altogether the Iberian section contains 371 pages, however more than 90 pages are blank.

which is the most interesting section and therefore will form the basic of our planned edition, we can find several excerpts of later published books. One of these is a very short alphabetical ordered register (*Ausszüg ettzlicher vornemer Hanndlungen auss dem Gifftjager von Wilhelmi Triphyllodacni*, fol. 145 r–150 v)[35]. This register refers to the *Gifftjager. Das ist: Ursach, Reinigung, Bewarung und Cur Pestilentzischer Lufft* (Klebitz, 1567), a book about poisons printed in Frankfurt am Main, Germany, with the German text[36] of the Zwinglian protestant theologian and mathematician Wilhelm Klebitz (around 1533–1568).[37] However, the accompanying folio numbers, which are given in this index, do not fit with the page numbers of the *Gifftjager*, of which Thurneysser made a register for himself.

In fact the folio numbers given here correspond with another part in the miscellany (fol. 209 r–240 v), which Thurneysser fills up with more than 60 pages on various treatises on poisons[38]: for example, recipes for the plague, lists of herbs to counter snakebites and the poison of other animals, or making comparisons between the medical device (humoralpathology) of the Antique physician and philosopher, Galen of Pergamon and his later critic by the Renaissance physician, botanist, alchemist, astrologer, and general occultist, Philippus Aureolus

---

35 The pseudonym *Wilhelmus Triphyllodacnus* refers to Wilhelm Klebitz, as already Degering, 1922: 344 reports.

36 Cf. Klebitz, 1567, forword: "Hab mich, sovil mueglich, der frembden woerter enthalten, vn[d] teutsche Namen der Kreuter gebrauchet, diweil es ein teutsch Buechlein seyn soll".

37 For Klebitz see Janse, 2001: This reformed theologian was born in Namitz (Brandenburg) around 1533. He became deacon in Heidelberg, where he bickered with his rival Tileman Heshusius (1527–1588), which led to his resignation in 1559. Later he dashed around Western Europe and published theological, philological and mathematical studies. Klebitz died 1568 in Paris. – As he mentions in the foreword of the *Gifftjager* (fol. 3) the excerpts from Dioscorides are taken from the spurious *Euporista. Ad Andromachum hoc est de curationibus morborum per medicamenta paratu facilia*, the first Greek edition of which was prepared by the Swiss botanist and physician Conrad Gessner (1516–1565) and Johannes Moibanus (1527–1562, a son of the Breslau humanist and reformer Ambrosius Moibanus (1494–1554), who studied in Wittenberg, wrote his dissertation in Italy in 1554 and later was physician, painter and musician in the Imperial city of Augsburg, cf. Konrad, 1891 and Siegel, 1928). This work was printed in Strasbourg in 1565, together with a Latin version (Gessner/Moibanus, 1565). On the *Gifftjager* cf. also Sudhoff, 1894: 128 f., Nr. 84 (who did not identify the pseudonym of Klebitz).

38 Cf. also the contemporary books on poisons, for example by the French dramatist Jacques Grévin (ca. 1539–1570), cf. Grévin, 1568 and Partington ²1969: 25 f.

Theophrastus Bombastus von Hohenheim (Paracelsus, 1493–1541)[39]. On the first page of this section (fol. 209 r), Thurneysser refers to Pedanius Dioscorides of Anazarbus, a Greek physician and pharmacologist, who served in Nero's armies as a botanist in the first century AD.[40] We have not studied this section of the miscellany until now in detail, but it seems that Thurneysser also used other contemporary sources[41], and some facts, for instance, when he reports that more than 100,000 people died in only four months, between May 15th and August 15th 1524, in the Dukedom of Milan (fol. 209 v), which corresponds directly with the above-cited *Gifftjager* (Klebitz, 1567: 2 f./cap. II)[42].

---

39 For an introduction on the life and work of Paracelsus see Goodman/Russell, 1991: 157–162; Golowin, 1993; Koyré, 2012; Letter, 2000; Meier, 1993; Moran, 2004 b and Webster, 2008. For general information see Hieronymus, 2005. – The chapter excerpted here on fol. 216 v–220 r (*Vnderscheidt zwischen den Galenis[ten] vnd Paracelsisten*) is nothing else than a short abstract of Klebitz, 1567: 52–75 (*Der ander Theil. Vnderscheydt zwischen den Galenisten vnd Paracelsisten in der Lehr von Artzney der Pestilentz*).

40 Dioscorides was the author of *De Materia Medica*, a five-volume encyclopedia about herbal medicine and related medicinal substance that was widely read for more than 1,500 years, cf. Aufmesser, 2000; Bertelli/Lilla/Cavallo, 1988/1992; Biedermann, ²1978; Dubler, 1952/1959; Lazaris, 2013 and Osbaldeston, 2000. For the plant names in his works cf. Wellmann, 1898. – The so-called *Wiener Dioscurides* (*Anicia-Juliana-Codex*), an illuminated Greek miscellany (dated around 512), which is preserved since 1569 in Vienna, Austrian National Library, med. Gr. 1 contains excerpts of Dioscorides, but also of other antique authors. With its depiction of 383 illuminated plants, it is a rare example of a late antique scientific text (Mazal, 1981; Nissen, ²1966: 17–20, here: 17 f.).

41 On fol. 220 r/v he refers to Conrad Gessner, who was working at the end of his life (1555–1565) on an extensive encyclopedia, the *Historia Plantarum* (with more than 1500 images of plants), which remained incomplete and was only published for the first time in 1751 (Schmidel, 1751); cf. Vogel, 2014: 23 and Zoller/Steinmann, 1987–1991. – For the life of Gessner and his naturalistic works, which are of most importance for zoology (for instance Gessner, 1551–1558: *Historiae animalium*, a 4,500-page encyclopedia of animals, see Gmelig-Nijboer, 1977; Kusukawa, 2010 and Kusukawa, 2015), cf. Freudenberg, 1999; Pyle, 2004; Leu, 1990 and the exhibition "Conrad Gesner. Physician, Scholar, Scientist 1516–1565" (1965). For his widespread library cf. Leu/Keller/Weidmann, 2008.

42 „Hieronymus Cardanus der beruempte vnnd hochgelehrte Mathematicus schreibet, daß im Jahr nach Christi geburt 1524. in der Stadt Mediolan, vnnd vmbligenden Doerffern vnnd Stedten des Herzogthum[b]s Mediolan, zwischen dem 15. tag Maij, vnnd dem 15. tag des Augustmonats, an der Pedstilentz gestorben hundert tausent Personen. Im vorgehenden Jar aber" (Klebitz, 1567: 2 r–v/cap. II) in comparison with Thurneysser, fol. 209 v: „Anno 1524 Inn der stat Mediolan, vnnd umbligenden dorfferen

Further extracts in this miscellany refer to the *Historia stirpium* (fol. 151 r–208 v: *Extractio oder Auszug Dodonei*) of the Flemish botanist Rembert Dodoens (Rembertus Dodonaeus, 1517–1585)[43], who had published his *Trium Priorum De Stirpium historia Commentariorum imagines* in Antwerp already in 1553 (Dodoens, 1553)[44], before Thurneysser's travel to Portugal. In the following year, the same author's vernacular book on herbs, the *Cruydeboeck*, appeared (Dodoens, 1554 b), which was illustrated with more than 700 woodcuts. Most of the botanic illustrations[45] in this work are copies of another herbal book (cf. Feldmann, 2010: 17, Nr. 5), which was printed in Thurneysser's home-town in Basel in 1542 (*De historia stirpium*)[46] and 1543 (*New Kreüterbuch*)[47], by the famous Swabian-born botanist Leonhart Fuchs (1501–1566)[48], who can be called, along with Otto

---

vnnd stetten des hertzogtums Mediolan, zwischen dem 15. Tag May vnnd dem 15 tag des Aug[u]stmonats seindt an der Pestilenntz gestorbenn hundert Tausendt Personen". – As a second example see the list on fol. 427 r of the Thurneysser manuscript, who lists the names of woods, which are good for cleansing the air. These names are exactly the same as specified in Klebitz, 1967: 9 r–10 v/cap. IIII: *Welche Hoeltzer sonderlich zur reinigung der lufft nuetzlich seind.*

43 For the life of Dodoens, who was later physician at the court of emperor Rudolph II in Vienna (1575–1578) and then received the Chair of Medicine at the University of Leiden in 1582, see the older studies by Van Meerbeeck, 1841/1980; Roentgen, 1842 and de Cock, 1890. – Dodoens was "a transitional figure between this and the next generation. […] the success of Dodoens' work points to the continued existence of a vernacular market for practical manuals. The same market existed in England, where John Gerard's plagiarized translation of Dodoens was immensely popular" (Ogilvie, 2006: 36 f.).

44 Dodoens later also dealt with cosmography, cf. his book *De Sphaera sive de astronomiae et geographiae principiis cosmographica Isagoge*, which appeared at the Antwerp publishing house of Christoffel Plantijn (Dodoens, 1584). In 1554 Dodoens also published his study *Posteriorum trium de stirpium historia commentariorum imagines* (Dodoens, 1554 a), cf. Baumann, 1998: 90–99.

45 For the botanic illustrations in the sixteenth century cf. Nissen, ²1966 and Zucchi, 2003. On the use of woodcuts cf. Treviranus, 1949.

46 This book with more than 900 pages and 511 woodcuts was a real de luxe edition cf. Meyer/Trueblood/Heller, 1999; Mägdefrau, 2013: 30 and Ogilvie, 2006: 113 f. and 194–197. More in general to medicinal plants cf. Kusukawa, 2012: Part 2: Picturing Medicinal Plants, 98–177.

47 Facsimile of the complete colored edition: Dobat/Dressendorfer, 2001.

48 For the life and work of Fuchs and the influence of his woodcuts cf. Baumann, 1998: 73–81; Baumann-Schleihauf, 2001; Brinkhus/Pachnicke, 2001; Dilg, 2010; Feldmann, 2010: 14 f., Nr. 3; Fichtner, 1968; Schlagbauer, 2002; Sprague/Nelmes, 1928–1931; Stübler, 1928 and Tillmann, 1988: 299–307. On the manuscript in Vienna, which was produced until 1564, cf. Baumann, 1998: 25–37; Baumann/Baumann, 2001 and Ganzinger, 1959.

Brunfels (ca. 1488–1534)[49] and Hieronymus Bock (1498–1554)[50] as one of the "fathers of botany"[51]. Dodoen's book was also translated to various languages and soon became very famous.[52]

It is even possible that Thurneysser read the herbal books by Leonhart Fuchs (Fuchs, 1542 and Fuchs, 1543) as a young boy. From these and other contemporary botanical works he seems to have copied a great deal of plant names, listed in his manuscript together with their Latin, Greek, German, Brabant and Gallic names, as one can see with the exemplary of "Ginciber carinum", which is the

---

49 For the life and work of Brunfels, who was also a theologian, physician and humanist, cf. Baader, 1978; Baumann, 1998: 63–72; Bautz, 1975; Mägdefrau, 2013: 24–27; Müller-Jahncke, 1982: 28–31; Roth, 1900; Sanwald, 1932; Sprague, 1928–1931; Telle, 1989 b; Tilmann, 1988: 174–180 and Treviranus, 1949. His *Contrafayt Kreüterbuch* (Brunfels, 1532–1537, Latin version: *Herbarum vivae eicones* in three parts [mostly a summary of ancient descriptions], Straßburg 1530, 1531 and [postum] 1536) was for the first time illustrated by the German Renaissance artist Hans Weiditz the Younger (ca. 1495–ca. 1537), with true-to-life illustrations of plants, which were artistically and scientifically of high quality, cf. Behling, 1957: 157–163 ("Exkurs: Das Kräuterbuch des Otto Brunfels und die Pflanzenaquarelle des Hans Weiditz"), here: 158; Ogilvie, 2006: 169 & 182 and Rytz, 1936. Some original watercolors from Weidlitz were acquired by the Swiss physician Felix Platter (1536–1614) and incorporated in his own herbarium, cf. Ogilvie, 2006: 172 (The earliest sheets in Platter's herbarium date from his days as Guillaume Rondelet's student in Montpellier in the 1550s), 180 and http://www.burgerbib.ch/platter-herbarium/. For the Platter family see the groundbreaking study of Le Roy Ladurie, 1998. For botany at the University of Basle in the 16th and early 17th centuries, cf. Reeds, 1991: 93–134.
50 For the life and work of Bock cf. Baumann, 1998: 81–85; Bergholz, 2005; Hoppe, 1969; Müller-Jahncke, 1982: 24–27; Staat, 1968; Telle, 1989 a and Tillmann, 1988: 243–247. The first edition of Bock's *Kreutterbuch* appeared in 1539 without illustrations. This book was very popular, because "his descriptions, which he composed himself, were much more precise than those of his contemporaries" (Ogilvie, 2006: 36). The version of 1546 (Bock, 1546) was illustrated by the Renaissance artist David Kandel (1520–1592, his monogram are his initials: DK), who was one of the pioneers of botanical art and science; he also copied Dürer's rhinoceros for a woodcut in the *Cosmographia* of Sebastian Münster, cf. Letkiewicz, 2009: 9, fig. 5. Münster's world map of 1569 was printed by Kandel. – Contemporary botanists as Adam Lonitzer (Adamus Lonicerus, 1528–1586) also have used Bock's work, cf. Lonitzer, 1557 (his first important work on herbs, the *Kräuterbuch* dealt for a large part with distillation), Baumann, 1998: 63–72, here: 69, Hoppe, 1969: 10–12 and 25 f. and Tilmann, 1988: 106 f.
51 Cf. Mägdefrau, 2013: 23–42 [Chapter 3: Die «Väter der Pflanzenkunde»].
52 See also the facsimile editions of his *Cruydeboeck*: Dodoens, 1971 and Gloning, 2005 (Online-Version of the extended edition of 1563).

Indian Pepper ("Inndianischer Pfeffer", fol. 197 r), or with the "Eberwurtz" (the Carline thistles, fol. 189 v).

In contrast, the source of another short extract (fol. 241 r–255 r) is immediately evident. Thurneysser provides us here with an index of the *Beschreibunng etlicher Kreütter auss dem Herbario Theophrasti Paracelsi Bombast, beider Artzney Doctoris*, as the title of his *Extractio oder Auszug der Beschreibung* tells us. Paracelsus, who may also have visited Portugal as a young man (maybe around 1517?)[53], is known for his revolutionary ideas: for insisting upon using observations of nature, rather than looking to ancient texts – and for his critique of the scholastic methods in medicine, science and theology.[54] We also know that one of his first essays, the *Herbarius*, written in the 1520's, dealt with herbs. It represents the curative effects of six medical substances: hellebore, persicaria, salt, angelic thistle, corals, and magnets (*von den Heilwirkungen der Nieswurz, der Persicaria, des Salzes, der Engel-Distel, der Korallen und des Magneten*, cf. Moran, 1993: 101). But this fragmentary text did not circulate long before it was printed.

The diverse sections of the Paracelsian *Herbarius* were printed separately: As part of *Herren Doctors Theophrasti Paracelsi Declaration, zuobereyten Hellebori* (Bodenstein, 1568)[55], the part concerning hellebore ("Nieswurz") was first edited by the German alchemist Adam von Bodenstein (1528–1577, an important translator of Paracelsus) in 1568. In the same year, the sections relating to persicaria ("Gemeiner Knöterich") and corals appeared anonymously at Augsburg as a supplement to the *Aphorismorvm Aliquot Hippocratis genuinus sensus & vera interpretatio* (Paracelsus, 1568)[56]. Two years later the part concerning the magnet

---

53 Paracelus reports in the foreword of his *Spitalbuch* (Nuremberg, around 1529), as well as, in his print *Die große Wundartzney* (Ulm, 1536 and other editions) that he had visited after his studies at the University of Ferarra (1513–1516), several places in Europe until 1524, among them Italy, Dalmatia, Spain and Portugal, France, Flanders, the Netherlands, England, Denmark, Sweden, Transylvania, Lithuania, Poland, Hungary, Wallachia, Croatia and Carniola. This part of his biography is not very well known because we do not have a lot of sources, cf. Benzenhöfer, ³2003: 31–34. – For his travels through the Iberian Peninsula cf. Lach, ²1994: 422, footnote 115; Schwedt, 1993: 87–100, here: 91 ("Daß Paracelsus auch in Portugal war, ist mit Sicherheit anzunehmen") and Zekert, 1963 (who dates the journey in 1517/1518, a wild guess). – Furthermore, Paracelsus mentions Portugal in two astrological divinations and tells us in his *Opusculum de Meteoris* that a special wind exists there. We thank Dr. Urs Leo Gantenbein from the "Zurich Paracelsus Project" for this information.
54 For early "Paracelsians" see Moran, 2005: 67–98 [Chapter 3] and Kühlmann/Telle, 2013.
55 Sudhoff, 1894: 149 f., Nr. 95.
56 Sudhoff, 1894: 159 f., Nr. 100.

was edited by the Bavarian ducal court doctor, the "Hofmedicus" Johannes Albert Wimpinäus (flor. 1563-1570) in his collection *Philippi Theophrasti Paracelsi von Hohenhaim, etliche Tractetlein zur Archidoxa gehörig* (Wimpinäus, 1570)[57]. Only in 1570, were all six sections of the *Herbarius* published in one collection for the first time, edited by the Tyrolean "poeta laureatus" Michael Toxites (Michael Schütz, 1514-1581, an editor of numerous medical and alchemical studies), as the second part of the *Ettliche Tractatus Des Hocherfarnen vnnd berümbtesten Philippi Theophrasti Paracelsi, der waren Philosophi vnd Artzney Doctoris* (Toxites, 1570: 286-532)[58]. As the Paracelsian *Herbarius* was printed much later, it seems that this part of the miscellany must have been written after 1570. This can be verified with the specific page numbers in Thurneysser's extract, which corresponds exactly with the Paracelsus edition of Toxites[59], and by a remark in a letter from Johannes Montanus (1531-1604) to Thurneysser (written in Striegau on 16th April 1574)[60].

The miscellany, probably composed and bound not until the end of the sixteenth century, contains a lot more material of great interest, which we have not studied yet. Among them, the various botanical descriptions (fol. 257 r-283 v, ordered from A to Z) with 27 illustrations of drawings of plants (fol. 275v-276 r; 279 r-283 v); furthermore, a register of herbs, animals, birds and other natural arts, histories and diseases (*Register der kreütern Thierenn Vogeln vnnd andere Naturliche[n] Kunstenn vnnd Historien vnd Kranckheitten*), with detailed comments (fol. 285 r-315 r) as well as an extensive catalogue of medicinal drugs (fol. 355 r-449 r)[61], that are indexed in alphabetical order according to their medical use (fol. 395 r-449 r).

---

57 Sudhoff, 1894: 213-215, Nr. 128. Cf. also 190-196, Nr. 119 and 215-217, Nr. 128.
58 Sudhoff, 1894: 196-198, Nr. 120.
59 The Bavarian State Library in Munich, which preserves the print (Toxites, 1570) also has a similar index (on the flyleaf) by an unknown scribe, who has interestingly chosen the same topics as Thurneysser.
60 Cf. Kühlmann/Telle, 2013: 383-391 (Nr. 113), here: 385, line 20 and 389.
61 Cf. the contemporary print by Clusius, 1567, who made a translation of the *Colóquios dos simples e drogas he cousas medicinais da Índia e assi dalgũas frutas achadas nella onde se tratam algũas cousas tocantes a medicina, pratica, e outras cousas boas pera saber* ("Conversations on the simples, drugs and materia medica of India and also on some fruits found there, in which some matters relevant to medicine, practice, and other matters good to know are discussed"; de Orta, 1563), a work of great originality published in Goa by the Portuguese Jewish physician and naturalist Garcia de Orta (1501-1563), who was a pioneer in tropical medicine, cf. Lopes, 2006: 20-24; Nobre de Carvalho, 2015; Rozeiro, 1970 and Vogl, 1887.

3. The Iberian Flora and Fauna described in the manuscript

After this short overview of the various contents of this exceptional miscellany containing mainly excerpts of books printed around 1570, we will now have a final look on the more original (Portuguese) part of the manuscript, where Thurneysser writes in four chapters (fol. 1 r–144 v and 317 r–354 v) about the natural scientific knowledge of the Iberian flora and fauna. This is of great importance, because the Renaissance craftsman Thurneysser provides us with a thorough description of Portuguese plants and animals, which he probably started to write while he was living in the house of the famous Portuguese humanist, Damião de Góis (1502–1574)[62] in Lisbon in 1555/1556, as the title of his first chapter tells us (fol. 1 r: *Zu Lysabon angefangen Anno* CHRISTI *1555 und 1556 Inn der behausung dess Edlen Herren vnnd Lusitanischen Ritters Herren Damiani de Goës umb die Zeitt des Solstitii Aestivi*[63], cf. fig. 2).

Fig. 2: Title page of the first chapter about the flora and fauna of Portugal (Staatsbibliothek zu Berlin, Ms. Germ. Fol. 97, fol. 1 r).

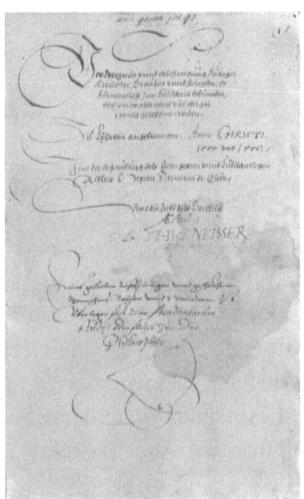

---

62 For the life and work of the famous Portuguese Humanist Damião de Góis, who also wrote about the city of Lisbon in his book *Urbis Olisponis Descriptio* (de Góis, 1554/ de Góis, 1996), cf. Bataillon, 1982; Beau, 1941; Hirsch, 1967; Marques, 1959; Matos, 2002–2006 and Rodrigues, 2002.

63 We also know that Thurneysser began his study around midsummer solstice in 1555, which was between 19th and 25th June 1555.

The first chapter (fol. 1 r–109 v) deals with Lusitanian herbs, bush plants and fruits (*Verzeichnus vnnd Beschreibung Etzlicher Kreütter, Stauden vnnd Früchten, so fürnemlich inn Lusitania befunden, bey vnns aber nicht viel oder gar wenig gesehen worden*) starting with an alphabetical index (referring to the red numbering of this chapter). After this, Thurneysser describes 32 plants in detail. Among them, one can find meticulous delineations of the "Malmakeiis" (in English: "Daisy", in German: "Gänseblümchen"), of which the description starts as follows:

> Was dieser plantæ generalem descriptionem belanget. Es ist <u>ein fruticulus oder Steüdlein, einer halben Elen, bißweilen</u> auch fast <u>dreij Viertheil einer Elen hoch,</u> wellches <u>etwas äschenfarbig</u> ist. <u>Hat</u> inn der <u>erst einen Geruch dem Fenchel nicht vngleich,</u> aber der <u>hernacher etwas stinckhet,</u> [...] <u>Inn den Hundstagen</u> kumbt es gemeinigleich vmb, vnnd <u>verdirbt vonn wegen grosser Hitz</u> (fol. 5 r).

This plant, which was depicted in various contemporaraneous drawings[64], is very common in Central Europe. This is why Thurneysser used contemporary plant books (here: Brunfels, 1532–1537: vol. 1: CXLIIII and CXLV) as his source; but his characterizations (fol. 5 r–10 r) are much more detailed than these models.

Conspicuously, the manuscript contains various plants, which can be found particularly or only on the Iberian Peninsula, such as the "Orches Lusitanicae" (fol. 57 r–59 r: "Lusitanische Hunndtshödlein", a form of the "Knabenkraut"[65] from the orchid family), various species of "Gladiolus" ("Schwertlilie"; fol. 11 r–12 r), moreover the "Medronho" ("Erdbeerbaum" or Strawberry Tree; fol. 13 r/v), the "Marmelos" ("Quitten"/Quince; fol. 17 r/v) or the "Verus Tamariscys" ("Tamarisken"/Tamarix; fol. 19 r). It still needs to be determined which sources Thurneysser used, and if these plants also play a role in any of his later publications, for instance in the *Historia sive descriptio plantarvm omnium*, which was printed in Berlin in 1578 in Latin and German (Thurneysser, 1578 a and b)[66] or in the *Magna Alchymia* (Thurneysser, 1583)[67].

Furthermore, the reader of the manuscript learns something about the use of fruits, when Thurneysser notices that the "Lusitani" made juice and jam of the "Marmeleiros":

---

64 Cf. the watercolor by Weiditz (1529) in Platter's herbarium (Burgerbibliothek Bern, Inv. Nr. Es 71, fol. 45), cf. Vogel, 2014: 14, Fig. 9 and Brunfels, 1532–1537: Vol. 1: CXLIIII (drawing) and CXLV (description).
65 Cf. Sauerhoff, 2001: 155.
66 For the Herbal books of Thurneysser cf. Spitzer, 1996: 84–89.
67 In Thurneysser, 1583: Liber Sextus: 86 he also refers to "*Lusitaniae*".

Die Lusitani richten zu vnd machen diese ire Marmellen ein mit dem succo oder Safft der Arantien, darvon sie dann gar durchsichtig vnnd lautter oder klar werden, vnnd wellche confectio vonn inen Marmolla[da] genanndt wirdt. [...] Es haben aber die Arantien fast einen solchen Geruch wie die wilde Feldt Rueben wann sie blüehen (fol. 17 v).

Unfortunately we do not have any illustrations of the Portuguese fruits and animals in this manuscript, but Thurneysser seems to have seen them in another book or manuscript, because he often refers to such figures with the words like "wie an derselbigen Figur so fol. 9 facie 1 deliniret ist" (fol. 22 r, beginning of the description of the plant "Numularia", and again on fol. 24 r).

Further species mentioned here are the "Lemtisco" (fol. 31 r; like the "willow" ["Weiden"] in Germany) or the "Darvera" ("rowan tree"; fol. 33 r–34 r). Thurneysser also tells us that the rowan berries ("Vogelbeeren") were used by the Lusitanian women for the eyes and to whiten their teeth:

DARVERA ist wie beij vnns die Ebereschen. [...] Letztlichen schmieren sie das Angesicht darmit, vnnd wirrdt vonn inen sehr lieb vnnd wertt gehalltenn. Mit desselbigen Holtze wetzen oder scherpfen sie auch ire Zehnen, dann wenn man dasselbige nach dem Essen inn den Mund nimbt, vnnd zerkauet, vnnd allso die Zenen darmit reibet, so werdenn sie gar schon weiß darvonn. Doher es vonn den Nigritis, den schwartzen Mohren oder ætijopischen Völckheren sehr gebraucht wierdt (fol. 33 r).

Elaborate is also item Nr. 17 with the description of the "coccus[68] vonn den Lusitanis GRAN genanndt, wellches sie gebrauchen anstatt der Gallöpfel", which was used in Lusitania instead of oak galls (fol. 43 r–50 r). Thurneysser informs us here that the "Grana Coccinea" was collected by Lusitanian women who lubricate their feet with garlic ("Die Weiber die dieselbige Grana Coccinea einsamlen vnd colligieren, die schmieren oder salben ire Füeß vnnd Stifel mit Knoblauch, vnnd essen denselbigen auch dazumahl", cf. fol. 45 v) and that they make powder of it, which was used to color their clothes (cf. fol. 46 v/47 r). The best Grana could be found in the Portuguese region Alentejo, which is romanticized by him here (fol. 48 v).

In the second chapter of his manuscript (fol. 111 r–127 v), Thurneysser describes Lusitanian animals and fishes, which were rarely seen in his own country ([*Verz*]*eichnus vnnd Beschreibung Ettlicher Tierenn, vnnd sunderlich Wasserthierleinn, so inn Lusitanis erfunden, bey vnns aber nicht viel gesehenn werden*). This part, signed by "Eines gela[e]rtem dieffsinnigen vnnd geschicktem Menschens Reysen vnnd Wannderer, Ist vberlegen fast aller Academiarum studijs vnnd fleiß Inn der Philosophia" starts with a motto by Plutarch (*Peregrinatio alit sapientiam*) and the German rimes

---

68 Cf. also Clusius, 1576: Cap. VI, 33–38: "De Ilice cocigera".

Wer Wannderenn thuet durch frembde Lanndt, Dem wierdt viel seltzam dinngs bekanndt, Erreicht dardurch Weissheit, Verstanndt, vnnd kumbt offt gros glick zuhanndt (fol. 111 r).

The short description of animals deals mostly with ichthyology. Fishes like the "Lamprea Lusitanica" (Lamprey/"Neunauge"; fol. 113 r–116 v) are mentioned. Thurneysser provides us here also with a list of species fished in the Tagus River (fol. 118 v–122 v: "Pisces in Lvsitania et praecipue ad Ostium Tagi Olispone. Das ist Erzehlunng oder Beschreibung der Fischen so inn Lusitania vnnd fürnemblich inn dem fluss Tago zu Lysabon gefanngen werden", cf. fig. 3) and a separate list with more than 80 fishes in Portugal (fol. 124 v–127v: PISCES IN PORTVGALLIA Etc., again referring to the folio pages of a book (or manuscript)[69] with images of these animals.

Fig. 3: The second chapter deals with lusitanic fishes (Staatsbibliothek zu Berlin, Ms. Germ. Fol. 97, fol. 118 v).

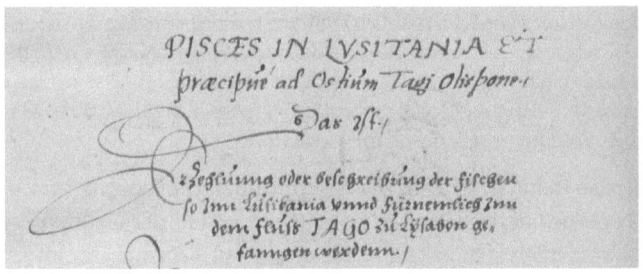

Further paragraphs deal with "Marinæ Conchæ" (slugs; fol. 122 v–124 v) and six names of „Monstris Marinis – Wunder thier des Meers" (sea monsters; fol. 127 v). It is interesting to see that the information given here again always refer to another manuscript, where Thurneysser probably excerpted his information from, although he sometimes refers directly to his Lusitanian wanderings and his stay in Lisbon, too:

Deßgleichen auch Anno 1555, do ich zu Lysabon gewesenn, seindt doselbst auch zween zimliche grosse Stier gefangen worden (fol. 119 r).

---

69 Contemporary is the great zoological work by Gessner, 1551–1558: the *Historia Animalium* (Description of animals) which was divided in four parts: quadrupeds, amphibians, birds and fishes; a fifth part about scorpions and snakes was published posthumously in 1587. A German translation (*Thierbuch*) of the four parts was published in Zürich (Forer, 1563). Cf. in general to zoology in the early Modern Europe: Guerrini, 2004.

or even directly to Damião de Góis:

> Es sagt der Edle Herr Damian de Goes, dass dieselbigen inn grosse mennge bey Bremoia gefanngen (fol. 115 v).

The third chapter (fol. 129 r–143 v) of the Portuguese part discusses mixed matters (*Miscellanea. Historica Geographica medica e[t] varie mixta*). Therein, Thurneysser describes with lively words the brutal rituals of the merchants and sailors from Norway (fol. 130 v–133 v). Furthermore, he gives also a detailed description of the black people he had probably seen on the *Rua Nova dos Marcadores*[70] in Lisbon (*Aethiopvm vel Nigritarum d[e]scriptio: Beschreibunng der Mohren Nigriten vnnd Äthiopier*; fol. 133 v–142 r, cf. fig. 4). From a modern perspective this is the most interesting and original section of the entire manuscript. Alongside one can also find here a very short essay about the ocean current called "Aestus Maris: Vonn dem Anlauff vnnd Ablauff des Meers, das ist vonn der Meeres fluet"(fol. 142 r–143 v).

Fig. 4: The manuscript also contains a description of the black people (Staatsbibliothek zu Berlin, Ms. Germ. Fol. 97, fol. 133 v).

In the last part (fol. 317 r–353 v, without title), Thurneysser gives further descriptions of animals, mainly of birds (fol. 317 r–322 v)[71], but also "De quadrupedibus. Von vierfüssigen Thieren" (of quadrupedal animals; fol. 323 r–328 v). He mentions monkeys, mules, snakes and rabbits ("*Cuniculi*", fol. 328 v). The Portuguese part ends with a description of Iberian plants, herbs and trees (fol. 329 r–354 v).

---

70 Jordan Gschwend/Lowe, 2015.
71 For the description of birds in the sixteenth century see especially the book of birds by Conrad Gessner (Volume 3 of his *Historia Animalium* from 1555) cf. Springer/Kinzelbach, 2009.

4. Perspective and Conclusion

The total transcription of the manuscript, which focuses on botanical knowledge[72], will allow us to discover also the sources used by Thurneysser. In particular, this text must be compared with other (later published) books, such as the book of herbs (*New Kreüterbuch* translated by Handsch, 1563) by the Italian naturalist Pietro Andrea Mattioli (1501–1577)[73], or the studies of the Flemish botanist Charles de l'Écluse (Carolus Clusius, 1526–1609)[74], which are also important for Portugal, because there Clusius "described about two hundred previously unknown plants" (Ogilvie, 2006: 52) in his book *Rariorum aliquot stirpium per Hispanias observatorum Historia* ("History of some rare species observed in Spain and Portugal"; Clusius, 1576)[75].

Moreover, the palaeography and the watermarks of the manuscript remain to be studied in detail (as well as linguistic characteristics), which will aid in the exact dating of the manuscript. Even if Thurneysser himself claims that he started to write the manuscript in 1555/1556 (fol. 1 r) it does not seem that

---

72 Kusukawa, 2006. Cf. also Smith/Findlen, 2002.
73 To Mattioli, who also wrote a commentary of Dioscurides (Mattioli, 1554) cf. Daxecker, 2004; Ferri, 1997 and Kusukawa, 2012: 162–177 ("The Authority of Pictures: Gessner, Mattioli and Jamnitzer").
74 Clusius' career as a naturalist started at the universities of Wittenberg and Montpellier. In 1557 he translated Dodoen's *Cruydeboeck* (Clusius, 1557) into French. In 1564/1565 he accompanied his pupil Jakob III Fugger (1542–1598, son of the German banker Anton Fugger) across the Iberian Peninsula cf. Barona, 2007; Ogilvie, 2006: 184 and Pardo-de-Santayana/Tardío/Morales, 2014: 32–34. Only in 1593, in the age of 67, was he honoured with a largely honorary professorship for botany at the University of Leiden, where he established the "hortus academicus" one of the oldest botanical gardens in the world. His *Exoticorum libri decem, quibus Animalium, Plantarum, Aromatum, aliorumque peregrinorum Fructuum historiae describuntur* ("Ten books of exotica: the history and uses of animals, plants, aromatics and other natural products from distant lands"; Clusius, 1605) is a famous illustrated zoological and botanical compendium, which contains translated and edited versions of earlier publications. Clusius, who never practiced as a physician, but had travelled extensively, died in Leiden in 1609. For his life and work (his description of plants are much more detailed than those provided by Bock and Dodoens) cf. Egmond, 2010; van Ommen, 2009. For his description of Portuguese and Spanish plants in particular cf. Ramón-Laca Menéndez de Luarca/Morales Valverde, 2005. – Ogilvie, 2006: 24 states that Iberian botanists "played little role in elaborating the Renaissance science of describing".
75 This work can be considered the first flora of the Iberian Peninsula. It included 323 plant species and 225 drawings.

this is the original text, because the handwriting does not compare with other documents written by him. In fact we can identify different hands throughout the miscellany, as well as the hand of his scribe, Adam Seidel[76], which means the manuscript is maybe a later copy.[77]

Furthermore, it is difficult to correspond the dates of his travel, with other dates of his biography: In 1555 he worked in Strasbourg and Konstanz (Moehsen, 1783: 57) and then he made a detour to Italy, about which he reports in the *Pison* (Thurneysser, 1572: CCXXXV: *Thurneysser saufft sich kranck*[78]), a voluminous work about rivers and their mineral waters[79]: On the 9th of September he drank water from a well in "Spriling"[80] between Lake Como and Graubünden, which made him ill. For more than five months he laid sick in Basle, until February 1556.[81] This means that if he had really started to write the manuscript around midsummer equinox (June 1555), he could not have stayed

---

76 Cf. Spitzer, 1996: 59 and 69. Schumacher, 2011: 224 claims that Adam Seidel is from Basle, but there is no proof for this. In a document (Ms. Germ. Fol. 423, Bl. 48r aZ and 65r nZ) mentioned by Spitzer, 1996: 53 he calls himself "Bürger und Buchdrucker zu Berlin jetziger Zeit des herrn Lienhart Thurneysser zum Thurn diener". Thurneysser's third wife, Marina Herbrott († 1610), returned to Switzerland in 1582 after her expulsion from Berlin (Schumacher, 2011: 259; Spitzer, 1996: 110). Daniel Seidel, the son of Adam, worked in Thurneysser's officine as a woodcutter from 1571 to 1583. For Seidel cf. also Spitzer, 1996: 24, 46, 54, 68–70, 74, 102 and 128.
77 Only the red marginal notes were written by Thurneysser, who corrected the text.
78 Exact text: "auff dem Spriling, welches ein hoher Berg, so zwischen dem Cumersee, vnd den GrawPüntern oder Curwalen gelegen ist, aus welchem Brunnen ich Anno 1555. den 9. Septembris, als ich aus Italia gen Feldkirch, vn[d] dem Bodensee zuzog, ein trunckh thet, vom dem ich erkrancket, vnd 22 Wochen zu Basel töedtlich kranck lag, da alle Artzte sagten, das ich weder Lung noch Leber mehr im Leib het, vnd nicht leben möcht" (Thurneysser, 1572: CCXXXV).
79 The *Pison* comprises ten books. Thurneysser describes the waters of the Danube (book 5, with affluents), the Rhine (book 6), the Elbe and the Spree. He also mentions rivers in Spain, cf. Thurneysser, 1572: II, VI and CCLIII. Interestingly in the same work he gives an incorrect date (22th July 1530?) for his baptism: "dem Ort, da unser zeit Sanct Leonharts Closter vnd Pfarrkirch stehet, in der ich von der gnaden Gottes Anno 1530. Den 22 tag Hewmonats getaufft bin" (Thurneysser, 1572: CCXXXVII). As Macco, 1934: 77 f. has demonstrated Thurneysser was baptized on 6th August 1531, probably born on 22th July of the same year.
80 This may be the Splügenpaß, identified by Schumacher, 2011: 52.
81 Moehsen, 1783: 59 reports that he was also acherontic in 1558 in Konstanz and healed himself ("welches ganz Kostniz in Verwunderung gesezt").

longer than ten weeks in Portugal.[82] Equally doubtful is also his alleged second journey through the Iberian Peninsula (supposedly in 1561), mentioned by some of his biographers (Bauer, 1893: 11; Harms, 1963 b: 302; Hofmeier, 2007: XXXV; Schumacher, 2011: 82)[83], often in relation with his "Grand Tour". The tour supposedly included England, Scotland and later Egypt, the Middle East, Greece, Italy and Hungary (cf. Spitzer, 1996: 17 and Boerlin, 1976: 15); but in fact there are no written sources which confirm this trip. Much more biographical research needs to be undertaken in the future (in particular finding facts regarding the end of his life). With this objective, a closer look at Thurneysser's extensive correspondence, which is also preserved at the Staatsbibliothek zu Berlin-Preußischer Kulturbesitz[84], will bring new insights about this Renaissance autodidact, who was portrayed by the Flemish painter, Frans Floris de Vriendt (1517–1570)[85]. Our planned edition of this rediscovered miscellany, Ms. Germ. Fol. 97, will at last provide access to this invaluable source which documents natural scientific knowledge of the Iberian flora and fauna in the mid-Renaissance.

Bibliography

Andresen, Andreas. 1973. *Jost Amman, 1539–1591. Graphiker und Buchillustrator der Renaissance. Beschreibender Katalog seiner Holzschnitte, Radierungen und der von ihm illustrierten Bücher. Mit einer biographischen Skizze und mit Registern seines Werkes und der Autoren illustrierten Bücher* (Scripta artis monographia 5). Amsterdam: G. W. Hissink, 100–448 [Reprint of the 1864 edition,

---

82 Hofmeier, 2007: XXXIV even thinks that Thurneysser was only in Basle from 1555 to 1558 (the year when his son Hans Jakob was born). – Mr. Schumacher kindly informed us that during the winter months (with a few hours of sunshine), a long travel would not have been sensible at that time.

83 His first biographer Moehsen, 1783 who wrote a whole chapter (55–66) about his journeys, only mentions his second (?) voyage to Iberia in 1561. He claims (Moehsen, 1783: 59) that these trips (and also the journey to Scotland in 1560) were undertaken upon the directive of Archduke Ferdinand II of Tyrol (1529–1595), who was married to Philipine Welser (1527–1580).

84 Staatsbibliothek zu Berlin, Preußischer Kulturbesitz, Ms. Germ. Fol. 420–426 (11 vols.), cf. Degering, 1925: 48, Nr. 420 a–426 and Spitzer, 1996: 26–31.

85 Cf. the image in the Öffentliches Kunstmuseum, Basel, as cited by Boerlin: 1976, 153, which was painted during his time in Münster in Westfalen (1569/1570). Thurneysser was also portrayed by Hermann tom Ring in the same year.

published by Danz, Leipzig, which was issued as part of vol. 1 of the author's *Der deutsche Peintre-Graveur*].

Aufmesser, Max. 2000. *Etymologische und wortgeschichtliche Erläuterungen zu „De materia medica" des Pedanius Dioscurides Anazarbeus* (Altertumswissenschaftliche Texte und Studien). Hildesheim et al.: Olms-Weidmann.

Baader, Gerhard. 1978. „Mittelalter und Neuzeit im Werk von Otto Brunfels". *Medizinhistorisches Journal* 13/3–4, 186–203.

Barona, Josep L. 2007. "Clusius' exchange of botanical information with Spanish scholars". In: Egmond, Florike et al. (eds.). *Carolus Clusius towards a cultural history of a Renaissance naturalist* (The history of science and scholarship in the Netherlands 8). Amsterdam: Koninklijke Nederlandse Akademie van Watenschappen.

Bataillon, Marcel (ed.). 1982. *Damião de Góis: humaniste européen. Études présentées par José V. de Pina Martins* (Centre de Recherches sur le Portugal de la Renaissance: Études 1). Paris: Touzot.

Bauer, Alexander. 1893. *Die Adelsdocumente österreichischer Alchemisten und die Abbildungen einiger Medaillen alchemistischen Ursprunges* (Monographien des Museums für Geschichte der Österreichischen Arbeit 3). Wien: Hölder.

Baumann, Brigitte/Baumann, Helmut (eds.). 2001. *Die Kräuterbuchhandschrift des Leonhart Fuchs*. Stuttgart: Eugen Ulmer Verlag.

Baumann, Susanne. 1998. *Pflanzenabbildungen in alten Kräuterbüchern. Die Umbelliferen in der Herbarien- und Kräuterbuchliteratur der frühen Neuzeit* (Heidelberger Schriften zur Pharmazie- und Naturwissenschaftsgeschichte 15). Stuttgart: Wissenschaftliche Verlagsgesellschaft.

Baumann-Schleihauf, Susanne. 2001. „Kräuterbücher und die Fuchsie erinnern an Leonhart Fuchs". *Pharmazeutische Zeitung* 146/6, 10–15.

Bautz, Friedrich Wilhelm. 1975. Art. „Braunfels (Brunfels), Otto (Otho)". *Biographisch-Bibliographisches Kirchenlexikon (BBKL)* 1, column 735 f.

Beau, Albin Eduard. 1941. *As Relações Germânicas do Humanismo de Damião de Góis*. Coimbra: Publicações do Instituto Alemão da Universidade de Coimbra.

Beck, Hanno. 1982. *Große Geographen. Pioniere – Außenseiter – Gelehrte*. Berlin: Dietrich Reimer Verlag.

Becker, C. 1838. „Leonhart Thurneisser zum Thurn. Mit besonderer Rücksicht auf seinen Aufenthalt in Münster und Berlin". *Zeitschrift für vaterländische Geschichte und Altertumskunde* 1, 239–264.

Behling, Lottlisa. 1957. *Das Kräuterbuch des Otto Brunfels und die Pflanzenaquarelle des Hans Weiditz*. Köln et al.: Böhlau.

Behrmann, Walter. 1954. „Hermann Wagner als akademischer Lehrer". *Die Erde. Zeitschrift der Gesellschaft für Erdkunde zu Berlin* 85/6, issue 3–4 [Festschrift für Otto Quelle zum 75. Geburtstag, 217–376], 362–368.

Benzenhöfer, Udo. ³2003. *Paracelsus* (Rowohlts Monographien 50595). Reinbek bei Hamburg: Rowohlt.

Bergholz, Thomas. 2005. Art. „Bock, Hieronymus". *Biographisch-Bibliographisches Kirchenlexikon (BBKL)* 25, column 81–86.

Bertelli, Carlo/Lilla, Salvatore/Cavallo, Guglielmo (eds.). 1988/1992. *De materia medica. Codex Neapolitanus, Napoli, Biblioteca nazionale, Ms. Ex. Vindob. Gr. 1.* 2 Bde. Roma: Salerno Editrice and Graz: Akademische Druck- und Verlagsanstalt.

Biedermann, Hans. ²1978. *Medicina Magica. Metaphysische Heilmethoden in spätantiken und mittelalterlichen Handschriften. Mit 30 Faksimile-Tafeln.* Graz: Akademische Druck- und Verlagsanstalt.

Bleker, Johanna. 1976. „Chemiatrische Vorstellungen und Analogiedenken in der Harndiagnostik Leonhart Thurneissers (1571 und 1576)". *Sudhoffs Archiv für Geschichte der Medizin und der Naturwissenschaften* 60/1, 66–75.

Bock, Hieronymus. 1546. *Kreütter Buͦch. Darin underscheydt, würckung und Namen der Kreiter so in Deutschen Landen wachsen. Auch der selbigen eigentlicher vnd wolgegründter gebrauch inn der Artznei fleissig dargeben. Leibs gesundeit zuͦ behalten und zuͦ fürderen seer nutzlich vnd tröstlich / Vorab dem gemeinem einfaltigen man. Von newem fleissig übersehen, gebessert, vnd gemehret, Dazuͦ mit hüpschen artigen Figuren allenthalben gezieret.* Straßburg: Wendel Rihel der Ältere.

Bock, Hans-Joachim. 1963. „Das Ibero-Amerikanische Institut". *Jahrbuch der Stiftung Preußischer Kulturbesitz* 1 (1962), 324–345.

Bodenstein, Adam von. 1568. *Herren Doctors Theophrasti Paracelsi declaration, zuobereyten Hellebori, inn sein arcanum, dardurch infectiones der vier Elementen außtriben werden: Darzuo getruckt ein caput von Perforata.* Basel: Samuel Apiarus für Peter Perna.

Böhm, Wolfgang. 1974. „Hermann Wagner und die Geographie an der Universität Königsberg". *Jahrbuch der Albertus-Universität zu Königsberg* 24, 196–201.

Boerlin, Paul H./Münster, L. 1960. „Der Alchemist Leonhard Thurneysser als Anatom auf einem Glasgemälde des 16. Jahrhunderts". *Ciba-Symposium* 8/1, 32–36.

Boerlin, Paul H. 1976. *Leonhard Thurneysser als Auftraggeber. Kunst im Dienste der Selbstdarstellung zwischen Humanismus und Barock.* Basel/Stuttgart: Birkhäuser Verlag.

Borsdorf, Axel/Leser, Hartmut. 2003. „Herbert Wilhelmy †". *Die Erde. Zeitschrift der Gesellschaft für Erdkunde zu Berlin*, Heft 1, 114 f.

Brauer, Adalbert. 1968. „Otto Quelle 1879–1959". In: *150 Jahre Rheinische Friedrich-Wilhelms-Universität zu Bonn 1818–1968. Mathematik und Naturwissenschaften*. Bonn: Bouvier et al., 215–222.

Breslau, Ralf. 1995. *Verlagert, verschollen, vernichtet ... Das Schicksal der im 2. Weltkrieg ausgelagerten Bestände der Preussischen Staatsbibliothek*. Berlin: Staatsbibliothek zu Berlin, Preussischer Kulturbesitz.

Brinkhus, Gerd/Pachnicke, Claudine (eds.). 2001. *Leonhart Fuchs (1501–1566). Mediziner und Botaniker. Ausstellung im Stadtmuseum Tübingen, 21. Juni bis 16. September 2001* (Tübinger Kataloge 59). Tübingen: Kulturamt.

Brunfels, Otto. 1530–1536. *Herbarum vivae eicones*. 3 vols. Argentoratum [Straßburg]: Schott.

Brunfels, Otto. 1532–1537. *Contrafayt Kreüterbuch. Vol. 1: Nach rechter vollkommener Art, vnud Beschreibungen der Alten, besstberümpten Ärtzt, vormals in Teütscher sprach, der masßen nye gesehen, noch im Truck außgangen. Sampt einer gemeynen Inleytung der Kreüter Urhab, Erkantnüsß, Brauch, Lob, und Herrlicheit. Vol. 2: Ander Teyl des Teütschen Contrafayten Kreuterbůchs*. Straßburg: Hans Schott.

Bugge, Günther. 1939. *Der Alchimist. Die Geschichte Leonhard Thurneyssers des Goldmachers von Berlin*. Berlin: Limbert [in total six editions].

Bulang, Tobias. 2013. „Zur Diskursivierung pflanzenkundlichen Wissens bei Leonhard Thurneysser zum Thurn". In: Burkard, Thorsten et al. (eds.). *Wissensdiskursivierungen. Themen, Medien und Räume des Wissens vom 14. bis zum 18. Jahrhundert*. Berlin: De Gruyter, 39–62.

Carreras, Sandra. 2004. „Die Quesada-Bibliothek kommt nach Berlin: zu den Hintergründen einer Schenkung". In: Carreras, Sandra/Maihold, Günther (eds.). *Preußen und Lateinamerika: im Spannungsfeld von Kommerz, Macht und Kultur*. Münster: Lit, 305–320.

Clusius, Carolus [L'Écluse, Charles de] (ed.). 1557. *Histoire des plantes, en laquelle est contenue la description entiere des herbes, c'est à dire, leurs especes, forme, noms, temperament, vertus & operations; non seulement de celles qui croissent en ce païs, mais aussi des autres estrangeres qui viennent en usage de medecine. Par Rembert Dodoens, medecin de la Ville de Malines; nouvellement traduite de bas Aleman en François par Charles de l'Escluse*. Anvers: de l'Imprimerie de Iean Loë.

Clusius, Carolus [L'Écluse, Charles de] (ed.). 1567. *Aromatum et simplicium aliquot medicamentorum apud Indos nascentium historia. Garcia ab Horto auc-*

tore. Nunc verò primùm Latina facta & in epitomen contracta a Carolo Clusio. Antverpiae: Plantin.

Clusius, Carolus [L'Écluse, Charles de]. 1576. *Rariorum aliquot stirpium per Hispanias observatarum Historia.* Antverpiae: Ex officina Christophori Plantini.

Clusius, Carolus [L'Écluse, Charles de]. 1605. *Exoticorum libri decem, quibus Animalium, Plantarum, Aromatum, aliorumque peregrinorum Fructuum historiae describuntur.* [Lugdunum Batavorum]: Raphelengius.

[s.n.] 1965. *Conrad Gesner. Physician, Scholar, Scientist 1516–1565. A quatercentenary exhibit held November-December 1965 in the National Library of Medicine.* Bethesda, Maryland.

Daxecker, Franz. 2004. „Der Botaniker und Arzt Pietro Andrea Matthioli". *Klinische Monatsblätter für Augenheilkunde* 221, 516 f.

Daveau, Suzanne. 1987–1991. *Geografia de Portugal por Orlando Ribeiro e Hermann Lautensach. Organização, comentários e actualização de Suzanne Daveau.* 4 vols. Lisboa: João Sá da Costa [six editions].

De Cock, Alfons. 1890. *Rembert Dodoens* (Volksboekjes 8). Gent: Vuylsteke.

Degering, Hermann. 1922. „Kleine Mitteilungen". *Zentralblatt für Bibliothekswesen* 39, 344 [about the excerpts by Thurneysser of the "Gifftjager" of Wilhelm Klebitz].

Degering, Hermann. 1925. *Kurzes Verzeichnis der germanischen Handschriften der Preussischen Staatsbibliothek (Berlin), volume I: Die Handschriften in Folioformat* (Mitteilungen aus der Preußischen Staatsbibliothek VII). Leipzig: Karl W. Hiersemann [unveränderter Nachdruck Graz: Akademisch Druck- und Verlags-Anstalt, 1970].

Delille, Maria Manuela Gouveia. 1969. „In Memoriam Albin Eduard Beau (1907–1969)". *Revista Portuguesa de Filologia* 15, 789–793.

Dilg, Peter. 2010. „Leonhart Fuchs: Arzt – Botaniker – Humanist". In: Köpf, Ulrich (ed.). *Die Universität Tübingen zwischen Reformation und Dreißigjährigem Krieg. Festgabe für Dieter Mertens zum 70. Geburtstag* (Tübinger Bausteine zur Landesgeschichte 14). Ostfildern: Thorbecke, 235–248.

Dobat, Klaus/Dressendorfer, Werner (eds.). 2001. *Leonhart Fuchs: The New Herbal of 1543.* Köln et al.: Taschen.

Dodoens, Rembert. 1553. *Trium Priorum De Stirpium historia Commentariorum imagines.* Antverpiae: Ex Officine Ioannis Loci.

Dodoens, Rembert. 1554 a. *Posteriorum trium de stirpium historia commentariorum imagines.* 1554 a). Antverpiae: Ex Officine Ioannis Loci.

Dodoens, Rembert. 1554 b. *Cruydeboeck: in den welcken die gheheele historie, dat es t gheslacht, t fatsoen, naem, natuere, cracht ende werckinghe van den Cruyden*

... *begrepen ende verclaert es, met dersel ver Cruyden natuerlick naer dat leven conterfeyt sel daer by gestelt.* Mechelen: s. n.

Dodoens, Rembert. 1584. *De Sphaera sive de astronomiae et geographiae principiis cosmographica Isagoge.* Antverpiae: Christophorus Platinus (Christofel Plantijn).

Dodoens, Rembert. 1971. *Cruydeboeck. Facs.-herdr. van de oorspronkelijke uitg. van 1554.* Nieuwendijjk: de Forel.

Dubler, César E. 1952/1959. *La 'Materia Medica' de Dioscórides. Transmisión medieval y renacentista.* 6 vols. Barcelona: Tipogr. Emporium.

Egmond, Florike. 2010. *The world of Carolus Clusius. Natural history in the making, 1550–1610* (Perspectives in economic and social history 6). London: Pickering & Chatto.

Eikermann, Erika. 2008. „Leonhard Thurneysser zum Thurn (1531–1596). Apotheker, Arzt, Alchemiker". *Geschichte der Pharmazie* 60, 29–38.

Eikermann, Diethelm. 2012. „Köln im Jahre 1596 und Leonhard Thurneysser zum Thurn (1531–1596)". In: Groten, Manfred/Kaun, Kulia/Soénius, Ulrich S. (eds.). *Jahrbuch 81 des Kölnischen Geschichtsvereins e. V. 2011/2012.* Wien et al.: Böhlau, 85–126.

Eikermann, Diethelm/Kaiser, Gabriele. 2012 a. *Die Pest in Berlin 1576. Eine wiederentdeckte Pestschrift von Leonhart Thurneisser zum Thurn (1531–1596).* Rangsdorf: Basilisken-Presse.

Eikermann, Diethelm/Kaiser, Gabriele. 2012 b. „Die Druckwerke von Leonhard Thurneysser zum Thurn (Basel 1531–Köln 1596)". *Gutenberg-Jahrbuch* 87, 171–198.

Feldmann, Reinhard (ed.). 2010. *Illustrierte Kräuter- und Pflanzenbücher der Frühen Neuzeit. Eine Ausstellung der Universitäts- und Landesbibliothek Münster in Zusammenarbeit mit dem Freundeskreis Propstei Clarholz im Museum in der Kellnerei (Klostermuseum Clarholz).* Münster: Universitäts- und Landesbibliothek.

Ferri, Sara (ed.). 1997. *Pietro Andrea Mattioli, Siena 1501–Trento 1578. La vita e le opere con l'identificazione delle piante.* Ponte San Giovanni, Perugia: Quattroemme.

Fichtner, Gerhard. 1968. „Neues zu Leben und Werk von Leonhart Fuchs aus seinen Briefen an Joachim Camerarius I. und II. in der Trew-Sammlung". *Gesnerus. Swiss journal of the history of medicine and sciences* 35, 65–82.

Fischer, Béat de. 1960. *Dialogue luso-suisse: Essai d'une histoire des relations entre la Suisse et le Portugal du 15e siècle à la Convention de Stockholm de 1960.* Lisbonne: Ramos Afonso & Moita.

Forer, Conrad (ed.). 1563. *Thierbuch, das ist ein kurze Beschreibung aller vierfüßigen Tieren.* […] *Erstlich durch den hochgeleerten herren D. Cûnrat Geßner in Latin beschriben yetzunder aber durch D. Cûnrat Forer zû mererem nutz aller mengklichem in das Teütsch gebracht vnd in ein kurtze komliche ordnung gezoge.* Zürich: bey Christoffel Froschower [Christoph Froschauer d. Ä.].

Freudenberg, Matthias. 1999. Art. „Ges(s)ner, Konrad". *Biographisch-Bibliographisches Kirchenlexikon (BBKL)* 15, column 635–650.

Fuchs, Leonhart. 1542. *De historia stirpium commentarii insignes.* Basileae: Isengrin.

Fuchs, Leonhart. 1543. *New Kreüterbuch, in welchem nit allein die gantz histori, das ist namen, gestalt, statt vnd zeit der wachsung, natur, krafft vnd würckung, des meysten theyls der Kreüter so in Teütschen vnnd andern Landen wachsen, mit dem besten vleiß beschrieben, sonder auch aller derselben wurtzel, stengel, bletter, blumen, samen, frücht, vnd in summa die gantze gestalt, allso artlich vnd kunstlich abgebildet vnd contrafayt ist, das deßgleichen vormals nie gesehen, noch an tag kom[m]en.* Basell: Isingrin.

Gantenbein, Urs Leo. 2011. „Leonhard Thurneysen [Thurneysser zum Thurn], 1531–1596". *Historisches Lexikon der Schweiz (e-HLS).* Basel: Schwabe [online: http://www.hls-dhs-dss.ch/textes/d/D14665.php].

Ganzinger, Kurt. 1959. „Ein Kräuterbuchmanuskript des Leonhart Fuchs in der Wiener Nationalbibliothek". *Sudhoffs Archiv für Geschichte der Medizin und der Naturwissenschaften* 43/3, 213–224.

Gessner, Conrad. 1551–1558. *Historiae animalium.* 4 vols. Zürich: Froschauer.

Gessner, Conrad/Moibanus, Johannes (eds.). 1565. *Euporista. Ad Andromachum hoc est de curationibus morborum per medicamenta paratu facilia.* Straßburg: Josias Rihel.

Gliech, Oliver C. 1990. „Das Ibero-Amerikanische Institut (Berlin) in der NS-Zeit: Grundprobleme einer Untersuchung". *Iberoamericana* 14/1, 5–16.

Gloning, Thomas (ed.). 2005. *Cruyde Boeck, Antwerpen 1563. Digitales Faksimile nach dem Exemplar der Universitätsbibliothek Marburg* (Schriften der Universitätsbibliothek Marburg 125). Marburg: Universitäts-Bibliothek [CD-Rom and online: http://archiv.ub.uni-marburg.de/dodoens/welcome.html].

Gmelig-Nijboer, Caroline Aleid. 1977. *Conrad Gessner's "Historia animalium": an inventory of renaissance zoology* (Communicationes Biohistoricae Ultrajectinae 72). Meppel: Krips [Dissertation].

Góis, Damião de. 1554. *Urbis Olisiponis descriptio. In qua obiter tractantur no[n]nulla de Indica navigatione, per Graecos, et Poenos et Lusitanos, diversis temporibus inculcata.* Eborae [Evora]: Andreas Burgensis.

Góis, Damião de. 1996. *Lisbon in the Renaissance. A New Translation of the Urbis Olisiponis Descriptio by Jeffrey S. Ruth*. New York: Italica Press.

Golowin, Sergius. 1993. *Paracelsus – Mediziner – Heiler – Philosoph. Die grosse Biographie zum 500. Geburtstag*. München: Goldmann.

Goodman, David/Russell, Colin A. (eds.). 1991. *The Rise of Scientific Europe 1500–1800*. Sevenoaks: Hodder & Stoughton.

Grévin, Jacques. 1568. *Deux livres des venins ausquels il est amplement discouru des bestes venimeuses, theriaques, poisons & contrepoisons*. Anvers [Antwerp]: Plantin.

Guerrini, Anita. 2004. "Zoology". In: Dewald, Jonathan (ed.). *Europe 1450–1789: Encyclopedia of the Early Modern World. Europe 1450–1789*. Vol. VI. New York et al.: Scribner, 259–261.

Handsch, Georg (ed.). 1563. *New Kreüterbuch Mit den allerschönsten und artlichsten Figuren aller Gewechß dergleichen vormals in keiner sprach nie an tag kommen*. Prag: Georg Melantrich von Aventin/Vincentius Valgriß.

Harms, Bruno. 1963 a. „Leonhard Thurneysser in Berlin. Leben und Wirken". *Der Bär von Berlin. Jahrbuch des Vereins für die Geschichte Berlins* 12, 28–49.

Harms, Bruno. 1963 b. „Leonhard Thurneysser in Berlin". *Berliner Medizin* 14, 301–309.

Hausmann, Frank-Rutger. 2001. „*Auch im Krieg schweigen die Musen nicht". Die deutschen Wissenschaftlichen Institute im Zweiten Weltkrieg* (Veröffentlichungen des Max-Planck-Instituts für Geschichte 169). Göttingen: Vandenhoeck & Ruprecht.

Hausmann, Frank-Rutger. ²2008. „*Vom Strudel der Ereignisse verschlungen". Deutsche Romanistik im ‚Dritten Reich'*. Frankfurt am Main: Klostermann.

[s.n.]. 1930. *Hermann-Wagner-Gedächtnisschrift. Ergebnisse und Aufgaben geographischer Forschung; dargestellt von Schülern, Freunden und Verehrern des Altmeisters der deutschen Geographie* (Petermanns Mitteilungen, Ergänzungsheft 209), Gotha: Perthes.

Herold, Bernardo Jerosch/Horst, Thomas/Leitão, Henrique. 2016a. *O manuscrito alemão de 1555 de Thurneysser com descrições de Portugal* [unpublished working paper, given at the Academia das Ciências de Lisboa on April 7th 2016 by Bernardo Jerosch Herold].

Herold, Bernardo Jerosch/Horst, Thomas/Leitão, Henrique. 2016b. *A "História Natural de Portugal" de Leonhard Thurneysser zum Thurn, ca. 1555–1556* [unpublished working paper].

Hieronymus, Frank. 2005. *Theophrast und Galen, Celsus und Paracelsus. Medizin, Naturphilosophie und Kirchenreform im Basler Buchdruck bis zum Dreissigjäh-

rigen Krieg. Ausstellungskatalog, 5 vols. (Publikationen der Universitätsbibliothek Basel 36). Basel: Universitätsbibliothek.

Hirsch, Elisabeth Feist. 1967. *Damião de Gois. The life and thought of a Portuguese humanist, 1502-1574* (Archives internationales d'historie des idées 19). The Hague: Nijhoff.

Hofmeier, Thomas. 2007. *Leonhard Thurneyssers Quinta essentia 1574. Ein alchemisches Lehrbuch in Versen* (Kleine alchemische Bibliothek 2). Berlin/Basel: Leonhard-Thurneysser-Verlag.

Hoppe, Brigitte. 1969. *Das Kräuterbuch des Hieronymus Bock. Wissenschaftshistorische Untersuchung. Mit einem Verzeichnis sämtlicher Pflanzen des Werkes, der literarischen Quellen der Heilanzeigen und der Anwendungen der Pflanzen.* Stuttgart: Hiersemann [Dissertation].

Janse, Wim. 2001. „Der Heidelberger Zwinglianer Wilhelm Klebitz (um 1533-1568) und seine Stellung im aufkommenden Konfessionalismus". In: Schindler, Alfred (ed.). *Die Zürcher Reformation: Ausstrahlungen und Rückwirkungen. Wissenschaftliche Tagung zum hundertjährigen Bestehen des Zwinglivereins 1997* (Zürcher Beiträge zur Reformationsgeschichte 18). Berlin et al.: Lang, 203-220.

Jordan Gschwend, Annemarie/Lowe, Kate J. P. (eds.). 2015. *The Global City: on the streets of Renaissance Lisbon.* London: Paul Hoberton.

Juntke, Fritz. 1980. „Über Leonhard Thurneisser zum Thurn und seine deutschen Kalender 1572-1584". *Archiv für Geschichte des Buchwesens* 19, column 1349-1400.

Juntke, Fritz. 1980. „Über Leonhard Thurneisser zum Thurn und seine Schriften nach der Flucht aus Berlin (1584)". *Archiv für Geschichte des Buchwesens* 21, column 679-718.

Kahn, Didier. 2014. "The Significance of Transmution in Early Modern Alchemy. The Case of Thurneysser's Half gold Nail". In: Beretta, Marco/Conforti, Maria (eds.). *Fakes?! Hoaxes, counterfeits, and deception in early modern science.* Sagamore Beach: Science History Press, 35-68.

Kaiser, Gabriele. 2016. „Leonhart Thurneysser zum Thurn (1531-1596) und sein Nachlass in der Staatsbibliothek zu Berlin" [in this volume], 121-132.

Kalwa, Erich. 2004. *Die portugiesischen und brasilianischen Studien in Deutschland (1900-1945). Ein institutionsgeschichtlicher Beitrag* (Beihefte zu Lusorama, 2. Reihe: Studien zur Literatur Portugals und Brasiliens 16). Frankfurt am Main: DEE Domus Ed. Europaea.

Kellenbenz, Hermann/Schneider, Jürgen. 1987. „Geschichte". In: Stegmann, Wilhelm (ed.). *Deutsche Iberoamerika-Forschung in den Jahren 1930-1980: For-*

schungsberichte über ausgewählte Fachgebiete (Bibliotheca lbero-Americana 32). Berlin: Colloquium Verlag, 43–80.

Klebitz, Wilhelm [Triphyllodacnus, Wilhelmus]. 1567. *Gifftjager. Das ist: Von vrsach, reynigung, bewahrung vnd Cur Pestilentzischer Luft, fürnemer Artzten rath und bedencken, mit angehencktem vnterschied der Schüler Paracelsi und Galeni, in der Cur gemelter Kranckheyt. Deßgleichen Wider allerley Gifft, so dem Menschen in Speiß und Tranck beygebracht werden mag, wider die Biss und Stich der gifftigen Thier, Auch wider gifftigen brand, krefftige, heylsame, vnnd bewerte artzneyen, auß den Schrifften der Hochgelehrten vnd weitberuempten Artzten: Dioscoridis Anazarbaei. Ferdinandi Ponzetti. Hieronymi Cardani. Conradi Gesneri. Theophrasti Paracelsi. Johannis Moibani. Vnd anderen mehr vertiert, zusamen gezogen, vnd in ein ordnung gebracht, vor nie in Teutsch gesehen.* Franckfurt [am Mayn: Getruckt bey Martin Lechler, in Verlegung Sigmumd Feyrabends und Simon Hüters].

Koch, Christine. 2003. *Das Bibliothekswesen im Nationalsozialismus. Eine Forschungsstandanalyse.* Marburg: Tectum Verlag.

Kohlhepp, Gerd (ed). 2004. *Herbert Wilhelmy (1910–2003). Würdigung seines wissenschaftlichen Lebenswerks* (Tübinger geographische Studien 141). Tübingen: Geographisches Institut der Universität.

Konrad, Paul. 1891. *Dr. Ambrosius Moibanus. Ein Beitrag zur Geschichte der Kirche und Schule Schlesiens im Reformationszeitalter* (Schriften des Vereins für Reformationsgeschichte 34/9,1). Halle: Verein für Reformationsgeschichte.

Koyré, Alexander. 2012. *Paracelsus (1493–1541).* Zürich: Diaphanes.

Kühlmann, Wilhelm/Telle, Joachim (eds.). 2013. *Corpus Paracelsisticum, Band III: Dokumente frühneuzeitlicher Naturphilosophie in Deutschland, Teilband 1* (Frühe Neuzeit. Studien und Dokumente zur deutschen Literatur und Kultur im europäischen Kontext 170), Berlin and Boston: De Gruyter.

Kulemeyer, Conrad Walter. 1942. *Glück und Ende des Leonhardt Thurneisser. Roman.* Berlin: Globus-Verlag.

Kusukawa, Sachiko. 2006. "The uses of pictures in the formation of learned knowledge. The cases of Leonhard Fuchs and Andreas Vesalius". In: Kusukawa, Sachiko/Maclean, Ian (eds.). *Transmitting knowledge. Words, images and instruments in early modern Europe.* Oxford et al.: Oxford Univ. Press, 73–96.

Kusukawa, Sachiko. 2010. "The source of Gessner's pictures for the Historia Animalium". *Annals of Science* 67/3, 303–328.

Kusukawa, Sachiko. 2012. *Picturing the book of nature. Image, text, and argument in sixteenth-century human anatomy and medical botany.* Chicago et al.: University of Chicago Press.

Kusukawa, Sachiko. 2015. "Drawing as an instrument of knowledge. The case of Conrad Gessner". In: Payne, Alina Alexandra (ed.). *Vision and its instruments. Art, science, and technology in early modern Europe*. University Park/Pennsylvania: Pennsylvania University Press, 36–48.

Lach, Donald F. ²1994. *Asia in the Making of Europe. Volume II: A Century of Wonder*, Chicago: University of Chicago Press.

Lautensach, Hermann. 1941. *Der Werdegang der portugiesischen Kulturlandschaft* (Schriftenreihe des Instituts für Portugal und Brasilien der Friedrich-Wilhelms-Universität Berlin 2). Berlin: Metzner.

Lautensach, Hermann. 1947/1982 *Bibliografia geográfica de Portugal*. 2 vols. Lisboa: Centro de Estudos Geográficos.

Lautensach, Hermann. 1964. *Iberische Halbinsel*. München: Keyser.

Lazaris, Stavros. 2013. "L'image paradigmatique: des Schémas anatomiques d'Aristote au De matéria medica de Dioscoride". *Pallas* 93, 131–164.

Le Roy Ladurie, Emmanuel. 1998. *Eine Welt im Umbruch. Der Aufstieg der Familie Platter im Zeitalter der Renaissance und Reformation*. Stuttgart: Klett-Cotta.

Letkiewicz, Ewa. 2009. "The Identification of a Print Study for a Woodcut in Hieronymus Köler's Album Amicorum in the British Library". *The Electronic British Library Journal*, 1–11 (http://www.bl.uk/eblj/2009articles/pdf/ebljarticle62009.pdf).

Letter, Paul. 2000. *Paracelsus. Leben und Werk*. Königsfurt: Krummwisch.

Leu, Urs B. 1990. *Conrad Gesner als Theologe. Ein Beitrag zur Zürcher Geistesgeschichte des 16. Jahrhunderts* (Schriftenreihe der Stiftung Franz Xaver Schnyder von Wartensee 55; Zürcher Beiträge zur Reformationsgeschichte 14). Bern et al.: Lang.

Leu, Urs B./Keller, Raffael/Weidmann, Sandra. 2008. *Conrad Gessner's Private Library*. (History of Science and Medicine Library 5). Leiden/Boston: Brill.

Liehr, Reinhard. 1992. „Geschichte Lateinamerikas in Berlin". In: Hansen, Reimer/Ribbe, Wolfgang (eds.). *Geschichtswissenschaft in Berlin im 19. und 20. Jahrhundert. Persönlichkeiten und Institutionen* (Veröffentlichungen der Historischen Kommission zu Berlin 82). Berlin et al.: de Gruyter, 633–656.

Liehr, Reinhard/Maihold, Günther/Vollmer, Günther (eds.). 2003. *Ein Institut und sein General. Wilhelm Faupel und das Iberoamerikanische Institut in der Zeit des Nationalsozialismus*. Frankfurt: Verlag Vervuert.

Lonitzer, Adam. 1557. *Kreuterbuch. New zugericht von allerhand Bäumen, Stauden, Hecken, Kreutern, Früchten, unnd Gewürtzen; eygentlicher Beschreibung der Gestalt, Underscheyd der Geschlecht, unnd leblicher Abconterfaytung, sampt jrem natürlichen Gebrauch, Krafft und Wirckung mit vilen newen Kreutern und Figuren [...] uber andere außgangene Edition gemehret; auch Distillierens Be-*

reytschafft und Bericht, allerley köstliche Wasser zubrennen [...] item der fürnembsten Gethier, Vögel und Fische, Metallen, Edelgesteinen, gebreuchlichen Gummi, und gestandenen Säfften, Beschreibung, und Nutzung. Franckfort am Mayn: Egenolff.

Lopes, Marília dos Santos. 2006, "A Revelação das Plantas. Garcia da Orta, Carolus Clusius e as espécies asiáticas na Europa". *Revista de Cultura* 20, 11–27.

Maar, Jürgen Heinrich. 2000. "Glauber, Thurneisser e Outros. Tecnologia Química e Química fina, conceitos não tão novos assim". *Química nova* 23/5, 709–713.

Macco, Hermann Friedrich. 1934. „Wann wurde Leonhard Thurneysser zum Thurn geboren?". *Zeitschrift des Vereins für die Geschichte Berlins* 51/3, 77–79.

Mägdefrau, Karl. 2013. *Geschichte der Botanik. Leben und Leistung großer Forscher*. 2. Aufl., 1992. Unveränderter Nachdruck. Berlin et al.: Springer Spektrum.

Marques, António Henrique R. de Oliveira. 1959. *Damião de Góis e os mercadores de Danzig* (Arquivo de Bibliografia Portuguesa 4,15/16). Coimbra: s. n.

Matos, Manuel Cadafaz de (ed.). 2002–2006. *Obras de Damião de Góis. Leitura diplomática e versão portuguesa por Miguel Pinto de Meneses*. 2 vols., Lisboa: Ed. Távola Redonda.

Mattioli, Pietro Andrea. 1554. *Petri Andreae Matthioli medici senensis Commentarii, in libros sex Pedacii Dioscoridis Anazarbei, de medica materia. Adiectis quàm plurimis plantarum et animalium imaginibus, eodem authore*. Venetia: Vincentius Valgrisi.

Mazal, Otto. 1981. *Pflanzen – Wurzeln – Säfte – Samen. Antike Heilkunst in Miniaturen des Wiener Dioskurides*. Graz: Akademische Druck- und Verlagsanstalt.

Mazzini, Giovanni. 1946. "»L'astrolabium« di Leonard Thurneisser zum Thurn". *Miscellanea Giovanni Mercati* (Studi e Testi 126). vol. VI: Paleografia, bibliografia, varia. Città del Vaticano. 414–431.

Meier, Pirmin. 1993. *Paracelsus. Arzt und Prophet. Annäherungen an Theophrastus von Hohenheim*. Zürich: Ammann.

Meyer, Frederick/Trueblood, Emily/Heller, John (eds.). 1999. *The Great Herbal of Leonhart Fuchs: De Historia Stirpium Commentarii Insignes, 1542 (Notable commentaries on the history of plants)*. 2 Vols. Stanford, Calif.: Stanford University Press.

Mittler, Elmar (ed.). 1986. *Bibliotheca Palatina. Katalog zur Ausstellung vom 8. Juli bis 2. November 1986, Heiliggeistkirche Heidelberg. Ausstellung der Universität Heidelberg in Zusammenarbeit mit der Bibliotheca Apostolica Vaticana. Textband*. Heidelberg: Edition Braus.

Moehsen, Johann Karl Wilhelm. 1976. *Leben Leonhard Thurneissers zum Thurn. Ein Beitrag zur Geschichte der Alchemie, wie auch der Wissenschaften und Künste in der Mark Brandenburg gegen Ende des 16. Jahrhunderts*. München: Fritsch [reprint of the edition Berlin/Leipzig: George Jakob Decker, 1783, with four copperplate engravings, which appeared in: *Beiträge zur Geschichte der Wissenschaften in der Mark Brandenburg von den ältesten Zeiten an bis zu Ende des sechszehnten Jahrhunderts*, 1–198.

Moran, Bruce T. 1993. "The Herbarius of Paracelsus": *Pharmacy in History* 35/3, 99–127.

Moran, Bruce T. 2004 a. "Alchemy". In: Dewald, Jonathan (ed.). *Europe 1450–1789: Encyclopedia of the Early Modern World. Europe 1450–1789.* Vol. I. New York et al.: Scribner, 32–35.

Moran, Bruce T. 2004 b. "Paracelsus (1493/94–1541)". In: Dewald, Jonathan (ed.). *Europe 1450–1789: Encyclopedia of the Early Modern World. Europe 1450–1789.* Vol. IV. New York et al.: Scribner, 392–394.

Moran, Bruce T. 2005. *Distilling Knowledge: Alchemy, Chemistry, and the Scientific Revolution.* Cambridge, Mass.: Harvard University Press.

Moran, Bruce T. 2013. "Art and artisanship in early modern alchemy". *The Getty research journal* 5, 1–14.

Morys, Peter. 1982. *Medizin und Pharmazie in der Kosmologie Leonhard Thurneissers zum Thurn (1531–1596).* (Abhandlungen zur Geschichte der Medizin und der Naturwissenschaften 43). Husum: Matthiesen.

Müller-Jahncke, Wolf-Dieter (ed.). 1982. *Biologica Marburgensia Illustrata. Eine Auswahl botanischer und zoologischer Abbildungswerke des 16. bis 19. Jahrhunderts in der Universitätsbibliothek Marburg* (Schriften der Universitätsbibliothek Marburg 14). Marburg: Universitätsbibliothek.

Ninhos, Cláudia. 2012. "«Com luvas de veludo». A estratégia cultural alemã em Portugal (1933–1945)". *Relações Internacionais* 35 (Setembero 2012), 103–118.

Nissen, Claus. ²1966. *Die botanische Buchillustration. Ihre Geschichte und Bibliographie.* 2 vols. in 4 parts, Stuttgart: Hiersemann.

Nobre de Carvalho, Teresa. 2015. *Os Desafios de Garcia de Orta. Colóquios dos Simples e Drogas da Índia.* Lisboa: Esfera do Caos Editores.

Ogilvie, Brian W. 2006. *The science of describing. Natural history in Renaissance Europe.* Chicago et al.: University of Chicago Press.

Orta, Garcia de. 1563. *Colóquios dos simples, e drogas he cousas medicinais da Índia e assim de algumas frutas achadas nella onde se tratam algumas coisas tocantes a medicina, pratica, e outras coisas boas pera saber.* Goa: por Ioannes de Endem.

Osbaldeston, Tess Anne. 2000. *Dioscorides: De Materia Medica. Being An Herbal With Many Other Medicinal Materials Written In Greek In The First Century Of The Common Era.* Johannesburg, South Africa: IBIDIS press [Online-Version:

https://oniehlibraryofgreekliterature1.files.wordpress.com/2015/09/94437559-dioscorides-de-materia-medica.pdf].

Paracelsus. 1536. *Die große Wundartzney*. Ulm: Hans Varnier [other editions: Augsburg: Haynrich Stayner; Frankfurt am Main: Georg Raben/Weygand Hanen].

[Anonymous]. Paracelsus, 1568. *Aphorismorvm Aliqvot Hippocratis genuinus sensus & vera interpretatio. Das ist Eygendtlicher verstandt un*[d] *warhafftige gegründte erklerung, uber etliche kurtze haupt sprüch Hippocratis; als nemlich vber alle XXV. Aphorismos primae sectionis, vnd vber die ersten VI. Aphorismos secundae sectionis. Neben dreyen hochnützlichen tractate*[n] *von sonderlicher verborgner kraft und würckung Coraliorum, Hyperici, & Persicariae. Durch den Hocherfarnen beyder Ertzney Doctorem, Herrn Theophrastu*[m] *Paracelsum von Hohenheym beschriben, vnd erst jetzt ans liecht kommen*. Augsburg: bey Mattheo Francken, in verleg Georg Willers.

Pardo-de-Santayana, Manuel/Tardío, Javier/Morales, Ramón. 2014. "Pioneers of Spanish Ethnobotany". In: Svanberg, Ingvar/Łuczaj, Łukasz (ed.). *Pioneers in European Ethnobiology* (Uppsala studies on Eastern Europe 4). Uppsala: Uppsala Universitet, 27–50.

Partington, James R. ²1969. *A History of Chemistry, vol. 2: 1500–1700*. London et al.: MacMillan.

Pehlivanian, Meliné (ed.). 2006. *Exotische Typen. Buchdruck im Orient – Orient im Buchdruck* (Ausstellungkatalog, Staatsbibliothek zu Berlin – Preußischer Kulturbesitz, Neue Folge 50). Berlin: Staatsbibliothek.

Peuckert, Will-Erich. 1956. *Der Alchymist und sei Weib. Gauner- und Ehescheidungsprozesse des Alchymisten Thurneysser* (Dokumente der Leidenschaft 1). Stuttgart: Fr. Frommanns Verlag.

Pinwinkler, Alexander. 2011. „ ‚Hier war die große Kulturgrenze, die die deutschen Soldaten nur zu deutlich fühlten…'. Albrecht Penck (1858–1945) und die deutsche „Volks- und Kulturbodenforschung". *Österreich in Geschichte und Literatur mit Geographie* 2011/2, 180–191.

Platter, Felix. 1552 ff. *Herbarium*. http://www.burgerbib.ch/platter-herbarium/

[s.n.]. 1939. *Portugal in Vergangenheit und Gegenwart. Ausstellung der portugiesischen Bibliotheken unter dem Protektorat der portugiesischen Regierung, April 1939 in der Staatsbibliothek zu Berlin*. Coimbra: Tip. da Atlàntida.

Pyle, C. M. 2004. "Gessner, Conrad". In: Dewald, Jonathan (ed.). *Europe 1450–1789: Encyclopedia of the Early Modern World. Europe 1450–1789*. Vol. III. New York et al.: Scribner, 60.

Quecke, Kurt. 1950. „Leonhard Thurneysser zum Thurn. Ein Scharlatan des 16. Jahrhunderts". *Neue Medizinische Welt* 19, 428–432.

Quelle, Otto. 1930. "Das Bonner Ibero-Amerikanische Forschungsinstitut und seine Geschichte". *Ibero-Amerikanisches Archiv* 4/1, 30–34.

Quelle, Otto. 1933. "Die verschollene Kolumbuskarte von 1498". *Ibero-Amerikanisches Archiv* 7/3, 309 f.

Quelle, Otto. 1939. "Die iberoamerikanischen Länder in Maunskriptatlanten des 17. und 18. Jahrhunderts der Wiener Nationalbibliothek". *Ibero-Amerikanisches Archiv* 13/2, 135–147.

Quelle, Otto. 1940. *Der spanisch-portugiesische Kulturkreis auf Wiener Gobelins. Eine kulturgeschichtliche Darstellung. Festschrift zum zehnjährigen Bestehen des Ibero-Amerikanischen Instituts Berlin. Mit 36 Tafeln.* Leipzig: Otto Harrassowitz.

Quelle, Otto. 1942. *Geschichte von Iberoamerika* (Die große Weltgeschichte 15: Geschichte Amerikas außer Kanada). Leipzig: Verlag Bibliographisches Institut.

Quelle, Otto. 1944 a. "Leonhard Thurneisser zum Thurn". *Revista do Instituto de Cultura Alemã. Zeitschrift des deutschen Kulturinstituts* 1/1, 99–102.

Quelle, Otto. 1944 b. "Deutsch-Portugiesische Kulturbeziehungen". *Zeitschrift für Politik* 34 (März/April 1944), 115–121.

Quelle, Otto. 1953 a. *125 Jahre Gesellschaft für Erdkunde zu Berlin 1828–1953*. Berlin: Selbstverlag der Gesellschaft für Erdkunde zu Berlin.

Quelle, Otto. 1953 b. *Portugiesische Manuskriptatlanten* (Abhandlungen des Geographischen Instituts der Freien Universität Berlin II). Berlin: Verlag von Dietrich Reimer.

Quesada, Ernesto. 1930. "Die Quesada-Bibliothek und das Lateinamerika-Institut". *Ibero-Amerikanisches Archiv* 4/1, 11–18.

Ramón-Laca Menéndez de Luarca, Luis/Morales Valverde, Ramón (ed.). 2005. L'Ecluse de Arras, Charles de: *Descripción de algunas plantas raras encontradas en España y Portugal. Rariorum aliquot stirpium per Hispanias observatarum historia. Amberes, C. Plantin, 1576* (Estudios de historia de la ciencia y de la técnica 26). [Valladolid]: Junta de Castilla y León, Consejería de Cultura y Turismo.

Reber, B. 1906. "Zwei neue Dokumente über Leonhard Thurneisser zum Thurn". *Mitteilungen zur Geschichte der Medizin und der Naturwissenschaften* 5, 431–439.

Reeds, Karen M. 1991. *Botany in medieval and Renaissance universities.* New York – London: Garland Publishing.

Richert, Gertrud. 1939. "Portugal in Vergangenheit und Gegenwart. Ausstellung in der Staatsbibliothek Berlin April 1939". *Ibero-Amerikanisches Archiv* 13/1, 13–21.

Rodrigues, Sónia Maria Correia (ed.). 2002. *Damião de Góis e o seu tempo 1502-1574. Actas do colóquio*. Lisboa: Academia Portuguesa da História.

Roentgen, Paul Ludwig. 1842. *Bemerkungen über Dodonäus Leben und Schriften. Nebst einem Commentar zu dessen Werke: Stirpium pemptades sex*. Würzburg: University [Dissertation].

Roth, Ferdinand Wilhelm Emil. 1900. „Otto Brunfels 1489-1534. Ein deutscher Botaniker". *Botanische Zeitung* 58/1, 191-232.

Rozeira, Arnaldo. 1970. *Garcia de Orta* (Publicações do Instituto de Botânica Dr. Gonçalo Sampaio da Faculdade de Ciências da Universidade do Porto, Serie 2). Porto: Instituto de Botânico Dr. Gonçalo Sampaio.

Rytz, Walther. 1936. *Die Pflanzenaquarelle des Hans Weiditz aus dem Jahre 1529. Die Originale zu den Holzschnitten im Brunfels'schen Kräuterbuch*. Bern: Haupt.

Sanwald, Erich. 1932. *Otto Brunfels 1488-1534. Ein Beitrag zur Geschichte des Humanismus und der Reformation. 1. Hälfte 1488-1524*. Bottrop i. W.: Postberg [Dissertation].

Sauerhoff, Friedhelm. 2001. *Pflanzennamen im Vergleich. Studien zur Benennungstheorie und Etymologie* (Zeitschrift für Dialektologie und Linguistik, Beihefte 113). Stuttgart: Franz Steiner Verlag.

Schaefer, Ingo. 1989. „Der Weg Albrecht Pencks nach München, zur Geographie und zur alpinen Eiszeitforschung". *Mitteilungen der Geographischen Gesellschaft in München* 74, 5-25.

Schindler, Hans-Georg, 1954. „Die Schriften Otto Quelles (chronologisch geordnet)". *Die Erde. Zeitschrift der Gesellschaft für Erdkunde zu Berlin* 85, Bd. 6, Heft 3-4 [Festschrift für Otto Quelle zum 75. Geburtstag, 217-376], 369-376.

Schlagbauer, Alfred. 2002. „Leonhart Fuchs (1501-10. Mai 1566)". In: Kavasch, Wulf-Dietrich (ed.). *Lebensbilder aus dem Ries vom 13. Jahrhundert bis zur Gegenwart*. Nördlingen: Verlag Rieser Kulturtage, 78-87.

Schmidel, Casimir Christoph (ed.). 1751. *Conradi Gesneri Philosophi Et Medici Celeberrimi Opera Botanica Per Dvo Saecvla Desiderata*. Norimbergae: Seligmannus.

Schmitz, Rudolf. 1988. „Medizin und Pharmazie in der Kosmologie Leonhard Thurneissers zum Thurn". In: Bergier, Jean-François (ed.). *Zwischen Wahn, Glaube und Wissenschaft. Magie, Astrologie, Alchemie und Wissenschaftsgeschichte*. Zürich: Verlag der Fachvereine, 141-166.

Scholz-Williams, Gerhild. 2013. "Pursuing the Inside Other: Thinking the Witch in Early Modern Print Media (16th and 17th Centuries) (Leonhard Thurneysser, Dr. Faustus, Pierre de Lancre, Theatrum Europaeum/Johannes Praetorius; Eberhard Werner Happel)". In: Manthripragada, Ashwin (ed.). *The Threat and Allure of the Magical: Selected Papers from the 17th Annual Interdisciplinary*

German Studies Conference, University of California, Berkeley. Newcastle upon Tyne: Cambridge Scholars Publ., 1–24.

Schröder, Karl-Heinz. 1970. „Herbert Wilhelmy zum 60. Geburtstag". In: Blume, Helmut (ed.). *Beiträge zur Geographie der Tropen und Subtropen. Festschrift zum 60. Geburtstag von Herbert Wilhelmy.* Tübingen: Geographisches Institut der Universität.

Schultz, Hans-Dietrich. ²2011. „Ein wachsendes Volk braucht Raum. Albrecht Penck als politischer Geograph". In: Nitz, Bernhard (ed.). *1810–2010. 200 Jahre Geographie in Berlin* (Berliner Geographische Arbeiten 115). Berlin: Geographisches Institut der Humboldt-Universität, 99–153.

Schultze, Joachim H. 1957. „Walter Behrmanns Wirken für die Geographie". *Berichte zur deutschen Landeskunde* 18/1, 46–59.

Schumacher, Yves. 2011. *Leonhard Thurneysser: Arzt – Abenteurer – Alchemist.* Zürich: Römerhof Verlag.

Schwedt, Georg. 1993. *Paracelsus in Europa. Auf den Spuren des Arztes und Naturforschers 1493–1541.* München: Diederichs.

Siegel, Karl. A. 1928. *Johannes Moibanus. Ein schlesischer Arzt und Künstler des 16. Jahrhunderts* (Schlesische Geschichtsblätter, Mitteilungen des Vereins für Geschichte Schlesiens 1). Breslau: Trewendt & Garnier.

Smith, Pamela H./Findlen, Paula (ed.). 2002. *Merchants & marvels. Commerce, science, and art in early modern Europe.* New York et al.: Routledge.

Spitzer, Gabriele. 1996. *... und die Spree führt Gold: Leonhard Thurneysser zum Thurn, Astrologe – Alchimist – Arzt und Drucker im Berlin des 16. Jahrhunderts* (Beiträge aus der Staatsbibliothek zu Berlin, Preußischer Kulturbesitz 3). Ausstellungskatalog. Wiesbaden: Dr. Ludwig Reichert Verlag.

Spitzer, Gabriele. 1997. „Leonhard Thurneysser zum Thurn – Arzt, Astrologe und Drucker im Berlin des 16. Jahrhunderts". In: Hoppmann, Jürgen G. H. (ed.). *Melanchthons Astrologie. Der Weg der Sternenwissenschaft zur Zeit von Humanismus und Reformation. Katalog zur Ausstellung vom 15. September bis 15. Dezember 1997 im Reformationsgeschichtlichen Museum Lutherhalle Wittenberg*, Wittenberg: Drei-Kastanien-Verlag, 80 f.

Sprague, Thomas Archibald. 1928–1931. "The herbal of Otto Brunfels". *Journal of the Linnean Society of London, Botany* 48, 79–123.

Sprague, Thomas Archibald/Nelmes, E. 1928–1931. "The herbal of Leonhart Fuchs". *Journal of the Linnean Society of London, Botany* 48, Nr. 325, 545–642.

Springer, Katharina B./Kinzelbach, Ragnar Kinzelbach. 2009. *Das Vogelbuch von Conrad Gessner (1516–1565). Ein Archiv für avifaunistische Daten.* Berlin: Springer Verlag.

Staat, Gerhard, 1968. „Der berühmteste badisch-pfälzische Botaniker, Hieronymus Bock". *Jahrbuch des Landkreises Kaiserslautern* 6, 115–119.

Stolberg, Michael. 2010. „Die Harnschau im 16. und frühen 17. Jahrhundert". In: Fuchs, Franz (ed.). *Medizin, Jurisprudenz und Humanismus in Nürnberg um 1500. Akten der gemeinsam mit dem Verein für Geschichte der Stadt Nürnberg, dem Stadtarchiv Nürnberg und dem Bildungszentrum der Stadt Nürnberg am 10./11. November 2006 und 7./8. November 2008 in Nürnberg veranstalteten Symposien* (Pirckheimer Jahrbuch für Renaissance- und Humanismusforschung 24). Wiesbaden: Harrassowitz, 129–143.

Stolberg, Michael. 2015. *Uroscopy in Early Modern Europe*. Translated by Logan Kennedy and Leonhard Unglaub. Farnham: Ashgate.

Strasen, Eduard August/Gândara, Alfredo. 1944. *Oito Séculos de História Luso-Alemã. Com 320 illustrações no texto, 3 iluminuras, índice de 1800 personagens, e 2 quadros genealógicos anexos*. Berlim: Instituto Ibero-Americano de Berlim.

Stübler, Eberhard. 1928. *Leonhart Fuchs. Leben und Werk* (Münchener Beiträge zur Geschichte und Literatur der Naturwissenschaften und Medizin 13/14). München: Verlag der Münchner Drucke.

Sudhoff, Karl. 1894. *Bibliographia Paracelsica. Besprechung der unter Theophrast von Hohenheims Namen erschienenen Druckschriften* (Versuch einer Kritik der Echtheit der Paracelsischen Schriften, I. Theil). Berlin: Georg Reimer.

Sudhoff, Karl. 1908. „Thurneissersche Kalender auf die Jahre 1591, 1594 und 1596". *Archiv für Geschichte der Medizin* 2/2, 129–135.

Telle, Joachim. 1989 a. Art. "Hieronymus Bock". In: Killy, Walther (ed.). *Literatur Lexikon. Autoren und Werke deutscher Sprache*. vol. 2. Gütersloh et al.: Bertelsmann Lexikon Verlag, 38 f.

Telle, Joachim. 1989 b. Art. "Otto Brunfels". In: Killy, Walther (ed.). *Literatur Lexikon. Autoren und Werke deutscher Sprache*. vol. 2. Gütersloh et al.: Bertelsmann Lexikon Verlag, 260 f.

Thurneysser zum Thurn, Leonhart. *Manuscripts in the Staatsbibliothek zu Berlin-Preußischer Kulturbesitz*:

- Ms. Bor. Fol. 680–687 und 691 [private Correspondence in 9 vols.: Letters to Thurneysser from 1570–1583].
- Ms. Germ. Fol. 97 [Miscellany, written in the second half of the sixteenth century].
- Ms. Germ. Fol. 176, fol. 195 r–204 v [Thurneysser's manuscript about libraries].
- Ms. Germ. Fol. 420–426 [Correspondence in 11 vols.: Letters to Thurneysser from 1564–1583].

Thurneysser zum Thurn, Leonhart. 1572. *Pison. Das erst Theil. Von Kalten, Warmen Minerischen vnd Metallischen Wassern, sampt der vergleichunge der Plantarum vnd Erdgewechsen 10. Buecher: Durch Leonhart Thurneisser zum Thurn, mit grosser m[ue]he vnd Arbeit, gemeinem nutz zu gut an tag geben. Mit Röm. Kay. Freyheit auff 10. Jar. 1572.* Gedruckt zu Franckfurt an der Oder: durch Johan Eichhorn.

Thurneysser zum Thurn, Leonhart. 1578 a. *Historia sive descriptio Plantarvm Omnium, tam domesticarum quam exoticarum: Earundem cum virtutes Influentiales, Elementares, & Naturales, cum Subtilitates, necnon Icones etiam veras, ad viuum artificiose expressas proponens: at*[que] *vnà cum his, partium omnium corporis humani vt externarum ita internarum picturas, & Instrumentorum Extractioni Chymicae servientium delineationem vsum*[que], *ac Methodos deni*[que] *Pharmaceuticas quasuis, ad curam valetudinis dextre tractandam necessarias complectens: Utilitatis vero publicae gratia. A Leonhardo Thurneissero zum Thurn, Medico ordinario Electoris Brandeburgici Conscripta.* Belin: Excudebat Michael Hentzske.

Thurneysser zum Thurn, Leonhart. 1578 b. *Historia Vnnd Beschreibung Influentischer, Elementischer vnd Natürlicher Wirckungen aller fremden vnnd heimischen Erdgewechssen, auch jrer Subtiliteten; sampt warhafftiger und Künstlicher Conterfeitung derselbigen, auch aller teiler, Innerlich und Eüsserlicher glider am Menschlichen Cörper, nebend fürbildung aller zu der Extraction dienstlichen Instrumenten, auch deren gebrauch, und alle[n] zu erhalten der gesundheit notwendigen Processen gemeine[n] nutz zu gut durch Leonhardt Thurneysser zum Tuhrn Churfürstlichen Brandenburgischen bestalten Leibs Medicum beschriben.* Berlin: bey Michael Hentzsken.

Thurneysser zum Thurn, Leonhart. 1583. *Megale Chymia* [Magna alchimia] VEL MAGNA ALCHYMIA. *Das ist ein Lehr vnd vnterweisung von den offenbaren vnd verborgenlichen Naturen, Arten vnd Eigenschafften, allerhandt wunderlicher Erdtgewechssen, als Ertzen, Metallen, Mineren, Erdsäfften, Schwefeln, Mercurien, Saltzen vnd Gesteinen.* [...] *Welches alles durch Leonharten Thurneissern zum Thurn von Basel, Churfürstischen Brandenburgischen bestalten Leibs Medicum, menniglichem zu nutz in 30. verscheidner Bücher, mit sonderlichem vnkosten, vleis vnd arbeit am tag geben.* Berlin: durch Nicolaum Voltzen.

Thurneysser zum Thurn, Leonhart. 1591. *Warhafftiger bericht Leonhardi Turneussers jetziger zeit zu Rom. VOn der Magia Schwartzen Zeuberkunst vnd was dauon zu halten sey etc.* Notopyrgen [i. e. Frankfurt am Main]: s. n.

Tillmann, Regine. 1988. *Neue Erkenntnisse zur Kräuterbuchliteratur des 16. Jahrhunderts.* Marburg: Universität [Dissertation].

Toxites, Michael (ed.). 1570. *Ettliche Tractatus Des Hocherfarnen vnnd berümbtesten Philippi Theophrasti Paracelsi, der waren Philosophi vnd Artzney Doctoris.* Straßburg: bey Christian Müllers Erben.

Treviranus, Ludolf Christian. 1949. *Die Anwendung des Holzschnittes zur bildlichen Darstellung von Pflanzen nach Entstehung, Blüte, Verfall und Restauration. Reprint der Ausgabe Leipzig 1855* (Classics in natural history 1). Utrecht: de Haan.

Van Meerbeeck, Philippe Jacques. 1841/1980. *Recherches historiques et critiques sur la vie et les ouvrages de Rembert Dodoens (Dodonaeus).* Utrecht: HES [Réimpression de l'éd. Malines 1841].

Van Ommen, Kaspar. 2009. *The exotic world of Carolus Clusius (1526–1609). Catalogue of an exhibition on the quatercentenary of Clusius' death, 4 April 2009* (Kleine publicaties van de Leidse Universiteitsbibliotheek 80), Leiden: Rijksuniversiteit Leiden, Bibliotheek.

Vogel, Gerd-Helge. 2014. „Wie kamen die Pflanzen in die Malerei? Zur botanischen Darstellung in der europäischen Kunst zwischen Spätgotik und Biedermeier". In: Idem (ed.). *Pflanzen, Blüten, Früchte. Botanische Illustrationen in Kunst und Wissenschaft.* Berlin: Lukas Verlag, 9–86.

Vogl, August. 1887. „Über Garcia de Orta und seine Bedeutung für die Pharmakognosie. Inaugurationsrede gehalten am 14. October 1887". Die feierliche Installation des Rectors der Wiener Universität 1887/1888, 23–70.

Voigt, Gudrun. 1995. *Die kriegsbedingte Auslagerung von Beständen der Preußischen Staatsbibliothek und ihre Rückführung. Eine historische Skizze auf der Grundlage von Archivmaterialien* (Laurentius, Kleine historische Reihe 8). Hannover: Laurentius-Verlag.

Wallich, Paul. 1934. „Leonhard Thurneißer". In: Hachel, Hugo/Papritz, Johannes/Wallich, Paul (eds.). *Berliner Großkaufleute und Kapitalisten. Erster Band: Bis zum Ende des Dreissigjährigen Krieges* (Veröffentlichungen des Vereins für Geschichte der Mark Brandenburg 32). Berlin: 1934, 311–319 [reproduction: Berlin: de Gruyter & Co., 1976].

Webster, Charles. 2008. *Paracelsus. Medicine, magic and mission at the end of time.* New Haven, Conn. et al.: Yale Univ. Press.

Wellmann, Max. 1898. „Die Pflanzennamen des Dioskurides". *Hermes. Zeitschrift für Klassische Philologie* 33/3, 360–422.

Wilhelmy, Herbert. 1952 [²1968]. *Südamerika im Spiegel seiner Städte. Heinrich Schmitthenner zum 65. Geburtstag in Verehrung und Dankbarkeit* (Hamburger romanistische Studien B: Ibero-amerikanische Reihe 23). Hamburg: Cram, de Gruyter.

Wilhelmy, Herbert. 1966. *Kartographie in Stichworten*. 4 Teile, Kiel: Hirt [altogether seven editions].

Wilhelmy, Herbert. 1975-1978. *Geomorphologie in Stichworten. Beiträge zur allgemeinen Geographie*. 4 Bände, Zug: Hirt [altogether seven editions].

Wilhelmy, Herbert. 1980. *Geographische Forschungen in Südamerika. Ausgewählte Beiträge. Zusammengestellt und mit einem einleitenden Lebensbild des Autors versehen von Gerd Kohlhepp* (Kleine Geographische Schriften 1). Berlin: Reimer.

Wilhelmy, Herbert (Mitarb.). 1982. *Lateinamerika* (Kohlhammer-Taschenbücher 1059). Stuttgart/Berlin: Kohlhammer.

Wilhelmy, Herbert/Engelmann, Gerhard/Hard, Gerhard. 1970. *Alexander von Humboldt. Eigene und neue Wertungen der Reisen, Arbeit und Gedankenwelt* (Erdkundliches Wissen 23). Wiesbaden: Steiner.

Wilhelmy, Herbert/Borsdorf, Axel. 1984/1985. *Die Städte Südamerikas*. 3 vols. Berlin et al.: Bornträger.

Wimpinäus, Johannes Albert (ed.). 1570. *Philippi Theophrasti Paracelsi von Hohenhaim, etliche Tractetlein zur Archidoxa gehörig*. München: bey Adam Berg.

Zaunick, Rudolph/Wein, Kurt. 1938. „Ein Brief von Johannes Thal an Leonhart Thurneysser zum Thurn aus dem Jahre 1582 in rebus botanicis". *Sudhoffs Archiv für Geschichte der Medizin und der Naturwissenschaften* 30/6, 401-405.

Zekert, Otto. 1963. *Die große Wanderung des Paracelsus (De Peregrinatione Paracelsis Magna). Von Einsiedeln nach Salzburg*. Ingelheim am Rhein: Boehringer.

Ziesche, Eva. 2002. *Verzeichnis der Nachlässe und Sammlungen der Handschriftenabteilung der Staatsbibliothek zu Berlin Preussischer Kulturbesitz* (Kataloge der Handschriftenabteilung, Zweite Reihe: Nachlässe 8). Wiesbaden: Harrassowitz.

Zoller, Heinrich/Steinmann, Martin (eds.). 1987-1991. *Conradi Gesneri Historia plantarum. Gesamtausgabe*. 2 vols. Dietikon-Zürich: Urs-Graf-Verlag [reprint].

Zucchi, Luca. 2003. "Brunfels e Fuchs. L'illustrazione botanica quale ritratto della singola pianta o immagine della specie". *Nuncius. Annali di Storia della Scienza* 18/2, 411-465.

Yves Schumacher

Basel – Fluchtpunkt der Humanisten und Alchemisten

Abstract: The first two Swiss people who reported about Lisbon are the physician, alchemist and philosopher Paracelsus and the polyhistor Leonhard Thurneysser zum Thurn. The biographies of the two adventurers show astonishing parallels. Both were at ware with traditional doctors who were adherers to the galenic system. Paracelsus had the audacity to give his lectures not in the time-honoured dog-Latin, but in good despicable German. The two scholars encountered serious difficulties. Paracelsus fall out with the academics and Thurneysser quarrelled with the authorities in connection with his divorce. After the latter had earned a fortune in Berlin he visited his hometown and intended to stay there for good. He reaped only contempt from his envious fellow citizens. In the final end both scholars had to leave the city hastily. Paracelsus fled to Alsace and Thurneysser to Rome. The latter became personal physician to Cardinal Markus Sittikus III of Hohenems (1533–1595).
Nevertheless, Basel was in the 16th century one of the most liberal cities of Europe and became the uncrowned capital of alchemy. It was therefore a place of refuge for many European scholars. The most famous and influential humanist of the Northern Renaissance, Erasmus of Rotterdam († 1536), lived there from 1514 to 1529 and was visited by his friend, the Portuguese humanist Damião de Góis (1502–1547).
The free spirit of the town with hardly more than 10.000 inhabitants lured also many typographers and printers to settle there. Johannes Frobenius († 1527) set up his press in 1491 and established standards for humanistic printing which were widely copied throughout the German-speaking world.
In 1529 the city was declared as reformed. Compared with Geneva or Zurich the Protestant reformation of Basel was rather mild. Thus in some churches were alternately held Catholic and Protestant services. Thurneysser religion beliefs were ambiguous. Although was baptized according to the Lutheran rite, he showed himself, depending on the situation, as a staunch Roman Catholic or Protestant.
While Thurneysser was often dismissed as a charlatan, is not to deny that he is one of the pioneers of modern chemistry.

Der gebürtige Basler Leonhard Thurneysser zum Thurn (1531–1596) war ein Grenzgänger zwischen Wissenschaft und Aberglauben. An Fürstenhöfen gefeiert, von seinen Feinden jedoch als Scharlatan verschrien, tanzte der Meister der Selbstdarstellung auf einem Hochseil zwischen Basel, Berlin und Rom. Er war Goldschmied, Söldner, Mineraloge, Alchemist, Arzt, Apotheker, Astronom, Geldverleiher, Schriftsteller und Verleger. Nach einem rastlosen Leben und ausgedehnten Forschungsreisen wurde er Leibarzt des Kurfürsten Johann Georg von Brandenburg (1525–1598) und baute seine Druckerei zum ersten Großunternehmen Berlins aus. Sein Verhältnis zu seiner Geburtsstadt Basel war indessen von Jugend an getrübt.

Über Thurneyssers Reise nach Portugal um das Jahr 1560/61 wissen wir leider wenig. Aber der junge Mann muss von Lusitanien tief beeindruckt gewesen sein, ansonsten hätte er wohl kein Manuskript über die Natur Portugals verfasst (s. Horst in diesem Band, 133-174).

Wenn wir Thurneysser mit anderen Ärzten, Alchemisten und Astrologen vergleichen, die an der Schwelle zwischen Mittelalter und Neuzeit standen, denken wir gleich an Philippus Theophrastus Aureolus Bombastus de Hohenheim (1493–1541), besser bekannt unter dem Namen Paracelsus. Beide waren Schweizer und beide waren mit Basel und Portugal verbunden. Die biografischen Analogien zwischen den beiden Gelehrten sind erstaunlich: Paracelsus reiste 1515 nach Portugal und war der erste Eidgenosse überhaupt, der Lissabon in seinen Schriften erwähnte. Die Reiserouten der beiden Männer weisen bis zu einem gewissen Punkt erstaunliche Parallelen auf. Es wäre deshalb nicht überraschend, wenn sich Thurneysser von Paracelsus' *Grand Tours* inspirieren ließ. Bemerkenswert ist auch, dass sich zu diesen ausgedehnten Fernreisen, die in der Renaissance ansonsten vom europäischen Adel unternommen wurden, zwei bürgerliche Schweizer hinreißen ließen. Paracelsus hatte zwar einen adeligen Vater, zählte aber zum minderen Stand, zumal seine früh verstorbene Mutter als Magd im Kloster Einsiedeln (Kanton Schwyz) gedient hatte (Jacobi, 1991: 239). Thurneysser gehörte, wie auch sein Vater Jakob, dem Handwerkerstand an, war also ein „Renaissance Craftsman". Er verehrte Paracelsus über alle Maßen. Ist es zudem ein Zufall, dass sich beide Männer in arge Schwierigkeiten mit der Basler Obrigkeit verwickelten?

Basel war in der Renaissance ein bevorzugter Treffpunkt namhafter europäischer Gelehrter. Ich beschränke mich auf wenige Beispiele: Der Buchdrucker Johannes Frobenius († 1527) aus Hammelburg in Franken, Desiderius Erasmus von Rotterdam († 1536) sowie der aus Württemberg stammende Reformator Johannes Oekolampad (1482–1531) lebten zeitweilig in Basel. Thurneysser war unter anderem mit dem Advokaten Basilius Amerbach (1533–1591) befreundet und verkehrte mit dem Rektor der Universität Basel, Johannes Huber (1507–1571), dem er in jungen Jahren als Famulus gedient hatte. Außerdem ging Thurneysser im Haus der bekannten Familie Platter ein und aus. Thomas Platter der Ältere (1499–1582) war ein humanistischer Gelehrter und Buchdrucker, sein Sohn Felix (1536–1614) ein hoch angesehener Mediziner.

Abb. 1 und 2: Leonhard Thurneysser und Paracelsus.

Quelle: Zentralbibliothek Zürich, Graphische Sammlung.

Die Frage ist, wie der junge Abenteurer Thurneysser zu Ehren kam, in Lissabon angeblich im Hause des Ritters Damião de Góis (1502–1574) aufgenommen zu werden (Schumacher, 2011: 51). Fakt ist, dass der Historiograph und Archivar Königs Dom João III. (1502–1557) sich in den Jahren 1533 und 1534 in Basel aufgehalten hatte, wo er sowohl Beziehungen zu Erasmus von Rotterdam als auch zum Humanisten und Komponisten Bonifazius Amerbach (1495–1562) pflegte (de Fischer, 1960: 79–87). Anzunehmen ist deshalb, dass Thurneysser mit einem gewichtigen Empfehlungsschreiben, von Basler Humanisten versehen, nach Portugal reiste.

Basel: Hauptstadt der Alchemie

Abb. 3: Basel 1493 in der Schedelschen Weltchronik.

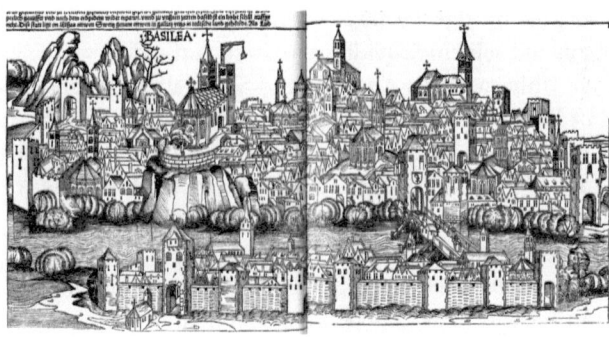

Quelle: Schedel, Hartmann. 1493. Liber chronicarum, Nürnberg: Anton Koberger, CCXLIIII.

Warum weilte eine Vielzahl von renommierten Humanisten ausgerechnet in Basel? Der Grund liegt darin, dass die Stadt vergleichsweise liberal war. Die Gelehrten genossen dort Redefreiheit und konnten von der Obrigkeit unbehelligt arbeiten. Basel galt im 16. Jahrhundert als ungekrönte Hauptstadt der Alchemie, wo zahlreiche alchemistische Laboratorien blühten. Davon zeugt unter anderem die beträchtliche Anzahl alchemistischer Handschriften, die im Nachlass des Druckers und Verlegers Johannes Oporinus (1507–1568) von dessen Schwager, dem Juristen und Kunstsammler Basilius Amerbach und seinem Neffen Theodor Zwinger der Ältere (1533–1588) aufgelistet wurden. Es waren 450 Schriften an der Zahl (Hofmeier, 2011: 24). Thurneyssers alchemistische Werke sind darin allerdings nicht erwähnt, zumal er sie nicht in Basel, sondern in Frankfurt an der Oder und in seinem Berliner Eigenverlag, dem Grauen Kloster, herausgegeben hatte.

Paracelsus und Thurneysser trugen in Basel die Folgen ihres aufbrausenden Charakters. Beide überschritten die Grenzen der gesellschaftlichen Konventionen und wurden deshalb diskriminiert und verachtet. Paracelsus brach auch ein Tabu: Er erdreistete sich als erster Mediziner, seine Vorlesungen nicht im damals üblichen Küchenlatein zu halten, sondern in schnödem Deutsch. Außerdem hätte er zur Erlangung der ordentlichen Mitgliedschaft zur Fakultät sein offenbar 1516 in Ferrara erworbenes Doktorat der Leib- und Wundarznei vorweisen müssen (Blaser, 1953: 33). Aber er scherte sich überhaupt nicht um akademische Übereinkünfte und beschimpfte die Basler Ärzte und Apotheker aufs Übelste (Schumacher, 2011: 28). Daraufhin wurde Paracelsus in einer in Basel angeschlagenen Schmähschrift als „Cacophrastus", das heißt „Scheißredner", tituliert (Classen, 2010: 167). In der Folge dieser Animositäten musste Paracelsus Ende 1528 die Stadt Basel fluchtartig verlassen.

44 Jahre später praktizierte Thurneysser, ohne akademische Qualifikationen, als Leibarzt des Brandenburgischen Kurfürsten Johann Georg von Brandenburg in Berlin. Der großspurige *Medicus* zerstritt sich mit der Basler Obrigkeit im Zusammenhang mit seinem Scheidungsprozess. Außerdem wurde er von seinen missgünstigen Mitbürgern als Angeber und Scharlatan diffamiert. Nachdem er sich einmal in Basel mit seinem gewaltigen Fleischkonsum in Berlin gebrüstet hatte, verspotteten ihn die Kinder auf der Straße und riefen ihm „Ochsenfresser" nach (Wieland, 1882: 1001). Offenbar war die wichtigtuerische Angabe Thurneyssers über den Fleischkonsum in fernen Städten für die einfachen Basler Bürger schlichtweg unvorstellbar und unglaubwürdig.

Thurneyssers unbescheidenes Selbstverständnis verdeutlichen zwei im Kunstmuseum Basel erhaltene Scheibenmalereien, die seinen Werdegang mit offensichtlich überhöhten Darstellungen aus seinem Leben zeigen. Mit diesem ursprünglich siebenteiligen Glasscheibenzyklus hatte Thurneysser sein Basler Haus geschmückt und wollte damit seine Mitbürger beeindrucken. Das brachte ihm allerdings keine Sympathien ein. Auch ein Holzschnitt aus dem 1581 in Nürnberg herausgegebenen Werk *Impletio* ist für Thurneyssers Selbstdarstellung charaktertypisch. In geradezu schizoider Weise zeigt er sich auf der linken Körperseite als geharnischter Krieger, derweil er sich in der rechten Körperhälfte als vornehmer Gelehrter mit einem Buch unter dem Arm präsentiert. Aufgrund der erwähnten Feindseligkeiten flüchtete Thurneysser 1584 Hals über Kopf nach Rom.

Bei der Textinterpretation der Biografien von Paracelsus und Thurneysser sind verschiedene charakteristische Aspekte der Stadt Basel im 16. Jahrhundert zu berücksichtigen. Bemerkenswert sind unter anderem die nicht standesgemäßen Wohnverhältnisse der beiden Männer. Paracelsus gab Vorlesungen in seinem Basler Wohnhaus (heute Leonhardstraße 1) im verrufenen Basler Quartier namens Kohlenberg, wo übrigens eine eigene Gerichtsbarkeit herrschte. In diesem Viertel hausten vor allem Angehörige der unehrlichen Berufe, die kein ständisches Ansehen hatten. „Unehrlich" bedeutete damals nicht etwa „betrügerisch", sondern „nicht ehrenwert". Neben dem Scharfrichter, vegetierten im Kohleberg unter anderem Totengräber, Prostituierte und Barbiere. Auch das fahrende Volk – also Nothausierer, Lumpensammler, Pferdetauscher, Kesselflicker oder Spielleute – zählten zu den Ehrlosen. Dass Paracelsus seinen Wohnsitz am Kohlenberg, inmitten von Bretterbuden, Zelten und Kloaken genommen hatte, wurde ihm von der akademischen Bürgerschicht übel genommen. Rund fünfzig Jahre später kaufte sich Thurneysser am Fuße des Kohlenberges ein stattliches, dreigeschossiges Haus für den übersetzen Preis von 4.500 Gulden (Boerlin, 1976: 125). Das Anwesen liegt just ein paar Gehminuten von Paracelsus' ehemaliger Wohnstätte entfernt.

Die Basler Herrschaftsverhältnisse

Basel bildet einen Sonderfall der Schweizer Geschichte. Um 1500 befand sich die Stadt in einer politisch äußerst prekären Lage. Aus kommerziellen Erwägungen nahmen die Bürger Rücksicht auf die nördlichen Landen. Gleichzeitig war Basel aber mit den eidgenössischen Orten verbündet. Im sogenannten „Schwabenkrieg", welchen die Eidgenossen gegen den habsburgischen Kaiser Maximilian I. (1449–1519) führten, blieb Basel deshalb vorerst neutral (Burckhardt, 1841: 12). Die Stadt gab dann dem Druck der Eidgenossen nach, und sie trat am 13. Juli 1501 dem damaligen Schweizerbund bei und wurde als vollberechtigtes Glied der Eidgenossenschaft anerkannt (Gittermann, 1949: 102). Manche Basler Patrizier betrachteten diesen Anschluss als Verrat am Kaiser und einige verließen die

Stadt. Auch schämten sich viele, nun zu den „Kuhschweizern" (Sieber-Lehmann, 1998: 7 f.) zu gehören (Buxtorf-Falkeisen, 1863: 5).

Die klassische Standesordnung „Kaiser, König, Edelmann, Bürger, Bauer, Bettelmann" mag für das Habsburgerreich des 16. Jahrhunderts charakteristisch sein, aber für die Stadt Basel galt sie nicht unbedingt. Es gab keinen König, kaum Edelmänner, wenig Bauern, aber sehr wohl Bürger und Bettelleute. Im Spätmittelalter bildete sich in Basel zwar ein Patriziat heran, das sich aus reich gewordenen Kaufmannsfamilien und aus dem stadtsässig gewordenem Landadel zusammensetzte. Die bürgerlichen Notabeln übernahmen die Lebensweise des Adels und bildeten ein Patriziat, das mit der italienischen Signoria vergleichbar ist. Dieses Basler Gentry-Patriziat war vor allem im Handel und im damals lukrativen Söldnerwesen tätig.

Zu Thurneyssers Zeiten gab es in Basel 15 verschiedene Handwerkerzünfte. Und die waren ziemlich mächtig. Thurneysser selbst gehörte als Goldschmied zwangsläufig der im Mittelalter gegründeten Korporation der „Hausgenossen" an, die auch Wechsler, Münzer, Kannen-, Hafen- und Glockengießer sowie die Büchsen- und Buchstabengießer vereinte. Auch Gelehrte verkehrten bei den „Hausgenossen", so etwa Thurneyssers Freund, der Medizinprofessor Felix Platter. Dieser wurde 1557 Mitglied der Zunft. Goldschmiede wie Leonhard und sein Vater Jakob Thurneysser († 1560) galten übrigens als absolut vertrauenswürdige Ehrenmänner, zumal ihnen der Besitz einer Silberwaage und der Handel mit dem Edelmetall gestattet war (Barth, 1978: 44). Doch schon in seinem 27. Altersjahr verspielte Leonhard seinen Ruf. Er ließ sich von einem dubiosen Freund zum Betrug eines jüdischen Pfandleihers in der nahegelegenen, zu Vorderösterreich gehörenden Kameralherrschaft Laufenburg bewegen, wo Juden das Niederlassungs- und Aufenthaltsrecht zugestanden wurde. Dort versuchte Leonhard, dem Pfandleiher eine mit Gold beschichtete Bleistange an Geldes statt unterzuschieben. Diese Jugendsünde blieb zeitlebens an Thurneysser haften (Schumacher, 2011: 53 f.).

Im Vorfeld der Reformation ging in Basel die Kritik am Papsttum mit der Forderung nach politischen und wirtschaftlichen Reformen einher. Der Rat[1] emanzipierte sich vom Einfluss der Patrizier und des Bischofs. 1515 hatten diese keine Privilegien mehr. Der Rat fasste die gesellschaftliche Nivellierung aber nicht als neue Unabhängigkeit der Bürger vom Klerus auf. Die Obrigkeit griff auch in kirchliche Angelegenheiten ein und unterstellte ihr die Klöster.

---

1 Der Kleine Rat war das oberste Organ der Stadt und wurde von Adligen und Patriziern beherrscht. Der Große Rat konstituierte sich hingegen aus Zunftleuten und konnte vom Kleinen Rat jederzeit einberufen werden (Alioth, 1981: 71).

Die reformatorischen Ideen Martin Luthers (1483–1546) fielen in der von humanistischen Gelehrten geprägten Stadt Basel auf fruchtbaren Boden. 1525 führte die konfessionelle Radikalisierung zu erheblichen Spannungen zwischen verschiedenen humanistischen Gruppierungen und reformatorischen Kreisen. Der Rat selbst war in Religionsfragen auch gespalten und beschränkte sich deshalb darauf, die persönliche Glaubensfreiheit zu propagieren. Seine schlichtende Politik vermochte die verstrittenen Anhänger der verschiedenen religiösen Auffassungen aber nicht zu versöhnen (vgl. Burckhardt, 1942: 9). Im Bildersturm von 1528/29 entluden sich die Spannungen in einer wilden Zerstörung kirchlicher Kunst. Im April 1525 suchte eine Jugendgruppe mitunter das Augustiner-Chorherren-Stift St. Leonhard heim. Unter dem Ruf „her, her, wir wollen die Pfaffen, die huren und das ruppennest (Raupennest) ganz zerschlagen" zerstörten die mit Spießen, Hellebarden und Büchsen bewaffneten Burschen und Mädchen die Kirche und die übrigen Gebäude fast vollständig (Wackernagel, 1959: 212).

Der Rat gab dem Druck der Aufständischen schließlich nach. 1529 erklärte die Basler Reformationsordnung die Stadt als reformiert. Der Rat konnte fortan auch über die Religionsausübung jedes Einzelnen bestimmen. Doch im Vergleich mit Genf oder Zürich verlief die Basler Reformation unter Johannes Oekolampad ausgesprochen mild. So wurden zum Teil in ein und denselben Kirchen abwechslungsweise katholische und protestantische Gottesdienste abgehalten. Thurneyssers Religionszugehörigkeit ist übrigens nicht eindeutig. Er war ein Opportunist. Er wurde zwar nach lutherischem Ritus getauft, aber je nach Lebenslage zeigte er sich als strammer Papist oder Protestant.

Kleinbürger und Kleingeister

Thurneysser war durch seine Aktivitäten im Grauen Kloster Berlin für Basler Verhältnisse unvorstellbar reich geworden. Die Vermarktung seiner Bücher war nur eine von vielen Einnahmequellen. Als Leibarzt des Kurfürsten von Brandenburg verdiente er jährlich die stolze Summe von 1352 Talern. Daneben brachten ihm seine im eigenen Laboratorium erzeugen Medikamente viel Geld ein. Unter anderem vermarkte er Arzneikästen mit 100 verschiedenen Medikamenten. Für jede dieser Apotheken verlangte er 386 Taler. Aber am meisten Einnahmen bescherten ihm zweifellos seine nach alchemistischen Prinzipien erzeugten Amulette und seine Horoskope. So bezahlte ihm zum Beispiel ein Graf von Oettingen für ein Horoskop die ansehnliche Summe von 100 Gulden (Wieland, 1882: 312).

Missgunst und Misstrauen gegenüber Menschen, die sich über gesellschaftliche Konventionen hinwegsetzten, zeichneten in Basel wie in allen anderen Städten den Kleingeist der Bürger aus. Basel war Mitte des 16. Jahrhundert ein Nest mit knapp 10.000 Einwohner, in dem fast jeder jeden kannte. Zu bedenken ist, dass natürlich nur die dünne Schicht der Obrigkeit und der Gelehrten von einem toleranten Geist beseelt war. Der Neid der biederen Basler gegenüber ihrem nach Brandenburg ausgewanderten Landsmann wurde in verschiedenen Episoden offensichtlich. Zum Beispiel im traurigen Schicksal eines Elchs:

Thurneysser baute in Berlin den ersten Zoo der Stadt auf, wo er seltene wilde Tiere zur Schau stellte. Neben seinem zweifellos echten wissenschaftlichen Interesse an der Tierwelt ging es ihm dabei wohl auch darum, es den Regenten gleichzutun und mit den seltenen Tieren Beziehungen zu fernen Ländern zum Ausdruck zu bringen. Eines Tages schenkte Fürst Nikolay Rudy von Radzivill (1512–1584), Grosshetman von Litauen, Thurneysser einen Elch, der wohl aus den Wäldern Polens stammte. Thurneysser kam auf die Idee, das Tier seinem Freund Felix Platter weiter zu schenken und in Basel der Öffentlichkeit zu präsentieren. In dieser Zurschaustellung kann man auch die Absicht Thurneyssers sehen, eine Promotion zum Absatz seiner Bücher zu veranstalten. 1559 brachte Thurneyssers Faktor Burkhard Speidel den Elch nach Basel, wo er am Kohlmarkt grosses Aufsehen erregte, denn Elche waren in der Schweiz schon im frühen Mittelalter ausgestorben. Diese anfängliche Begeisterung der Basler für das unbekannte Tier schlug bald in Feindseligkeit um. Weil Thurneysser aufgrund seiner alchemistischen Schriften und Horoskope als Schwarzmagier verschrien war, galt der harmlose Elch bald einmal als Zauberteufel aus seiner Hand. Um diesen Dämon unschädlich zu machen, warf ihm eines Tages eine alte Frau einen mit Nadeln bespickten Apfel zum Frass hin, worauf das arme Tier natürlich qualvoll verendete (Schumacher, 2011: 115 f.).

Ein turbulenter Scheidungsprozess

Thurneyssers Scheidungsprozess in Basel ist ein Kapitel für sich. Der Volkskundler und Hochschullehrer Will-Erich Peuckert (1895–1969) verarbeitete diesen Prozess in seinem Roman *Der Alchymist und sein Weib* (Peuckert, 1957). Es handelte sich hierbei um einen erbarmungslosen Zwist zwischen Thurneysser und seiner dritten, um 20 Jahre jüngeren Frau Marina Herbort (um 1553–1610) aus Konstanz, der Ehebruch vorgeworfen wurde. Dieser Prozess wirft nicht nur ein Licht auf Thurneyssers ungezügeltes Temperament mit fatalen Auswirkungen, sondern gibt uns auch eine Ahnung von den für diese Zeit erstaunlichen Rechte der Frauen in Basel (Wunder, 1995: 173 f.).

Von 1529 bis 1875 gab es in Basel ein sogenanntes Ehegericht, das als Klageinstanz fungierte. Interessant, dass diese Instanz von den Frauen viel häufiger in Anspruch genommen wurde als von Männern. 1582 wurde Thurneysser vom Gericht aufgefordert, Berlin zu verlassen und unverzüglich nach Basel zu kommen, um mit seiner getrennt lebenden Frau ehelich zu haushalten. Ein paar Tage nach dem Urteil des Ehegerichts teilte das Stadtgericht der jungen Marina Herbot das gesamte Hab und Gut Thurneyssers zu. Auch Kapitalerträge aus Thurneyssers Immobilien in Basel wurden ihr zugesprochen. Eine Intervention des Kurfürsten von Brandenburg zugunsten seines Leibmedikus beeindruckte die Basler Gerichtsbarkeit in keiner Weise. Der Ausgang dieses Prozesses veranlasste Thurneysser zur Herausgabe von Pamphleten, in denen er seine Frau und die Basler Regierung in vulgärster Weise verunglimpfte.

Die in der früheren Literatur kolportierte Aussage, wonach Thurneysser bei diesem Prozess seine letzten Gulden und Taler verloren hätte, ist beim heutigen Wissenstand stark zu bezweifeln. Wie hätte er sonst später den prachtvollen Landsitz „Belvedere" in Frascati bei Rom für die horrende Summe von 11.300 Scudi-d'Oro – was in etwa 33.900 Gulden entsprach – erwerben können? In Rom zeigte sich Thurneysser natürlich als guter Katholik und verbrachte zehn Jahre als Leibmedicus des Kardinals Markus Sittikus III. von Hohenems (1533–1595) (Schumacher, 2011: 281).

Mit einer Rückkehr Thurneyssers nach Basel oder Berlin war nicht zu denken. Seine Spott- und Drohbriefe, die er an die Basler Regierung gerichtet hatte, ließen die Wogen seines Prozesses nie glätten. Über Thurneyssers Ende gibt es zahllose Spekulationen. Fakt ist, dass Thurneyssers letzter Wunsch in Erfüllung ging. Er wurde 1596 tatsächlich in unmittelbarer Nähe seines Vorbildes Albertus Magnus (1200–1280) in Köln bestattet. Das hat Thurneysser-Forscher Diethelm Eikermann († 2015) aus Köln nachgewiesen (Eikermann, 2012: 69–77).

Die erwähnten Zusammenhänge lassen mich zu einer kühnen Behauptung hinreißen: Ohne Reformation und ohne Persönlichkeiten wie Thurneysser wäre Basel nicht das, was die Stadt heute ist: eine Metropole der chemischen Industrie. Zur Erläuterung dieser These schlage ich nun einen großen Bogen: Chemie lag bis im 17. Jahrhundert im Zuständigkeitsbereich der Mediziner und Alchemisten, namentlich der Paracelsisten. Den ersten Chemie-Kurs in Basel gab Theodor Zwinger III. (1658–1724), dessen Großvater, der Mediziner und Paracelsist Theodor Zwinger I., mit Leonhard Thurneysser befreundet war. Drei Jahre später veröffentlichte der Basler Theologe Samuel Werenfels (1657–1740) die Abhandlung *Meditatio de atomis*, in der er mit Hilfe von mathematisch-philosophischen Erwägungen die Existenz von Atomen zu beweisen suchte. Seine Anschauungen ebneten dem englischen Naturforschers John Dalton (1766–1844) den Weg zu seinen grundlegenden Untersuchungen zur Atomtheorie, die ihn als Begründer der modernen Chemie und Kernphysik gelten lassen.

So gesehen, gehören Paracelsus und Thurneysser nicht nur zu den Wegbereitern der Beziehungen zwischen der Schweiz und Portugal; sie sind auch als Schrittmacher für die Entwicklung der chemischen Industrie Basels zu würdigen.

Literatur

Alioth, Martin. 1981. *Basler Stadtgeschichte. Vom Brückenschlag 1225 bis zur Gegenwart*, Basel: Reinhardt.

Barth, Ulrich. 1978. *Zur Geschichte des Basler Goldschmiedhandwerks (1261–1820)*. Muttenz: Rolf Mayer.

Blaser, R[obert] H[enri]. 1953. *Neue Erkenntnisse zur Basler Zeit des Paracelsus*. Einsiedeln: Schweizerische Paracelsus-Gesellschaft.

Boerlin, Paul-Henry. 1976. *Leonhard Thurneysser als Auftraggeber*. Basel: Birkhäuser.

Burckhardt, L[udwig] A[ugust]. 1841. *Der Kanton Basel, historisch, geographisch, statistisch geschildert. Beschreibung seiner Lage, natürlichen Beschaffenheit, seiner Bewohner, politischen und kirchlichen Verhältnisse und Ortschaften. Ein Hand- und Hausbuch für Kantonsbürger und Reisende*. St. Gallen: Huber.

Burckhardt, Paul. 1942. *Geschichte der Stadt Basel von der Zeit der Reformation bis zur Gegenwart*. Basel: Helbling & Lichtenhahn.

Burghartz, Susanna. 1999. *Zeiten der Reinheit – Orte der Unzucht. Ehe und Sexualität in Basel während der frühen Neuzeit*. Paderborn: Schöningh.

Buxtorf-Falkeisen, Karl. 1863. *Baslerische Stadt- und Landgeschichten aus dem sechzehnten Jahrhundert*. Bd. 1. Basel: Schweighauser.

Classen, Albrecht (Hrsg). 2010. *Paracelsus im Kontext der Wissenschaften seine Zeit: Kultur und mentalitätsgeschichtliche Annäherungen*. Berlin/New York: Walter de Gruyter.

Eikermann, Diethelm. 2012: „Leonhard Thurneysser zum Thurn (1531–1596): Neue Quellen zu einem Ende in Köln". *Geschichte der Pharmazie. Zeitschrift der Deutschen Gesellschaft für Geschichte der Pharmazie e. V. und Mitteilungsblatt der Internationalen Gesellschaft für Geschichte der Pharmazie e. V.* 64/4 (2012), 69–77.

Fischer, Béat de. 1960. *Dialogue Luiso-Suisse, Essai sur les relations Luso-Suisses à travers les siècles*. Lisbonne: Ramos Afonso & Moita.

Gittermann, Valentin. 1949. *Geschichte der Schweiz*. Zürich: Büchergilde Gutenberg.

Hofmeier, Thomas: 2011. *Basel – Hauptstadt der Alchemie*. Berlin: Leonhard-Thurneysser-Verlag.

Jacobi, Jolande (Hrsg). 1991. *Paracelsus – Arzt und Gottsucher an der Zeitenwende. Eine Auswahl aus seinem Werk*. Olten: Walter-Verlag.

Peuckert, Will Erich. 1957. *Der Alchymist und sein Weib. Gauner- und Ehescheidungsprozesse des Alchymisten Thurneysser* (Dokumente der Leidenschaft 1). Stuttgart: Fr. Frommanns Verlag.

Schollenberger, J[ohann Jakob]. 1920. *Das Bundesstaatsrecht der Schweiz. Geschichte und System*. Berlin/Heidelberg: Springer.

Schumacher, Yves. 2011. *Leonhard Thurneysser. Arzt – Alchemist – Abenteurer*. Zürich: Römerhof Verlag.

Sieber-Lehmann, Claudius u. a. (Hrsg). 1998. *In Helvetios – wider die Kuhschweizer: Fremd- und Feindbilder von den Schweizern in antieidgenössischen Texten aus der Zeit von 1386 bis 1532* (Schweizer Texte/Neue Folge 13). Bern u. a.: Haupt.

Wackernagel, Hans Georg. 1959. *Altes Volkstum der Schweiz. Gesammelte Schriften zur historischen Volkskunde*. Basel: Schweizerische Gesellschaft für Volkskunde.

Wieland, Karl. 1882. *Leonhard Thurneysser zum Thurn. Beiträge zur vaterländischen Geschichte*. Band I, 1901. Basel: Historische Gesellschaft zu Basel.

Wunder, Heide (Hrsg.) et al. 1995. *Eine Stadt der Frauen. Studien und Quellen zur Geschichte der Baslerinnen im späten Mittelalter und zu Beginn der Neuzeit (13.–17. Jh.)*. Basel/Frankfurt am Main: Helbing & Lichtenhahn.

Annemarie Jordan Gschwend

Anthonio Meyting: Artistic Agent, Cultural Intermediary and Diplomat (1538–1591)[1]

> *Sy wellen unns ettwas selzamen uund ihr Lannds frembden Sachen [...] zuekommen lassen, auch mit ettwan zur steur in unnser Chunnstchamer begaben thue.*[2]
> Albrecht V, Duke of Bavaria (r. 1550–1579), to the Queen of Spain, 1574

Abstract: Fürstliche Sammler der Renaissance wie die Wittelsbacher oder Habsburger bauten spektakuläre Sammlungen in Form von recht ansehnlichen Kunst- und Wunderkammern auf. Dies gelang ihnen vor allem aufgrund der weit verstreuten Handelsnetzwerke, welche oberdeutsche Handelsgemeinschaften mit Sitz in Augsburg, Nürnberg und Regensburg seit dem ausgehenden 15. Jahrhundert auf der Iberischen Halbinsel errichtet haben. Dadurch wurde überhaupt erst ein globaler Markt für den Export von Luxusgütern und Exotica aus Portugal und Spanien eröffnet.
Aber wie erwarben die fürstlichen Sammler diese Raritäten, die sie stolz in ihren Kunstkammern präsentierten? Und wie waren die fürstlichen Höfe mit den einschlägigen Marktplätzen in Afrika, Asien und der Neuen Welt im 16. Jahrhundert vernetzt?
Der vorliegende Beitrag hebt die Rolle von Zwischenhändlern bei der Errichtung solcher Sammlungen nördlich der Alpen hervor: Dies wird am Beispiel der vom bayerischen Herzog Albrecht V. (1528–1579) errichteten Münchener Kunstkammer eindrucksvoll demonstriert: Im Jahre 1579 hatte Albrecht für seine Schatz- und Wunderkammer bereits mehr als 3.500 kunstvolle Objekte erworben, die auf verschiedene Räumlichkeiten im Marstall der Münchener Residenz verteilt und dort ausgestellt waren.
Seit den 1560-er Jahren hatte Albrecht seine Kunstkammer zügig ausgebaut, indem er hier ausgewählte Objekte miteinander vereinte, die ihm von seinen habsburgischen Verwandten an den Höfen in Lissabon, Madrid und Wien als Geschenke überreicht wurden. Zudem halfen ihm auch die auf der Iberischen Halbinsel angesiedelten Faktoren und Händler der Fugger bei der Suche nach außergewöhnlichen Objekten. So wurde Albrecht etwa über den Augsburger Fugger-Faktor Nathaniel Jung, der mehrere Jahre in Lissabon gelebt

---

1   A shorter version of this essay presented in Jordan Gschwend, 2012.
2   Bayerisches Hauptstaatsarchiv, Munich, Kurbayern, Äusseres Archiv, 4853, fol. 315 v (May 24, 1574). In a letter written to his niece, Anna of Austria (1549–1580), Albrecht V begged her to find for him suitable rarities available in marketplaces in Spain and Portugal, and to please forward him "strange, foreign objects for his Kunstkammer", he created in the Munich Residence.

hatte, mit zahlreichen kunstvollen Handelsgütern versorgt, die aus Asien oder Brasilien importiert wurden.

Von großer Bedeutung ist hierbei auch der Kunstagent und Diplomat Anthonio Meyting (1538–1591), ebenso ein Augsburger, der Karriere am spanischen Hofe machte und einer der engsten vertrauten Diener des bayerischen Herzogs auf der Iberischen Halbinsel war. Meyting war eine kultivierte Persönlichkeit des Renaissancezeitalters, der nicht nur in mehreren Sprachen bewandert war. Er repräsentiert zugleich eine neue Art eines Händlers im ausgehenden 16. Jahrhundert, der sich insbesondere als hochqualifizierter Kunstagent, Diplomat, Unternehmer und Bankier einen Namen machte.

Meyting war für seinen bayerischen Kunden somit ein geeigneter Mittelsmann: Er sandte ihm die besten Exotica aus der Neuen Welt und aus Asien, die schließlich Eingang in die Münchener Kunstkammer fanden. Seine Vermittlung ist bemerkenswert und jüngst aufgefundene Archivdokumente belegen eindrucksvoll seine internationale Karriere. Somit können in der Münchener Residenz bis heute außergewöhnliche Exotica bestaunt werden, welche Meyting für Albrecht V. im 16. Jahrhundert aus Lissabon erwarb.

Royal Consumption in Renaissance Portugal, Spain and Central Europe: a brief introduction

A great deal of scholarship has focused in the past twenty-five years on the creation and formation of *Kunst and Wunderkammern* in the sixteenth and seventeenth centuries.[3] The bibliography on royal, aristocratic and patrician collectors, their multifarious collections and diverse collecting habits and tastes is comprehensive, with many studies targeting specific areas of collecting within the context and framework of the *Kunstkammer*. Although it may seem the investigation of *Kunstkammers* and their origins in the Renaissance has been exhausted, many open questions still require further research. In particular, how the regular flow of luxury goods, jewels and precious stones from overseas was channeled and procured for royal collectors in Iberia and Central Europe.

Renaissance *Kunstkammers* simply did not appear. Nor can such distinct collections, many formed between the early and the late sixteenth century, be regarded as "heirloom" treasuries passed down from generation to generation, upon which later descendants expanded. Sixteenth-century *Kunstkammers* were novel in conception, some collections dictated by the Age of Discovery, whereby vast distances circumscribed by new means of travel and communication played a determinant role. Price and cost mattered little: royal and patrician indulged in their passion(s) for collecting even if debts were incurred.

---

3   The most significant monographs are von Schlosser, 1908 and Impey/Macgregor, 2001.

One question that begs to be addressed by the present line of *Kunstkammer* scholarship, is how Renaissance collectors came to acquire their global objects from distant continents, especially imported luxury goods and animals of an exotic nature? If such objects and exotica were not inherited, where did the shopping elite of Europe buy? Did royal or aristocratic collectors shop at stores? The answer is yes they did. Major metropolises in Renaissance Europe: Venice, Genoa, Antwerp, Seville and Lisbon were marketplaces where luxury goods, exotica, foreign commodities, plants and animals were offered for sale and purchased. Did such elite shoppers, however, go shopping themselves? From extant archival documents and correspondence, we can deduce that the answer is no. Royal and princely collectors turned to trusted, competent intermediaries, such as resident ambassadors based abroad at strategic European courts in Milan, Venice, Genoa, Naples, Rome, Madrid and Lisbon, or relied on courtiers who traveled between courts and cities whom they recruited to fulfill wish lists. Collectors also resorted to highly inventive specialists, individuals who marketed themselves as art brokers-merchants-dealers-financiers, who made it their business and livelihood to procure exquisite *Kunstkammer* objects for a royal clientele. It is precisely this aspect of the history of collecting in the Renaissance, which needs better clarification and understanding.

The recent discovery of two paintings depicting the principal shopping street of Lisbon, the *Rua Nova dos Mercadores*, – the Fifth Avenue or Bond Street of its day –, offers a unique window into the past.[4] We are confronted, in this late sixteenth-century view, with a bustling street, lined with stores underneath a protective arcade, convenient for shopping in inclement weather (Fig. 1). The *Rua Nova* was the largest, most frequented street of Lisbon, and a financial center where a currency and stock exchange operated. Reputable German, Italian and Flemish merchants and financiers based their headquarters and offices in and around the *Rua Nova*: the trading families and employees of the Welser, Fugger, Imhof, Herwart, Affaitaidi, Rovalesco and Giraldi. Eleven booksellers and nine apothecaries selling drugs and spices from Portuguese Asia were counted here in a 1552 census, and by 1580 six shops alone specialized in selling Chinese blue and white Ming porcelain imported from Macau and Malacca. Travelers visiting Lisbon in the last decades of the sixteenth century wrote with amazement of the quantities of luxury goods, wild animals and exotica from Asia and the Americas for sale here: textiles and diamonds from India, ivories and crystal buttons from Ceylon, pearls from Ormuz, mounted coconuts, rhinoceros horns and nautilus

---

4   The *Rua Nova*, and the paintings of this street, were the subject of a recent monograph. See Jordan Gschwend/Lowe (eds.), 2015.

shells, lacquer furniture and chests from China and Japan, small mother-of-pearl and tortoiseshell caskets from Gujarat, even live turkeys from the New World. By the 1580s, Lisbon's reputation as the commercial center of Europe was well established, and for insiders, this hub was a shopper's haven much appreciated by Habsburg and Wittelsbach collectors in Central Europe and Bavaria.[5]

Fig. 1: Anonymous Netherlandish artist, View of the Rua Nova dos Mercadores: Rua Nova dos Ferros with a corner view of the Largo do Pelourinho Velho. oil on canvas, c. 1570–1619, 65 x 95.5 cm, The Society of Antiquaries, Kelmscott Manor, Oxfordshire, inv. no. KM 186.1.

© The Society of Antiquaries of London (Kelmscott Manor).

But, if these royal collectors could not travel themselves to Lisbon, or even further abroad to India or to the Americas, how did they access their luxury goods and exotica? Who helped them buy the outstanding objects they avidly sought, and who helped them fulfill their dreams of possessing the best, the bizarre and the rarest? More study must be undertaken on patronage and clientage networks linking Europe to Africa, Asia and the Americas in the sixteenth century. To what degree were dealers, agents and merchants directly involved with centers of production in Asia or in the Far East? How were personal directives from royal patrons in Europe transmitted to merchants based both in Europe and abroad? How were wish lists carried out and met with satisfaction? Complex trade networks were established between Europe, Asia, Africa and the New World which

---

5   Jordan Gschwend/Pérez de Tudela, 2003 and Pérez de Tudela, 2011.

permitted great distances, time and space to be bridged with incredible speed, while allowing for a global transfer of peoples, goods, commodities, plants and animals in dimensions never experienced before. Novel patterns of collecting were set into motion and the players behind the scenes – the silent personages behind a number of Renaissance *Kunstkammers* – assumed a new level of importance and dimension for the first time as entrepreneurs, adventurers and dictators of taste and fashion.[6]

This essay will take a closer look at one individual, Anthonio Meyting (1524–1591), a businessman and patrician from Augsburg, who played a leading role as financier, intermediary, art dealer, and at various stages in his career,[7] as courtier and diplomat for the Bavarian court.[8] Meyting played an influential role in the material formation and development of the renowned Munich *Kunstkammer* of his patron,[9] Duke Albrecht V (1528–1579), while serving as an agent and artistic consultant to Philip II of Spain (1527–1598).[10] Through research undertaken in archives in Portugal, Spain, France (Besançon), Austria, Germany, and Brno in the Czech republic, it has been possible piece together through correspondence and other related documents Meyting's international career. Many outstanding art works and pre-eminent exotica once in Renaissance *Kunstkammers* in Lisbon, Madrid, Munich, Vienna, Prague and Innsbruck, extant today in major European museums, would not have entered these collections without Anthonio Meyting's directives.

Meyting (better known as Anton Meuting) was a German patrician, born and raised in Augsburg.[11] Best described as an early modern agent, Meyting embarked on a successful career as an entrepreneur and cultural mediator. At the start of

---

6 See Lieb, 1952; Backmann, 1997; Neuwirth, 2000; Meadow, 2002; Strohmeyer, 2003; Burkhardt/Karg (eds.), 2007 and Johnson, 2008.
7 Mention of Meyting first made by Stockbauer, 1874: 89 and 99; Bäader, 1943.
8 Meyting is often referred to in Spanish documents as, "Antonio Mayting un criado del Duque," or servant of the Duke of Bavaria. Cf. Pérez de Tudela, 2011: 1787, n. 53.
9 For Lorenz Seelig's complete bibliography regarding the Munich Kunstkammer consult Seelig, 2008.
10 Research regarding Meyting, his career in Spain and the cultural and artistic exchange he generated between the Spanish court, Bavaria and Central Europe first published in Pérez de Tudela/Jordan Gschwend, 2001. Subsequently in Jordan Gschwend/Beltz (eds.), 2010: 97–104 and Pérez de Tudela, 2011: 1769–1836.
11 Meyting invariably signed his letters, *Anthonio Meyting* and the author has opted for this spelling. Meyting was also given the sobriquet of "the Spaniard" or *der Spanische*, because of the long years he lived and worked in Spain (approximately 43 years).

his career in Spain as an apprentice,[12] he quickly advanced to a junior business partner. Subsequently, his most prestigious responsibilities were those he carved out for himself as an independent merchant-dealer and art broker at the Spanish and Bavarian courts, where he exploited and re-defined commercial and artistic networks between Iberia and Central Europe.

Anthonio Meyting's Career in Spain

Meyting's international career started in Ausgburg, where he was born in 1524.[13] He moved to Northern Spain in 1538 at the age of fourteen, to Valladolid where his elder sister, Ursula Meuting, resided. One year prior, her husband, Ulrich Ehinger, a prominent German merchant at Emperor Charles V's court since 1518, had died in August of 1537. Meyting may have gone to help his widowed sister with her definitive return to Augsburg in 1539. At this juncture, Meyting took up an apprenticeship with the South German merchant-financier from Nuremberg, Alberto Cuon (Albrecht Khun or Kuen, 1502–c. 1558), who like Ehinger was based in Valladolid. Cuon quickly recognized Meyting's hidden talents,[14] and from 1543 to 1546, Meyting worked as his junior business partner.[15] Although Meyting established a base in Valladolid, he still maintained commercial contacts with his natal city, Augsburg, where he married in the early 1550s.[16] He continued to live and travel between Augsburg and Spain until

---

12 Häberlein/Bayreuther, 2013: 66–80 posit Meyting arrived in Spain before 1537 as a young teenager, conjecturing he lived in the Valladolid residence of his elder sister, Ursula Meuting (1507–1588) and her husband, the Augsburg banker Ulrich (Enrique) Ehinger (1485–1537). Ehinger, a member of the Ehinger family from Ulm, was active in Constance and Zurich, becoming the foremost German merchant in Spain (based in Seville and Valladolid). He began his career as a partner in Iberia for the Augsburg bankers and merchants, the Welsers, was subsequently elected a member of Emperor Charles V's council, made a knight of the Order of Santiago and became a respected personage at the Spanish court. There is no documented evidence Meyting lived in his sister's house before 1542, where according to Häberlein/Bayreuther, 2013: 79 he supposedly completed his education and mercantile training, while gaining insights into life at the Spanish court. Ehinger's published will written in Valladolid on August 13, 1537 does not mention Meyting, although all members living in his household at this date were named. For Ehinger's testament witnessed by Alberto Cuon, Hieronymous Sailer, Christoph Peutinger and Ugo (Hugo) Angelo see Pascual Molina, 2007.
13 Steiner, 1978: 25 and 90 where Meyting's close ties with Albrecht V are underscored.
14 For Cuon, who originated from Baden-Württemberg, see Kellenbenz, 1961.
15 Häberlein/Bayreuther, 2013: 66–84, especially 76 f.
16 Häberlein/Bayreuther, 2013: 85–94.

his death in 1591. During the regency of Juana of Austria (r. 1554–1559), the younger sister of Philip II of Spain (1527–1598), the Spanish court resided in Valladolid, and it was likely during these interim years Meyting cultivated ties with Juana, Princess of Portugal (1535–1573) and her nephew and Philip's son, Infante Carlos (1545–1568).

In Spain, Meyting learned to move with ease in aristocratic and court circles, soon gaining the respect and admiration of Philip II, who had taken over the Spanish throne in 1556, after his father, Emperor Charles V, abdicated. Meyting was not only well educated, but also well read, evidently influenced by a humanist cultural background. It is well known that the Fugger merchant-banking family educated their sons in mathematics, law and languages, the younger generations even attending universities, which underscores how the Fugger world of commerce was clearly intersected by cultural interests and the arts.[17] Meyting's early education surely followed similar paths.

One of Meyting's favorite diversions was music, theatrical performances and comedies.[18] He had a command of languages, having learned to speak Castilian fluently, probably also Portuguese since he traveled to Lisbon frequently, collaborating with the commercial agent-merchant residing there, Nathaniel (or Nathanael) Jung († 1578),[19] who was often commissioned to seek out for Meyting, high-quality exotica, Indian diamonds, Columbian emeralds, foreign animals and animal by products (rhinoceros horns, bezoar stones and animal skins). Meyting's financial activities took him back and forth between Spain and Augsburg with frequent regularity, and his alacrity in resolving business matters, his familiarity with current political issues, his love of the arts, coupled with his superb eye in buying superior objects and exotica soon caught the attention of Duke Albrecht V in Munich, who equally placed his trust in him as a courtier and official intermediary. One of Meyting's roles was to act as translator both at the Spanish and Bavarian courts.[20]

---

17 Meadow, 2002: 187.
18 Meyting's knowledge of contemporary plays and actors was admired by his contemporary and colleague, Hans Fugger (1531–1598). Cf. Karnehm, 2003: vol. 2/1, 154, letter 357. Also 211, letter 506, in which Hans Fugger equally praises Meyting's eloquence and aptitude in writing well.
19 For Jung's career in Lisbon consult Jordan Gschwend/Beltz, 2010: 98 f.
20 On several occasions, Meyting acted as translator at the Bavarian court for Spanish envoys sent with messages and letters written in Castilian for Duke Albrecht V, and in Spain, both Meyting and his nephew, Peter Renz (active 1590s), would translate letters written in German for Philip II.

In a short time span, Meyting soon became the intimate of a royal elite, acting as the principal link between the Madrid and Munich courts, relaying letters sent between the two cousins, Philip II and Albrecht V, and when necessary acting as their personal courier and messenger. Before long, Meyting was advising both rulers about artistic matters, intervening in their acquisitions of exotica from Portuguese Asia and the Americas for Albrecht V's *Kunstkammer* in Munich, while buying quantities of luxury goods made in Augsburg, among them intarsia furniture, caskets and writing desks so beloved by the Spanish king, his family, and other aristocratic consumers in Iberia (Fig. 2).[21] During lengthier stays in Augsburg, Meyting would act there as Philip II's agent and factor, personally advising the Spanish king on artistic, political and financial matters, receiving a salary for his services while at the same time supervising purchases of luxury goods made on Philip's behalf.[22]

Fig. 2: South German Portable Writing Desk (with front fall lid), Augsburg, fruitwood and marquetry with an ivory Virgin and Child (inside), late sixteenth century, private collection. Photo credit: Pedro Lobo.

21 For inlaid boxes, caskets, chests and portable writing desks see Berger, 2000. For a study of Augsburg marquetry woodwork exported to Spain in the sixteenth century see Aguiló, 1990.
22 When Meyting died, Philip II raised the question of who would replace his trusted agent in Augsburg, and Meyting's nephew, Peter Renz (Pedro Ranz), who worked with Meyting in Spain and was a fluent Spanish speaker, was chosen as the king's replacement with 150 ducats *per annum* as his salary. Archivo General de Simancas, Estado, leg. 2450, letter written by Philip II to his ambassador in Prague, Guillén de San Clemente (June 24, 1593). I thank Almudena Pérez de Tudela for this reference.

In the Service of Albrecht V's Kunstkammer

> Hyer au soir venit vers moy Anthonie Meyting, me mostrant une lettre du duc de Baviere a luy, allendroit de alcunes curiosites quil vouldroit bien avoit pour cabinet quil faict...[23]

Meyting functioned on multiple levels for Duke Albrecht V and the Munich *Kunstkammer* during the years he was engaged in his service. Typically, Meyting was given direct orders to seek out in Spain rare, strange and exotic things for the Duke's collection. On other occasions, Albrecht V's commissions for Meyting were indirect, transmitted through correspondence with his close Fugger contacts in Augsburg, Marx Fugger (1529–1597) and Hans Jakob Fugger (1531–1598), both of whom equally worked in the Duke's service as an antiquarian and a librarian respectively, and who sent these requests on to Madrid via the Fugger postal network.[24] Othertimes, Meyting acted on his own initiative, sending Albrecht luxury goods as personal gifts, or offering him exclusive objects, clothes, sword blades and jewels for sale in Spain, where Meyting worked closely with German merchants based there. One merchant-dealer with whom Meyting had a close working relationship was Sebastian Renner (active 1560s), who owned a shop in Madrid where luxury goods and intarsia furniture from Augsburg were sold.[25] Meyting cultivated a wide, extensive network, which kept him informed of goods, textiles, commodities and exotica for sale in Lisbon, Seville, Valladolid and Medina del Campo.

Alone, and sometimes with Sebastian Renner as a partner, Meyting took advantage of auctions (*almonedas*) of royal collections broken up after the deaths of their illustrious owners. Meyting is documented buying exclusive exotica and luxury objects at the estate sales of Queen Mary of Hungary (1505–1558), the former regent of the Netherlands, in Valladolid in 1558, in Madrid in 1571 when the collection of Philip II's son, Infante Carlos (1545–1568), was sold off to pay outstanding debts, and, in 1574, when the collection of Ruy Gómez da Silva (1516–1573), one of Philip II's most trusted courtiers, who had amassed a substantial collection of Chinese Ming porcelain and other luxury Asian wares, went

---

23 Excerpt of a letter addressed to Philip II of Spain, in 1572, which the *Kunstkammer* (*cabinet*) of Albrecht V and the curiosities Anthonio Meyting acquired for him in Spain are mentioned. Published in Pérez de Tudela/Jordan Gschwend, 2001: 35.
24 Meadow, 2002: 189–193. For Marx's role as agent, patron and collector see the recent study by Wölfle, 2009.
25 Pérez de Tudela, 2011: 1772; Häberlein/Bayreuther, 2013: 94–115.

on the block.[26] Meyting made significant purchases at this particular auction on behalf of Albrecht V, spending the large sum of 3,000 crowns. Cost and expense mattered little for the Bavarian Duke, who had given Meyting *carte blanche* to buy as much as he could and only the best.[27]

Albrecht V was always pleased with the curiosities Meyting brought back to Munich, just as the Duke was delighted with gifts the Spanish court frequently sent him. In a thank you written to the Spanish king in 1574, Albrecht raved about an emerald mineral stone (*handstein*) from Peru he received from Philip II, confessing how crazy he was for exotica: *quia exoticis eiusmodi plurimum delectamur*.[28] Between 1558 and 1577, one year after Albrecht V died, Meyting is documented supplying Munich with exotica of diverse origins: lemon trees for the palace gardens, woven mats from Asia, probably from Japan (the first Tatami mats documented in a Renaissance German collection), five painted cloths from India, and a "painting" of St. John the Baptist, now lost, made from "parrot" feathers, an Amerindian object manufactured in Michoacán, Mexico.[29] Albrecht V cultivated interests in foreign cultures and clearly his tastes were global.

Meyting's Diplomatic Mission to Portugal: 1573

faisant [...] le d. Anthoine Meyting son voyage de Portugal que le duc de Baviere luy a enchargé...[30]

To conclude, one of the highlights of Meyting's career was a diplomatic mission Albrecht V entrusted him at the height of Meyting's service to the Bavarian court. Meyting was sent to Spain and Portugal on an extended visit from late 1572 until 1574, where he was expected to procure rarities and exotica for the Munich Kunstkammer, oversee business for the Fuggers, as well as undertake a delicate mission for Albrecht at the Lisbon court.[31] For this purpose, Albrecht appointed

---

26 I am grateful to Cinta Krahe for sharing with me her research on Gómez da Silva's Asian collection. Consult Krahe, 2015.
27 Jordan Gschwend/Beltz, 2010: 98.
28 This mineral stone brought to Munish by Meyting. Cf. Pérez de Tudela, 2011: 1786; Seelig, 2000; Bujok, 2007 a; Wappenschmidt, 2008 and Bujok, 2009.
29 Pérez de Tudela/Jordan Gschwend, 2001: 45. Cf. Bujok, 2007 b.
30 Excerpt of a letter to Philip II, dated November 23, 1573 cited in Pérez de Tudela/Jordan Gschwend, 2001: 35 f.
31 According to the Austrian Imperial ambassador in Spain, Hans Khevenhüller (1538–1606), Meyting visited Portugal and Spain for a period of 17 months, during which interval Meyting not only took care of business for Albrecht but also for Jakob Fugger. Consult Österreichisches Staatsarchiv, Haus, Hof-, und Staatsarchiv, Vienna,

Meyting as Ambassador Extraordinary.[32] Meyting's purpose was to conduct secret negotiations between Albrecht V and the Portuguese queen, Catherine of Austria (1507–1578), in order to arrange a marriage between Albrecht's youngest daughter, Maximiliana Maria (1552–1614), and the Portuguese king, Catherine's grandson, Sebastian (1554–1578). Copies of Maximiliana's portrait no doubt executed by Albrecht's court painter Hans Mielich (1516–1573) were duly dispatched to the Lisbon and Madrid courts for royal inspection and Catherine's approval,[33] who hoped to bring the young Bavarian bride to Lisbon to educate and prepare her for her future role as queen of Portugal. Aside from shopping for exotica in Lisbon, Meyting was expected to finalize the marriage contract between Bavaria and Portugal. Albrecht counted on this Lusitanian match, and Catherine speculated on the successful outcome of Meyting's secret visit and his astute mediation upon his return to Munich.

Meyting reached Lisbon in late 1573, and closeted meetings were held between Catherine and the Duke's envoy, who was presented with a rare illuminated genealogy of the Kings of Portugal for the Bavarian court, probably painted by the queen's former court painter, Antonio de Holanda (active 1518–1556).[34] During Meyting's visit, the Portuguese queen also selected for Albrecht V exclusive exotica from her collection and *Kunstkammer* as personal gifts for her nephew: two Ceylonese caskets stored in leather cases, three combs and almost certainly other Ceylonese items, among them crystal utensils (spoons and forks) and a gold thimble mounted in gold and set with rubies (Fig. 3). This rare, sixteenth-century thimble from Ceylon (present day Sri Lanka) resembles the one Catherine selected for her nephew. Catherine intentionally chose prestigious objects she had received from the King of Kotte, Bhuvaneka Bahu VII, in 1542, when the first Asian embassy to ever visit Europe arrived in Lisbon.[35]

Spanien, Diplomatische Korrespondenz, Karton 8, konv. 20, fol. 104 v (letter written by Khevehüller in 1574): "Ihn ansehung Anthony Meyting (welcher in der Portugesischen heyrat tractation auch Herrn Jacoben Fuggers particular handlungen wegen des Herzogen von Bayern bis in die sibenzehen monat alhie gewest) abraisen." (Translation: "Anthony Meyting has just departed. He was here [in Spain and Portugal] for almost 17 months negotiating the Portuguese marriage for the Duke of Bavaria, while also taking care of personal business for Jakob Fugger").

32 Karnehm, 2003: vol. 1, 476–479, letter 1088.
33 Biblioteca da Ajuda, Lisbon, Ms. 49-X-4, fols. 182–183.
34 Österreichisches Staatsarchiv, Haus, Hof-, und Staatsarchiv, Vienna, Spanien, Diplomatische Korrespondenz, Karton 8, konv. 20, fols. 56, 85v, 95, 105.
35 Jordan Gschwend/Beltz, 2010: 33–45.

Fig. 3: Ceylon, gold thimble set with rubies and sapphires, second half sixteenth century, private collection. Photo credit: Pedro Lobo.

Despite Catherine's calculated dealings with Meyting, and the presents she sent to Bavaria, this marriage never took place. However, for Albrecht V and his *Kunstkammer*, Catherine's gifts proved in time to be a windfall; the most significant Asian ivories to have ever been given a Renaissance queen arrived in Munich in Meyting's baggage in 1574. Two magnificently carved ivory caskets and three ivory combs have survived the vicissitudes of time, and are on view today at the Schatzkammer in the Munich Residenz.[36] For this we can thank Anthonio Meyting who surely impressed upon the Portuguese queen the merit and quality of Albrecht V's *Kunstkammer* in the Munich Residenz.

Manuscript Sources

Archivo General de Simancas, Estado, leg. 2450

Bayerisches Hauptstaatsarchiv, Munich, Kurbayern, Äusseres Archiv, 4853

Biblioteca da Ajuda, Lisbon, Ms. 49-X-4.

Österreichisches Staatsarchiv, Haus, Hof-, und Staatsarchiv, Vienna, Spanien, Diplomatische Korrespondenz, Karton 8, konv. 20.

Bibliography

Aguiló, María Paz. 1990. *El mueble en España en los siglos XVI–XVII*. Madrid: CSIC-Anticuaria.

---

36 Bayerische Verwaltung der staatlichen Schlösser, Gärten und Seen, Residenz München, Schatzkammer, Inv. nos. 1241, 1242, 1243, 1244, 1245. Cf. Jordan Gschwend/Beltz, 2010: 88 f., cat. 32.

Bäader, Berndt. 1943. *Der bayerische Renaissancehof Herzog Wilhelms V. (1568–1579). Ein Beitrag zur bayerischen und deutschen Kulturgeschichte des 16. Jahrhunderts*. Leipzig: Heitz.

Backmann, Sibylle. 1997. "Kunstagenten oder Kaufleute? Die Firma Ott in Kunsthandel zwischen Oberdeutschland und Venedig (1550–1650)". In: Bergdolt, Klaus/Brüning, Jochen (eds.). *Kunst und ihre Autraggeber im 16. Jahrhundert. Venedig und Augsburg im Vergleich*. Berlin: Akademie, 175–197.

Berger, Ewald. 2000. *Prunkkassetten. Europäische Meisterwerke aus acht Jahrhunderten. Hanns Schell Collection, Graz*. Stuttgart: Arnold.

Bujok, Elke. 2007 a. "Die frühe Sammeltätigkeit der Wittelsbacher. Ethnografica in der Münchner Kunstkammer um 1600". In: Müller, Claudius/Stein, Wolfgang (eds.). *Exotische Welten. Aus den völkerkundlichen Sammlungen der Wittelsbacher 1806–1848*. Munich: Staatliches Museum für Völkerkunde, 21–52.

Bujok, Elke. 2007 b. "Kunstkammerbestände aus portugiesischen Seereisen". In: Kraus, Michael (ed.). *Novos Mundos – Neue Welten: Portugal und das Zeitalter der Entdeckungen*. Berlin: Deutsches Historisches Museum, 240–255.

Bujok, Elke. 2009. "Ethnographica in early modern Kunstkammern and their perception". *Journal of the History of Collections* 21/1, 17–32.

Burkhardt, Johannes/Karg, Franz (eds.). 2007. *Die Welt des Hans Fugger (1531–1598)*. Augsburg: Wißner-Verlag

Häberlein, Mark/Bayreuther, Magdalena. 2013. *Agent und Ambassador. Der Kaufmann Anton Meuting als Vermittler zwischen Bayern und Spanien im Zeitalters Philipps II*. Augsburg: Wißner-Verlag.

Impey, Oliver/Macgregor, Arthur (eds). 2001. *The Origins of Museums: The Cabinet of Curiosities in Sixteenth- and Seventeenth-Century Europe*. London: House of Stratus.

Johnson, Christine. 2008. *The German Discovery of the World. Renaissance Encounters with the Strange and the Marvelous*. Charlottesville, Va.: University of Virgina Press.

Jordan Gschwend, Annemarie. 2012. "Exotica for the Munich Kunstkammer. Anthonio Meyting: Fugger agent, Art dealer and Ducal Ambassador in Spain". In: *Exotica. Kunstkammer Georg Laue*. Munich: Georg Laue, 8–28.

Jordan Gschwend, Annemarie/Pérez de Tudela, Almudena. 2003. "Exotica Habsburgica. La Casa de Austria y las colecciones exóticas en el Renacimiento temprano". In: Alfonso Mola, Marina/ Martínez Shaw, Carlos (eds.). *Oriente en Palacio. Tesoros asiáticos en las colecciones reales españolas*. Madrid: Patrimonio Nacional, 27–44.

Jordan Gschwend, Annemarie/Beltz, Johannes Beltz (eds.). 2010. *Elfenbeine aus Ceylon. Luxusgüter für Katharina von Habsburg (1507–1578)*. Zurich: Museum Rietberg.

Jordan Gschwend, Annemarie/Lowe, Kate J. P. (eds.). 2015. *The Global City: on the streets of Renaissance Lisbon*. London: Paul Holberton.

Karnehm, Christl. 2003. *Die Korrespondenz Hans Fuggers von 1566 bis 1594. Regesten der Kopierbücher aus dem Fuggerarchiv* (Quellen zur Neueren Geschichte Bayerns). 3 vols. Munich: Kommission für Bayerische Landesgeschichte.

Kellenbenz, Hermann. 1961. "Alberto Cuon. Auf den Spuren eines Nürnberger Kaufmanns in Valladolid". In: Goldmann, Karlheinz (ed.). *Norica. Beiträge zur Nürnberger Geschichte. Bibliotheksdirektor a. D. Dr. Friedrich Bock zu seinem 75. Geburtstag*. Nürnberg: Stadtbibliothek, 21–27.

Krahe, Cinta. 2015. *Chinese Porcelain in Habsburg Spain*. Madrid: Centro de Estudios Europa Hispánica.

Lieb, Norbert. 1952. *Die Fugger und die Kunst im Zeitalter der Gotik und der frühen Renaissance*. Munich: Verlag Schnell & Steiner.

Meadow, Mark. 2002. "Merchants and Marvels. Hans Jakob Fugger and the Origins of the Wunderkammer". In: Smith, Pamela/Findlen, Paula (eds.). *Merchants and Marvels: Commerce, Science and Art in Early Modern Europe*. London: Routledge, 182–200.

Neuwirth, Markus. 2000. "*Portugal, die süddeutschen Fernhandelshäuser und Erzherzog Ferdinand II*". In: Seipel, Wilfried (ed.). *Exotica. Portugals Entdeckungen im Spiegel fürstlicher Kunst- und Wunderkammern der Renaissance*. Vienna: Kunsthistorisches Museum, 49–53.

Pascual Molina, Jesús Félix. 2007. "Testamento e inventario de bienes de Enrique Ehinger, agente de los Welser, en Valladolid". *Boletín de la Academia Puertorriqueña de la Historia*. Vol. II. *Homenaje a Manuel Ballesteros*, XXIV–XXV, January 2004–July 2005, 15–78.

Pérez de Tudela, Almudena. 2011. "Relaciones artísticas de los Duques de Baviera con España en el reinado de Felipe II". In: Martínez Millán, José/González Cuerva, Rubén (eds.). *La dinastía de los Austria: las relaciones entre la Monarquía Católica y el Imperio*, Madrid: Polifemo, 1769–1836.

Pérez de Tudela, Almudena/Jordan Gschwend, Annemarie. 2001. "Luxury Goods for Royal Collectors: Exotica, princely gifts and rare animals exchanged between the Iberian courts and Central Europe in the Renaissance (1560–1612)". In: Trnek, Helmut/Haag, Sabine (eds.). *Exotica. Portugals Entdeckungen im Spiegel fürstlicher Kunst- und Wunderkammern der Renaissance. Die Beiträge des am 19. und 20. Mai 2000 vom Kunsthistorischen Museum Wien veranstal-*

teten Symposiums (Jahrbuch des Kunsthistorischen Museums Wien 3), Mainz: Philipp von Zabern, 1–127.

von Schlosser, Julius. 1908. *Die Kunst- und Wunderkammern der Spätrenaissance.* Leipzig: Klinkhardt & Biermann Verlag.

Seelig, Lorenz. 2000. "Exotica in der Münchner Kunstkammer der bayerischen Wittelsbacher". In: Seipel, Wilfried (ed.). *Exotica. Portugals Entdeckungen im Spiegel fürstlicher Kunst- und Wunderkammern der Renaissance.* Vienna: Kunsthistorisches Museum, 144–161.

Seelig, Lorenz. 2008. "Die Münchner Kunstkammer". In: Diemer, Dorothea and Peter/Sauerländer, Willibald/ Seelig, Lorenz/Volk, Peter/Vork-Knüttel, Brigitte (eds.). *Die Münchner Kunstkammer*, 3 vols., Munich: Bayerische Akademie der Wissenschaften, vol. 3, 1–124.

Steiner, Robert. 1978. *Die Meuting in Augsburg* (Genealogia Boica 3/1), Munich: self-published.

Stockbauer, Jakob. 1874. *Die Kunstbestrebungen am Bayerischen Hofe unter Herzog Albrecht V. und seinem Nachfolger Wilhelm V. Nach den in K. Reichsarchiv vorhandenen Correspondenzacten.* Vienna: W. Braumüller.

Strohmeyer, Arno. 2003. "Kulturtransfer durch Diplomatie: Die kaiserlichen Botschafter in Spanien im Zeitalter Philipps II. und das Werden der Habsburgermonarchie (1560–1598)". In: Schmale, Wolfgang (ed.). *Kulturtransfer. Kulturelle Praxis im 16. Jahrhundert.* Innsbruck: Studienverlag, 205–230.

Wappenschmidt, Fridericke. 2008. "Selzame und Hir Lander Fremde Sachen. Exotica aus Fernost im Münchner Kunstkammer inventar von 1598". In: Diemer, Dorothea and Peter/Sauerländer, Willibald/Seelig, Lorenz/Volk, Peter/Vork-Knüttel, Brigitte (eds.). *Die Münchner Kunstkammer*, 3 vols., Munich: Bayerische Akademie der Wissenschaften, vol. 3, 293–309.

Wölfle, Sylvia. 2009. *Der Kunstpatronage der Fugger, 1560–1618.* Augsburg: Wißner-Verlag.

Samuel Gessner

Lost Between Centuries: a Celestial Globe (1575) from Augsburg in the Portuguese Royal Collections

Abstract: Christoph Schissler der Ältere (1531–1608) war einer der herausragendsten Instrumentenbauer astronomischer und mathematischer Geräte in Augsburg in der zweiten Hälfte des sechzehnten Jahrhunderts. Estácio dos Reis zeigte bereits 1990 das Vorhandensein eines Himmelsglobus Schisslers in Portugal an, dessen Existenz bislang nicht bekannt war. Dieser Globus von etwa 40 cm Durchmesser ist um 1575 datiert. Von der sorgfältigen Fertigung her (mit gestochenen Figuren in vergoldetem Kupfer) bietet er alles Nötige für ein Sammlungsstück eines hohen Adligen. Heute wird er im Museum Palácio Nacional de Sintra aufbewahrt. Er gehörte einst zu den Sammlungen des portugiesischen Königshauses, spätestens ab dem frühen zwanzigsten Jahrhundert. Dieser Aufsatz folgt den Spuren dieser Himmelskugel, von ihrer Herstellung bis zur heutigen Aufstellung in Sintra.
Daraus entsteht eine bruchstückhafte Lebensgeschichte, welche den wechselnden epistemologischen und kulturellen Status von Himmelsgloben allgemein zum Vorschein bringt. Gleichzeitig ergeben sich daraus neue Fragen zum Gebrauch und der Verbreitung von solch aufwändigen Instrumenten zwischen den Deutschen Landen und Portugal während des Entdeckungszeitalters.

This is the 'life story' of a particularly precious celestial globe. Its life is special because it is the only sixteenth-century globe that travelled from Germany to Portugal to have survived today. It is now preserved at the Museum of the *Palácio Nacional de Sintra* (National Palace of Sintra). While this fact is exceptional, the trajectory of this globe through the centuries is fairly typical of many a Renaissance celestial globe.

The sixteenth century saw a significant rise in globe making comprising both printed and manuscript globes. This is true for celestial as well as for terrestrial globes or, as they were rather called, cosmographical globes. However, while cosmographical globes had never been described before the fifteenth century, celestial globes were an old idea. A celestial globe is an astronomical instrument and as such its structure was already described in Ptolemy's Μεγάλη συντάξις or *Almagest* (c. 130 AD), the influential work on mathematical astronomy. The oldest extant metallic celestial globes preserved today stem from the Islamic world and are inscribed in Arabic.

A celestial globe is like a model of the heavenly vault: it allows one to represent the star positions as seen by an observer placed at the centre. Moreover, it records the star names and the constellation names. There were times in history when, in addition to that, the globe was used to explain the principal concepts of spherical astronomy: the celestial equator, the ecliptic, the meridian, the local horizon, the colures, the tropic circles etc. One could operate with a globe to readily make ('approximate') calculations. One could, for instance, determine the rising times of stars, the conversion of ecliptical to equatorial coordinates, or predict the height of the sun at noon for a given day and so on. By using a globe it was also possible to set up the astrological house divisions for a horoscope.

Over time celestial globes seem to have undergone varying popularity. While they were frequent in Arabic culture, and known to the medieval scholars in Europe, they reached large dissemination when printed globes were invented in the 16th century. Today they are almost forgotten. The ones on sale at museum shops are of deplorable quality. It is normal then that all globes produced in the past, as any instrument, feel the ups and downs and change in status that concern all pieces of their kind. In the course of their 'life span' they become part of various contexts and are used in different ways and acquire a different status in each context. Their last stage, very often, is oblivion, when they might even be considered trash.

This is exactly what occurred with the globe we are presenting here (see figure 1): had it not been for the pioneer of globe studies in Portugal, António Estácio dos Reis, and his persistence, this Schissler globe would still be largely unknown even today. We must credit Estácio for litterally 'rediscovering' the globe at the *Palácio Nacional de Sintra* around 1987, and for alerting the scientific community of its existence. The first detailed description (with photos) is also due to Estácio.[1]

Fig. 1: *Christoph Schissler the Elder, celestial globe, Augsburg (1575), gilt copper and brass, diameter 42 cm. National Palace of Sintra, inv. nr. PNS53457 (Credits: DGPC/ADF, Luisa Oliveira, 2002).*

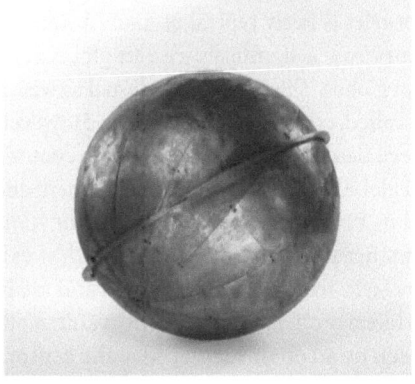

---

1   Estácio dos Reis, 1990: 57–62; plates 8 and 9.

This may sound unbelievable given the Schissler globe's rarity, size, its historical importance, the precious craftsmanship it represents as well as the renown of its maker. The globe consists of a large gilt body. The sphere is engraved with stars and expressive, figurative constellation images. It measures 42 cm in diameter. The globe turns on an axis attached to a graduated meridian ring. The inscriptions indicate its date and provenance: it was made by the famous instrument maker Christoph Schissler the Elder (1531–1608) in Augsburg in 1575. The globe is entirely metallic: inside, completely invisible to the observer, it consists of a riveted brass skeleton. The latter is covered by twelve copper gores, i.e. spherical segments. This is a truly unusual construction technique. These copper parts were engraved and then fire gilded.

Some parts are missing today that once belonged to the Schissler globe.[2] One can see that there must have been a chapter ring for an hour pointer near the northern pole. This feature was typical of the sixteenth century globes. It allowed one to reckon the position of celestial bodies at certain times. Moreover, for a sixteenth-century globe to be fully functional it needed to sit in a horizon ring. Typically the horizon ring would be marked with a 'theory of the sun', i.e. scales that would yield the ecliptical position of the sun as a function of the date. At any time, half of the globe exactly will show above that ring, the other half stays below. This ring is prepared on the southern and northern part to receive the meridian ring in a recess. It is very probable that the Schissler globe had its stand with a horizon ring. The inclination of the globe's axis would have been adjustable to the desired latitude. The fact that this stand is missing today indicates a rather dramatic event in the past during which it was destroyed or lost. Currently the globe is safe. It is now on permanent exhibition at the museum of the Palace of Sintra in the so-called King Sebastião's bedchamber. It rests on a donut shaped cushion on a period table.

On the following pages I will bring together many known but scattered pieces of information about the Schissler globe. The only novel finding here concerns the exact date of the globe's arrival at Sintra. The larger part of its existence remains still in the dark. So for the seventeenth to the nineteenth centuries I will succinctly indicate the general status of celestial globes in European culture; a status that inevitably also affected the Schissler globe, wherever it may have been during these years. In telling the life story of this globe I keep to a tradition in instrument studies. In a seminal article, Samuel Alberti proposed to use this 'biographical' perspective for the purpose of understanding the epistemological change happen-

---

2 For convenience, we will designate this globe in Sintra, inv. PNS53457, as 'Schissler globe'.

ing to objects when they enter a museum.³ He pointed out that the 'life' or 'career' makes an object different from other, similar objects. So he proposed to study the shifting relationships between it and the surrounding objects and people, the human relationships realised by producing, exchanging and using objects. The observed trajectory may point to patterns in the political and social climate. Accordingly, this meant for Alberti borrowing various methods from a number of research areas: instrument studies, historical epistemology, ethnology and archaeology. He then distinguished three key moments of an object's biography. The first moment concerns the movement of the object from its manufacture through collecting or exchange to a museum or cabinet of curiosity. The second moment examines the use of the object when it has become part of a collection. And a third moment analyses the nature of the relationship between the object and its viewer in a museum setting today, and the various roles it may play according to various visitor categories. Although the record that comes with the celestial globe we focus on here is slight, we propose to follow it through these moments.

It is important to note that the full biography in Alberti's sense involves a number of inquiries and questions that are of fundamentally different type. That universe of questions addressed to an object is strongly heterogeneous. As you will notice on the following pages, I have tried to bring all possible aspects together: including the globe's material and archival record, as well as its cultural and scientific background. For a global view of these questions it has proved useful to map them onto four 'quadrants' (I to IV) with each quadrant representing a specific type of questions.⁴ In the case of our globe, the strictly biographical questions (quadrant IV) would be those that concentrate narrowly on the singular globe we have now before us and follow it from Schissler's workshop, in 1575, onwards down to the showcase at the Palace of Sintra. To reach the full potential of biographical questions, and Alberti is fully aware of this, other questions need to be asked. First of all, we need to inspect the object carefully as it sits before us (quadrant I). For this we might even use modern techniques of imaging or material analysis. But then, of course, it becomes important to make comparisons with other globes. And the comparisons on all levels raise a number of specific questions about how globes are made and what functions they allowed to perform, namely a form of analogue computation (quadrant II). Finally, the cultural history of celestial

---

3   Alberti, 2005.
4   These are the four quadrants, which I have described in an article published with Marta Lourenço, cf. Lourenço/Gessner, 2014.

globes and, more generally of mathematical instruments at large, form the natural background (quadrant III) for understanding the Schissler globe's own biography.

Within the scope of the present chapter all the questions from these quadrants will not be addressed. A great part of them has been answered in previous studies: Estácio (1994) has published a thorough description of the globe.[5] This initial description has been taken further in my first publication on the globe (2010) in which Schissler's use of Caspar Vopelius's printed globe (1536) is established.[6] Many of these findings will be here presented again in chronological form, this time, however, with a special attention given to the changing status of Schissler's 1575 globe.

From Augburg to the Iberian Peninsula: patronage and middlemen

The inscription on the globe informs us of its maker, the place where it was made and the date:

CHRISTOPHORVS SCHISSLERVS AVGVSTANVS GEOMETRICVS ET ASTRONO-MICVS FABER GLOBVM HVNC CÆLESTREM FACIEBAT ET DESCRIBEBAT ANNO DOMINI 1575.
Christoph Schissler, citizen of Augsburg, geometrical and astronomical craftsman made and inscribed this celestial globe in the year of the Lord 1575.

At that time the imperial city of Augsburg was a major European centre of metalwork production. Not only its weapons and armours were famous. Innumerable reliquaries and other kinds of goldsmith works, clocks and automata, mathematical instruments and sundials came from Augsburg. These pieces were often commissioned for the highest-ranking nobility. Many of them served as diplomatic gifts.[7] This was the city where Christoph Schissler the Elder ran one of the most notable and most prolific workshops of mathematical instruments.[8] He stood out among the numerous compass and sundial makers who were his fellow citizens. One of the reasons for which he was exceptional is evident from the range of instruments that Schissler manufactured. This went far beyond the common sundials and astronomical *compendia* by masters of the guild of metalworkers. He proposed not only to produce larger (and hence more expensive and more precious) instruments, but also more complex ones. Schissler's son, Christoph

---

5   Estácio dos Reis, 1994.
6   Gessner, 2010.
7   Seelig, 1995.
8   Bobinger, 1954 and Bobinger, 1966. Cf. also Gessner, 2012 a.

the Younger, was trained as a clockmaker but later also crafted sundials clocks and table globes; nevertheless he did not reach his father's level of production.

Celestial globes were produced more frequently in Schissler's time than in the fifteenth century.[9] The development of wood-cut and soon copper-plate printing was one factor in making the globes more common. Three names belong to the early stage of print globe making: Johannes Schöner (1477–1547), Caspar Vopell (Vopelius, 1511–1561) and Gerard Kremer (Gerhard Mercator, 1512–1594). It is also through the action of these men that it became common to conceive globe as pairs: each including a celestial and a cosmographical one. The cosmographical globes were essentially terrestrial globes with features that allowed using them in a similar way as celestial globes. The ecliptic, the tropics and a series of bright stars would be inscribed beside the geographical information. The production included printed globes that used spheres with wooden structure that was covered by a mixture of plaster and glue. This support could then be turned in a loath to become perfectly spherical before the paper gores with the printed images would be pasted on them. The astronomical writer Johannes Schöner would offer such printed globes for sale starting with a terrestrial globe in 1515 and a celestial one in 1517. Another teacher of mathematics who produced such globes was the Caspar Vopelius from Medebach (near Cologne). His printed celestial globe (dated 1536) was used by Schissler as a model for the star and constellation names as well as for the iconography.[10] It is very interesting to note how the production of printed globes made models for further globe production much more readily available.

At that time, celestial globes belonged to the category of mathematical instruments. Indeed, Schissler calls himself a *geometricus et astronomicus faber* in the text of a cartouche on the surface of the Sintra globe. This translates as 'astronomical and geometric instrument maker'. Globe makers then were usually makers of other mathematical instruments, like astrolabes, quadrants or sundials, too. The contemporaneous literature on the use of the globe, or the solid sphere as it was often called, informs us about the ways such an instrument could be put to use. Here are some of the typical operations that may be performed as presented in Johannes Schöner's seminal text on the use of the celestial globe[11]:

---

9   Dekker, 1995.
10 Gessner, 2010.
11 Cf. Schöner, 1516/1548: Chapters IV ("Latitudinem regionis cuiuscunque siue eleuationem poli nobis ignotam, adiutorio huius globi demensam computare"), VII ("Zenith ortus et occasus solis, ac stellae uel puncti coeli cuiuscunque addiscere"), VIII ("Altitudinem Solis et stellarum meridianam atque horariam utrunque computare") and

- To calculate an unknown latitude (i.e. elevation of the pole) of any place what so ever measured by using this globe.
- To learn the direction of the rising and setting of the sun, of a star or of any point in the sky whatsoever.
- To calculate the height of the sun and of the stars at noon and at any hour.
- To determine in how much time a sign or whatever signs of the zodiac will rise completely according to their right ascension or taken on the ecliptic.

A celestial globe, beside all the things it symbolizes, the record of star positions, constellation names, therefore is also a tool. One of the conditions for the globe to work properly is, obviously, that the stars are marked at the correct positions. One peculiarity of these is the slow change of their longitude over the centuries due to a phenomenon known as 'the precession of the equinoxes'. Therefore the star positions on any globe are always set for one particular year, the so-called epoch. It is conspicuous that Schissler adds an extra cartouche to account for the updated star positions:

> STELLÆ HVIVS GLOBI NVMERATAE AC DISTRIBVTÆ SVNT SECVNDVM CVRSVM SPHÆRÆ OCTAVÆ AD NOSTRVM TEMPUS ANNVMQUE ACCOMMODATÆ 1575.
> The stars of this globe are measured and arranged according the course of the eighth sphere to our time and adapted to the year 1575.

This shows that the maker, as any competent instrument maker, was aware of this fundamental fact. If we check the position of a couple of stars close to the ecliptic, it turns out that Schissler indeed used updated positions, that correspond to Ptolemaic longitudes plus 21° 10'. Such values point to the use of star coordinates that go back to Copernicus's publication and theory of precession. His *De revolutionibus* was published in 1543 and very soon gave rise to the production of astronomical tables using some parameters slightly different from the traditional (Ptolemaic and Alfonsine) values.[12] This is all just to say that Schissler not only produced a globe that was especially valuable because of its size and expensive materials. Its value lay also in that it could be put to use as a geometrical device where the positions of the stars were adapted for the year it was made.

To appreciate the status of our celestial globe, we must bear in mind the following cultural traits of European mid-sixteenth century within which it emerged. Mathematical instruments of this costly kind, i.e. executed as luxury items with precious metals, attracted a variety of potential collectors. Such objects were sought

---

XII ("In quanto tempore quodlibet signum uel signa quaelibet zodiaci peroriantur in sphaera recta et obliqua perscrutari").
12 The most important such tables were Erasmus Reinhold's *Tabulae Prutenicae* (1551).

for as representative of high-level craftsmanship. They simultaneously demonstrated the intellectual culture of their owners. That period saw the foundations of numerous *Kunstkammern* boasting rare and curious objects including both *naturalia* and *artificialia*. Moreover, operations performed with mathematical instruments in a courtly setting would require the tools to be of appropriate material and sumptuous elaboration. For instance: if equating the celestial houses when casting a horoscope could be done by tables, astrolabes or globes, those codices, astrolabes and globes used in courtly settings would typically be the decorated and rare pieces. Schissler could provide such type of artefacts. He was well connected and managed to furnish instruments to the most high-ranking noblemen across Europe. He was able to sell a large geometric square, a waywiser and other instruments to the *Kunstkammer* of the emperor Rudolf II (r. 1576–1612; Prague in the beginning; later Vienna). Schissler also had a correspondence with the prince elector August I of Saxony (r. 1553–1586) himself in which he advertised instruments he could make. Quite a number of Schissler's instruments were delivered at Dresden. The core of August's *Kunstkammer* including some of those instruments forms part, today, of the collection still kept in Dresden.[13] A few of Schissler's letters to the prince elector have been preserved at the Sächsisches Hauptstaatsarchiv in Dresden. The inventory of the collection of the German banker Anton Fugger (1493–1560) once located at the castle of Oberndorf in Biberbach near Augsburg shows that Schissler also provided instruments to this rich, culturally alert merchant collector.[14] His oldest son Marx Fugger (1529–1597) also acted as a middleman to send Schissler's instruments to such far away places as the Iberian peninsula. This is mentioned in a letter by the Augsburg mathematics teacher Johannes Maior († 1615) who wrote to the famous astronomer Tycho Brahe in 1576:

> [...] Porro te ignorare nolo, mi Domine Tycho, Schislerum γνωμονοποιον nostrum confecisse Globum ex cupro fabrefactum secundum omnes dimensiones perfectae, veluti tuum Automaton, rotunditatis cuius diameter quinquepedalis. Globus autem tali arte est fabrefactus, ut in 8 partes dißolvi et rursus in integrum restitui poßit, neceßitate exigente, quem in Hispaniam unâ cum Astrolabio quantitatis eiusdem per MARCUM FUGGERUM est misurus. Cum itaque viderem Globum, de tuis commodis cogitans quaerebam, num etiam maiorem conficere poßit? qui respondit ita, seque ad tale corpus conficiendum motum eße ex eo, quod audivißet D[ominationem] tuam hîc artificem

---

13 The collection of the Mathematisch Physikalischen Salon, part of the Staatliche Kunstsammlungen Dresden (SKD) still includes a series of small and large instruments signed by Schissler.
14 Bobinger, 1966: 336 f.; cf. about the instrument collecting by the Fuggers, see Meadow, 2002: 184 and 192.

quaesivißs, qui id præstaret, seque poße aeque conficere eiusmodi corpus, cuius diameter eßet 8 vel 10 pedum, et ea arte, ut in 40 circiter partes resolutus facilime vasique inclusus, commodißime vel per mare vel per terras transportari poßit. Cum autem ulterius quaererem, quantum cuperet pro tali corpore conficiendo, me ad D[ominationem] tuam hac de re scripturum, respondit, non minus quam 1000 aureos, circulis exceptis et pede. || Quare cum hîc Globum viderim ligneum, qui tamen propter ligni naturam suis ex aequo partibus non respondere videretur, hac de re certiorem facere volui, ut si forte eiusmodi corpus globosum D[ominatio] tua habere cuperet, quod et ad durationem et perfectionem magis accederet, sciret, hîc esse Artificem, qui hoc se praestare posse diceret.[15]

Johannes Maior and Marx Fugger were middlemen for the diffusion of Schissler's works. We see Maior acting here by recommending Schissler to Tycho and explaining why his globes were desirable. Tycho does not seem to have purchased the large globe of 8 to 10 foot (2.5 to 3 meters!). The one thousand *aurei* might have been too expensive even for him. We have no knowledge about where the mentioned five-foot globe went to in *Hispania* (Spain or more generally the Iberian Peninsula). But we see that around Schissler, a network of connoisseurs with their scholarly or commercial correspondences is available. The network provides the channels through which such luxury goods were advertised and distributed, first to the patricians in Augsburg and then all over Europe. While distributing the beautifully crafted instruments, the network also helped diffuse the values of the cultivated collector: the letters implicitly (sometimes explicitly) contain the message of what is desirable, suitable and appropriate for a prince, for a man of culture: to possess globes, to be knowledgeable about what they contain in information and about how they could be used. Sharing the same type of predilection for certain objects and activities (mathematical exercise for instance) could be not only a way of claiming a similar status (as the emperor or the king). Simultaneously, it is a sign of allegiance and is socially felt as an obligation. Conspicuously, Maior tells Tycho of Fugger's shipment to Spain insinuating perhaps, although there is no specific mention of that, that the Spanish court or King Philip II himself were eager to receive large instruments by Schissler. Tycho will have got this message and considered whether he should not emulate the cultural gesture of this renowned prince.

Such networks can be understood as being instrumental not only for the dissemination of cultural objects. Information about the practices and cultural behaviours were also broadcast through such networks. In this way they contributed to the propagation of fashions and values.[16] Schissler as the maker of the mathematical

---

15 Brahe/Dreyer, 1924: 36 who transcribes the original Austrian National Library, Vienna, Handschriften und Alte Drucke, Cod. 10686/66, fol. 74r–75r.
16 Cf. Smith/Findlen, 2002.

instruments would benefit from the high esteem for the objects of that type. He obviously counted on the network he was connected to by direct correspondence and cultural middlemen. Schissler's particular globe that is now in Sintra has left the Augsburg workshop after 1575 when it was finished. It could very naturally have been sent to Lisbon back then. There is, however, no evidence so far to document it.

There has been no notice of the particular circumstance in which the globe came to Lisbon. Many possibilities are open. One may be particularly worth considering: the particular date of the globe's making, 1575, coincides with the year of the opening of the pepper trade to period contracts. A very active financial speculator from Augsburg, Konrad Rott, maintained the best relations with the Portuguese king D. Sebastião (r. 1557–1578) and Cardinal infante D. Henrique (r. 1578–1580). He eventually secured the monopoly of the pepper trade starting in 1575, but soon went bankrupt.[17] Alongside such a business deal it could have been appropriate to bring a gift from Augsburg, which would certainly be appreciated for its preciousness and prestige. Also, in 1576 a commercial agent, Hieronymus Kramer, travelled to Lisbon on behalf of Prince Elector August with the mission to achieve several deals. One was about the sale of copper. Would the copper globe by Schissler not be an appropriate gift to incline the King or any other princely collector to strike a deal? It will take much further historical research to examine the documents of the business agents from Germany that could involve the exchange of precious mathematical instruments.[18]

This is as far as we can go in order to characterize Alberti's first key moment in the Schissler globe's biography. We can understand that its status assured it an easy travel through the networks of cultivated middle men that connected makers and collectors. Its first and original owner remains still unknown however. That's why the next stage of the globe's biography will be written without knowledge of what specifically occurred to it and is no more than a short account of the cultural change that would have affected its status, wherever it was actually kept.

Obsolescence and Oblivion: the shifting status of celestial globes within collections

There might be a simple reason for the absence of any notice of the Schissler globe during the 17th century: it quickly became obsolete. This is not to say that celestial

---

17 Haebler, 1895 and Schirmer, 2001: 152–155.
18 The above-mentioned agents Hieronymus Kramer and Konrad Rott actually purchased also rare items in Lisbon, like portulan charts and exotic specimens for the Prince Elector in Dresden, cf. Haebler, 1895.

globes were no longer sought for, collected, exhibited or used. On the contrary, the first part of the 17th century saw a rise of globe production as an editorial business. Dutch globe makers like Blaeu and Hondius dominated the scene. It became standard to equip a library with a pair of globes, celestial and terrestrial, and an armillary sphere. For Schissler's celestial globe, obviously, the star positions, their ecliptical longitudes, would become outdated. Moreover, over the decades more stars and constellations, especially on the southern hemisphere, became conventional. So it would be natural that the globe was seen as deficient and it would be replaced.

The seventeenth century also saw the emergence of large-scale globes. The general of the Franciscan order Vicenzo Coronelli (1650–1718) conceived the most famous ones. These globes included encyclopaedic information inscribed on their surface. Not only were the star and constellation names given in many languages. The observed positions of comets or new stars are also mentioned. In general, these globes turned into encyclopaedias and paintings and their mathematical aspect became a marginal aspect. For Coronelli's largest globes, like the 4 m-diameter-globes offered to king Louis XIV (r. 1674–1692), their size turned them utterly useless for calculation.

In the eighteenth century, the age of spectacular science, globes continued to be an important mode of representing the sky and the earth. The famous Paris mathematical instrument maker and author Nicolas Bion (1652–1733) published a lavish book with the title *L'usage des globes célestes et terrestres, et des sphères, suivant les differens systemes du monde* in 1700 (cf. Bion, 1700). Its third part still included a variety of problems to be solved by globes; quite similar to those described two centuries earlier by Schöner. However, globes were no rarity anymore but a commodity that was routinely produced and available in most libraries and schools. Internationally, the Dutch and French globes dominated the market. English globes were also on the rise in the 18th century.

This dynamics also meant that more up-to-date editions would replace older globes, wherever the funds were available to acquire a more recent product. It must have been a time when many of the globes of the first two centuries of printed globe making were trashed. Obsolescence is a dangerous state: the fact that Schissler's large globe has survived signals particular circumstances, of which, however, we know nothing. All that we can say is that these circumstances allowed it to outlast this dangerous period of its life and reach happily the safe place of a new stage of its career: the stage of 'historical value'.

Rediscovery on the attic of a palace: the era of 'historical value'

The man who rediscovered Schissler's globe around 1900 was Alberto Girard (1860–1914), a trained oceanographer. He was a close collaborator of the penultimate king, Carlos I (1863–1908) of Portugal. The king had entrusted him to organise his personal library in one of the royal palaces of Lisbon, in the so-called Necessidades Palace, before 1907.[19] The Necessidades Palace lies on a slight elevation overlooking the Tejo estuary. In that place there had been already a church, *Igreja da Nossa Senhora das Necessidades*, when in the 1740s, under King João V (r. 1706–1750), royal order was issued to renovate the church, to build in the adjacent space a religious congregation, a royal palace and a college.[20] Later the Oratorians would occupy the monastery buildings and run a college that achieved some fame for the modernity of their teaching methods.[21]

If we believe Girard's own testimony it was he who found the Schissler globe on the attic of this palace 'many years' before he gave his account in 1913. He points it out as one of the valuable ornamental pieces, alongside a ceramic piece by Bernard Palissy, he had personally chosen for the adornment of the king's personal library in the former monastery wings.[22]

The interesting question here is: Where could the globe have come from? As long as no other evidence appears the field is open for speculation and there are many hypotheses. Here I will only evoke one. The college run by the oratorians included an impressive cabinet with 'mathematical instruments'. We may read this expression with the broad meaning it received during the 18th century: to include all instruments of natural sciences, astronomy but also surveying, and other practical arts. A globe like Schissler's could well have been part of such a cabinet, although we have not the slightest evidence that this was so. A description dated 1756 mentions the following:

> Returning to the corridor leading to the Palace, as I said, it continues underneath a staircase of the tower, with glass pane windows in niches, and it ends in a beautiful room called [room] of the instruments. It is of square plant and proportionally high. The ceiling is painted with extraordinary decorations representing allegories of the sciences. The floor is tiled, and there are three glass pane windows in niches. In this room, there are the mathematical instruments within cupboards along the walls. The cupboards are shut with glass fronts so the instruments are to be seen and still are protected from dust.[23]

---

19 Girard, 1907.
20 Ferrão, 1994: 15–45 and 327.
21 Carvalho, 1982: 56–62.
22 Gessner, 2012 b: 274.
23 'Tornando ao transito que dá serventia como disse para o Palacio, este vay continuando por debayxo de hum lanço da torre com janellas rasgadas todas com vidraças e termina com huma formoza caza chamada dos instromentos. He ella quadrada, e à proporção a altura. O teto he de singular pintura, e figuras de sciencias, o chão ladrilhado, com tres

Obviously, many other possibilities are open for the globe's provenance. It can not even be entirely excluded that it came with a 19th century purchase. The only clue Girard's declaration yields is that the globe ended its period of obsolescence in the attic of a palace.

The personal library of King Carlos, where in room III the Schissler globe would be presented on a counter ('contador'), boasted not only many volumes with literature and scientific topics, with an emphasis on oceanography, but contained collections of natural history and medals. The globe does not appear on the photographic image of the library that I have found. The image gives an idea of how the library of a monarch would look like in 1908 (see figure 2). I suppose that the original stand of the globe was then already lacking, but there is no proof of that. We may safely assume that it served as a 'conversation piece', still another proof – if need be – of the king's great care for and knowledge of the arts and sciences, according to a carefully crafted image of a modern monarch, and simultaneously as a decoration underlining the ancient lineage of the Bragança dynasty.

Fig. 2: Private library of King Carlos at the Necessidades Palace, Lisbon. Taken from Collaço/Palhares/Torralba, 1908: 92.

The globe had moved from an attic to the representative place of a king's private library! Alas, it should benefit from this enormous promotion for a few months

---

janellas rasgadas com vidraças. Nesta caza estão em armarios encostados as paredes os instrumentos mathematicos. Tem os armarios vidraças para se verem os instromentos, e para estarem guardados do pó.' Father Manuel do Portal, congregation of the oratory described the whole abbey in 1756, after the earthquake, in a manuscript ('Historia da ruina da cidade de Lisboa cauzada pelo espantozo terramoto e incendio que reduzio a pó e cinza a melhor, e mayor parte desta infeliz cidade'), which was published by Sousa, 1928: 739–748 and reproduced from that source in Ferrão, 1994: 299.

only. In a context of extreme political tension King Carlos fell victim to an attack in 1908 and soon the first Portuguese Republic was declared in 1910. Within a month the newly constituted Republican assembly decided to make inventories of all belongings of the Crown. It was stipulated in the process that all items of national interest and being of artistic, archaeological or historical value would become public property. The inventories, the so-called 'Arrolamentos', in the form of large bound volumes still exist and among the items listed we find Schissler's globe.[24] It is the earliest extant document referring the artefact:

> nr. 6962: A celestial sphere of gilt copper, with engravings, dated 1575.

Most significantly, the globe was singled out in the inventory in one of the columns labelled 'classification' and got the mention as being 'of historical value', as one of very few items. This comment appears to be the judgement of the commission in charge of the inventory. This mention would imply that the property of the globe would be transferred from the Crown to the Portuguese State.

As a consequence the globe would remain for over thirty years in the same palace, although not in the same room. The documents show that it was moved to the safe room, the so-called 'Casa Forte' or treasure vault, where the important silver and gold ware from the royal household were stored. The safe place, always in the same building complex of Necessidades Palace, located at the basement of the church tower, could be visited on special request, but it was not a public place. The Schissler globe was now being kept on behalf of the Republican State together with a large number of other precious metal ware all considered of historical or artistic value. It is the third clearly documented stage in the globe's 'life' after the years on the attic and the brief time of glory in the King's private library.

Before I turn to the next stage in the globe's biography, it is useful to describe the historical context that would determine the scenery. For the globe to move out again from that hidden place a transformation was needed: the former possessions of the crown needed to become integrated into the picture of 'national' heritage. This evolution was promoted by a long lasting effort of the *Estado Novo*, the authoritarian one party rule in Portugal presided by Salazar, to turn the former royal palaces into public museums. The government of the *Estado Novo* would support all kinds of initiatives that helped forging national pride and patriotic history. The Secretariat of National Propaganda (*Secretariado de Propaganda Nacional*, SPN) together with the administrative service in charge of the public buildings and

---

24 APNA – Arquivo do Palácio Nacional da Ajuda, DGFP – Direcção da Fazenda Pública (fondo). [ca. 1910]. *Arrolamento dos Paços, Edifício do convento* [das Necessidades] (Arrolamentos judiciais 3), 968.

monuments (*Direcção Geral dos Edifícios e Monumentos Nacionais*, DGEMN) were institutional instruments of this process. A crucial actor within the latter institution was Raúl Lino (1879–1974).[25] Trained as an architect, he was appointed from the 1940s onwards to reorganize the royal palaces and adapt them to their use as museums. In this connection the program Lino devised for the Palace of Sintra is particularly significant:

> In fact, the architect Raul Lino as to this palace defined a historical and stylistic programme, in which the building would be taken as of earlier origin but with an image fixed in a period around the end of the gothic era (manueline), and in the Renaissance, with only a few hints at the 17th and 18th centuries. The way he organized the circuit for visitors and some of the decorations of the rooms shows already signs that were clearly of museographic nature.[26]

The space, conceived as a 'palace-museum', would offer only guided tours to visitors in those days. The guards led the tours. The information provided, some would note, depended on the guides who actually inhabited the palace as residents. Moreover, the palace would be used for a variety of diplomatic receptions and cultural events.[27] Now the stage is set for the Schissler globe's reappearance.

Casimiro Gomes da Silva (director of the palace of Sintra from 1944 to at least 1971) seconded the efforts of Lino. He prolonged and fixed the orientation given by the architect. In the 1950s when the refurbishing of several parts of the palace had been concluded, Gomes da Silva was looking for adequate furniture for the palace. The obvious source for it would be 'internal', i.e. among the vast heritage managed by the administrative body (DGEMN) Gomes da Silva worked in. The documental record shows that the director of the Sintra Palace visited the *Casa Forte* of the Necessidades Palace on 31st December 1956. There he selected several among the precious items to be transported to Sintra.[28]

Correspondence of those days reveals that Gomes da Silva was concerned about the value of the various silver pieces. As I've only recently discovered, the

---

25 Soares 2010: 27.
26 Ibid., 34, note 158: 'De facto, o Arquitecto Raul Lino na sua acção neste palácio definiu um programa histórico e estilístico, em que este edifício se assumiria como um palácio de origem anterior mas com uma imagem fixada no período correspondente ao final do gótico (manuelino) e ao período da Renascença, entrando um pouco pelos séculos XVII e XVIII. No modo como organizou o percurso de visita e alguns dos arranjos de salas denota também já preocupações claramente museográficas.' See also Lino, 1948.
27 Soares, 2010: 91.
28 Ibid., Annexo 25, citing correspondence dated 6th January 1957, *Correspondência Ofícios P.N.S.*, documents conserved at Sintra Palace.

letters included a list of objects 'to be transferred' to Sintra: '[nr] 6.962 – Uma esfera celeste de cobre dourada gravada, datada de 1575'. The globe came together with a Chinese porcelain terrine, dozens of silver candlesticks, silver writing utensils and silver basins. It does not seem, however, that it was immediately exhibited in the premises of Sintra Palace. Nevertheless, this move can be equated with Alberti's second key moment mentioned in the introduction of this chapter. For the first time, as far as we know, the Schissler globe became part of a public collection.

There is the curious fact that the Palace of Sintra would serve as 'Presidential Palace' of the fictitious country 'Manchuria' in the movie comedy *The secret of my success* by Andrew L. Stone that came out in 1965.[29] Quite a few pieces of furniture remained on the movie set. I watched the film with the hope of discovering the first visual record of the Schissler globe. However, it is not to be seen. Only from the 1970s onwards there is a declared goal by the directors that Sintra Palace must become a museum in the full sense of that word, and the whole estate became part of the Ministry of Culture. The director then was committed, beyond openness to visitors, also to research, to education and conservation, and outreach events.

In his account António Estácio dos Reis, whom I mentioned at the beginning of this chapter, attributes the merit of presenting the globe to the visitors to the then director of the Palace, Maria Matilde Pessoa de Magalhães Figueiredo de Sousa Franco. She held the office of director from 1984 to 1990.[30] Estácio summarized the re-discovery reporting that Matilde de Sousa Franco took the globe out of a cupboard 'giving it the due importance' and putting it on display.[31] Estácio knew about the existence of the globe since around 1989. At the instigation of the Coronelli Society, he had started to draw up a list of old globes held by Portuguese institutions. Rubem Amaral Júnior, 'consul-geral' of Brazil in Lisbon, as a friend of Estácios' was aware of this interest. He had received Estácio's draft globe list. When taking illustrious visitors to the Sintra Palace, Amaral Júnior sees the globe lying on a table upon which he alerted Estácio.

The latter published the first description with photo of the Schissler globe in the international journal *Der Globusfreund* in 1990. A large photo (265 x 375 mm) was again published for a wider public in an issue of the magazine *Océanos* (Estácio dos Reis, 1992). Now the globe began the normal life of a museum object, and reached – in Alberti's terms – the third key moment: it was not only on display at its home institution in Sintra (see figure 3). From there it also travelled to national

---

29 Stone, 1965: see 1h 38 m and onwards.
30 Franca, 1987.
31 'dar a este último globo a devida importância, tirando-o de um armário onde estava encafuado, e expondo-o ao público)', see Seruya/Pereira, 2004: 20.

exhibitions, for instance, to one at the *Biblioteca Nacional* where it was part of *Pedro Nunes (1502–1578): nouas terras, nouos mares, e o que mays he: nouo ceo e nouas estrellas* (23th April to 7th September 2002) and more recently the exhibition at the Gulbenkian gallery *360° Ciência Descoberta* (2nd March to 2nd June 2013) both curated by the eminent historian of science Henrique Leitão.

In these varying contexts Schissler's globe of 1575 is presented as an illustration of the material culture of science of its time. The titles of the exhibitions it is part of suggests that it is made to carry messages about the prestigious history of astronomical knowledge, or about the high status science enjoyed with the nobility and the rich merchants during the Renaissance.

Fig. 3: Sala das pegas. Photo: Francisco d'Almeida Dias (Luzes da Ribalta). Taken from Romão/Correia, 1995: 15.

Conclusion

Over the centuries, the cultural position and epistemological status of celestial globes changed. While they were mathematical instruments that could replace cosmographical calculations by instrumental approach for approximate results, in a procedure that put to use the geometrical configuration, the globes later became simply decorative items that would suitably ornate a library. Being representative of Renaissance culture, astronomical lore and metal working skills, the globe from the sixteenth century became of 'historical importance' and turned eventually into a genuine museum object in Alberti's sense during the twentieth century.

For the large fire gilded copper globe by Schissler in the Sintra Palace, we may suspect that it formed part of the exchange of prestigious luxury objects between Northern Europe and the Iberian Peninsula during Renaissance. While exotica, specimen of plants or animals and maritime charts were taken northwards, metalwork (weapons and instruments), tapestries etc. flowed southwards to Spain and Portugal. The Schissler globe, quite probably, was among those. These flows of intriguing objects require still a lot further study.

Bibliography

Alberti, Samuel J. M. M. 2005. "Objects and the Museum". *Isis* 96, 559–571.

APNA – Arquivo do Palácio Nacional da Ajuda, DGFP – Direcção da Fazenda Pública (fondo). [ca. 1910]. *Arrolamento dos Paços, Edifício do convento* [das Necessidades] (Arrolamentos judiciais 3).

Bion, Nicolas. 1700. *L'usage des globes célestes et terrestres, et des sphères, suivant les differens systemes du monde.* Amsterdam: Halma.

Bobinger, Maximilian. 1954. *Christoph Schissler der Ältere und der Jüngere* (Schwäbische Geschichtsquellen und Forschungen 5). Augsburg: Die Brigg.

Bobinger, Maximilian. 1966. *Alt-Augsburger Kompaßmacher. Sonnen-, Mond- und Sternuhren,* astronomische und mathematische Geräte, Räderuhren (Abhandlungen zur Geschichte der Stadt Augsburg 16). Augsburg: Rösler.

Brahe, Tycho/Dreyer, Johan Ludvig Emil (ed.). 1924. *Tychonis Brahe Epistolae astronomicae* (Dani Opera omnia 7), København: Gyldendal.

Carvalho, Rómulo de. 1982. *A Física Experimental em Portugal no Século XVIII* (Biblioteca breve: Série pensamento e ciência 63). Lisboa: Instituto de Cultura e Língua Portuguesa.

Collaço, Jorge/Palhares, António/Torralba, Roiz. 1908. *Sua Magestade El-Rei Dom Carlos I e a sua obra artística e científica.* Lisboa: António Palhares.

Dekker, Elly. 1995, "Conspicuous features on sixteenth century celestial globes". *Der Globusfreund. Wissenschaftliche Zeitschrift für Globen- und Instrumentenkunde* 43/44, 77–97.

Estácio dos Reis, António. 1990. "The oldest existing globe in Portugal. Der älteste in Portugal erhaltene Globus". *Der Globusfreund. Wissenschaftliche Zeitschrift für Globen- und Instrumentenkunde* 38/39 (1990), 57–65.

Estácio dos Reis, António. 1992. "Globos antigos em colecção imaginária". *Océanos* 9, 36–49.

Estácio dos Reis, António. 1994. "Old globes in Portugal", Boletim da Biblioteca da Universidade de Coimbra 42, 281–298.

Ferrão, Leonor. 1994. *A real obra de Nossa Senhora das Necessidades*. Lisboa: Quetzal Editores.

Franco, Matilde Sousa. 1987. *Palácio Nacional de Sintra: Residência Querida de D. João I e D. Filipa de Lencastre*. Sintra: Palácio Nacional de Sintra/The British Historical Society of Portugal/Lloyds Banc plc.

Gessner, Samuel. 2010. "The Vopelius-Schissler connection: transmission of knowledge for the design of celestial globes in the 16th century". *Bulletin of the Scientific Instrument Society* [SIS] 104, 32–42.

Gessner, Samuel. 2012 a. " 'Geometricus et astronomicus faber'. Chr. Schissler aus Augsburg als Hersteller eines wenig bekannten großen Himmelsglobus (1575)". In: Hamel, Jürgen/Korey, Michael (eds.). *Weiter sehen. Seeing further. Beiträge zur Frühgeschichte des Fernrohrs und zur Wissenschaftsgeschichte Augsburgs. In Memoriam Inge Keil* (Acta Historica Astronomiae. Beiträge zur Astronomiegeschichte 45). Frankfurt am Main: Harri Deutsch, 123–154.

Gessner, Samuel. 2012 b. "O globo de Schissler em Portugal – a história silenciosa de uma raridade quinhentista". In: Semedo de Matos, Jorge (ed.), *António Estácio dos Reis. Marinheiro por vocação e historiador com devoção. Estudos de Homenagem*. Lisboa: Comissão Cultural de Marinha, 269–282.

Girard, Alberto Alexandre. 1907. *Bibliotheca particular de sua magestade El-Rei*, [S.l.: s.n.].

Haebler, Konrad. 1895. "Conrad Rott und die Thüringische Gesellschaft". *Neues Archiv für sächsische Geschichte und Alterthumskunde* 16, no. 3–4, 177–218.

Lino, Raúl. 1948. *Quatro palavras sobre os Paços Reais da Vila de Sintra*. Lisboa: Valentim de Carvalho.

Lourenço, Marta C./Gessner, Samuel. 2014. "Documenting Collections: Cornerstones for More History of Science in Museums". *Science & Education. Contributions from History, Philosophy and Sociology of Science and Mathematics* 23, 727–745.

Meadow, Mark A. 2002. "Hans Jacob Fugger and the Origins of the Wunderkammer". In: Smith, Pamela/Findlen, Pauls (eds). *Merchants & Marvels. Commerce, Science and Art in Early Modern Europe*. New York/London: Routledge, 182–200.

Romão, Ana Maria de Arez/Correia, Brito. 1995. *Sintra – Palácio Nacional*. Lisboa: IPPAR.

Schirmer, Uwe. 2001. "Öffentliches Wirtschaften in Kur-Sachsen (1553–1631). Motive-Strategien-Strukturen". In: Schneider, Jürgen (ed.). *Öffentliches und privates Wirtschaften in sich wandelnden Wirtschaftsordnungen*. Stuttgart: Franz Steiner, 121–157.

Schöner, Johann. 1516/1548. "Sphærici ac solidi Corporis sive Globi Astronomici Canones Vniuersum eiusdem vsum ordinatirssima triginta propositionum seu capitum serie explicantes". In: Phrysius, Gemma (ed.). *De principiis Astronomiae et Cosmographiae, deque usu Globi Cosmographici ab eodem editi. De Orbis diuisione & Insulis, rebusque nuper inuentis. Eiusdem De Annuli Astronomici usu. Ioannis Schoneri De usu Globi Astriferi opusculum.* Antwerp: J. Steelsius.

Seelig, Lorenz. 1995. *Silver and gold: courtly splendour from Augsburg.* Munich/ New York: Prestel-Verlag.

Seruya, Ana Isabel/Pereira, Mário (dir.). 2004. *Globos Coronelli/Globes Coronelli. Sociedade de Geografia.* Lisboa: Instituto Português de Conservação e Restauro.

Smith, Pamela/Findlen, Paula (eds). 2002. *Merchants & Marvels. Commerce, Science and Art in Early Modern Europe.* New York/London: Routledge.

Soares, Luís Filipe da Silva. 2010. *Palácio Nacional de Sintra. Circuito Expositivo. Análise da sua evolução.* MA dissertation in Museology. Universidade Nova de Lisboa. (Online: https://run.unl.pt/handle/10362/4432, accessed 29th June 2016).

Sousa, Franzisco Luiz Pereira de. 1928. *O terramoto do 1º de Novembro de 1755 em Portugal e um estudo demográfico.* III – Distrito de Lisboa. Lisboa.

Stone, Andrew L. 1965. *The secret of my success.* Los Angeles: Metro Goldwyn Meyer [movie].

Wolfgang Köberer

"The Right Foundation of Seafaring"

German-Portuguese Connections in the Sixteenth Century with Regard to Nautical Science

Abstract: Im Zeitalter der Entdeckungen waren die Handelsbeziehungen zwischen den deutschen Territorien und Portugal wahrscheinlich enger als auf dem Gebiet der Wissenschaft, obwohl es genügend Beispiele dafür gibt, dass etwa die astronomische und kosmographische Literatur wechselseitig von den daran interessierten Kreisen rezipiert wurde. Dies trifft allerdings nicht zu in Bezug auf die nautische Wissenschaft: So wurde vor mehr als 100 Jahren die Behauptung, dass Martin Behaim die nautischen Tafeln und Instrumente, die für die astronomische Navigation der portugiesischen Seefahrer benötigt wurden, aus Nürnberg nach Portugal gebracht hätte, als Mythos entlarvt. Andererseits gibt es auch keine Anhaltspunkte dafür, dass die nautischen Kenntnisse aus Portugal den deutschen Seefahrern bekannt waren und die nautische Literatur in der deutschen Sprache beeinflusst hätten.

Es gibt allerdings vereinzelte Ausnahmen dieser allgemeinen Beobachtung: In einer der großen fürstlichen Bibliotheken Deutschlands – der Herzog-Albrecht-Bibliothek in Wolfenbüttel – befindet sich ein portugiesisches Manuskript, in dem das Standard-Material der portugiesischen nautischen Manuskripte und Bücher des frühen 16. Jahrhunderts enthalten ist. Und das älteste Handbuch der Navigation in (nieder)deutscher Sprache aus dem Jahr 1578 beruht auf diesen Methoden und Tafeln in seinen Kapiteln zur astronomischen Navigation. Der Beitrag beschäftigt sich mit den Übereinstimmungen in den Informationen in diesen Dokumenten mit ihren portugiesischen Vorbildern und versucht nachzuvollziehen, auf welchen Wegen diese Dokumente nach Deutschland gekommen sind.

Although the German territories[1] were farther from Portugal than France or Italy resulting in travelling time of at least a couple of weeks[2], economical and cultural exchange between both regions in the Renaissance took place in many ways. We can see the information about the Portuguese voyages of discovery in the contemporary literature and in artefacts, for instance the Schöner-Globus of 1515 in the

---

1 There was no unified German state until 1871; before that the German territories consisted of a multitude of kingdoms, duchies, archbishoprics and other principalities as well as a handful of free cities, among them the towns of the Hanse.
2 Travelling on foot or by cart one could expect to cover about 30–40 km a day, which meant 4 to 6 weeks for a journey from southern Germany to Lisbon.

"Historisches Museum" of Frankfurt am Main[3]. Also there was a steady influx of objects and books from the German territories to Portugal[4].

On the other hand German contributions to the Portuguese discoveries appear to have been rather marginal. As Hermann Kellenbenz once put it: participation in the discoveries was rather modest[5]. Some South German families (Fugger[6] and Welser among others) were involved in financing a few voyages but that did not amount to much[7]. There may also have been some trade in scientific instruments from Nuremberg, which at the time was a center of production and trade in scientific instruments in Europe[8] although nautical instruments proper clearly were not produced in Nuremberg. A major contribution to the cultural impact of the discoveries may finally be seen in the work of printers hailing from Germany: Valentim Fernandes[9], Herman de Campos, who published the earliest known manual of navigation[10] and Jakob Cromberger. Not only did they publish the narratives of the expeditions but they also published the texts used by the pilots.

Traces of the influence of German science on nautical science in Portugal in the Age of the Discoveries?

The question whether German science – especially cosmographical and astronomical science (which was flourishing in the German territories in the late 15th and early 16th centuries and can be connected with the names of Regiomontanus, Waldseemüller and Schöner just to name a few) – had any influence on the Portuguese (and Spanish) discoveries was hotly debated in the 19th and early 20th centuries.

In Germany a catalyst of this debate was Alexander von Humboldt who quoted Barros (1778) mentioning that Martin Behaim[11] had been a member of the "Junta dos Matematicos"[12]. Arthur Breusing, head of the nautical school at Bremen since 1858 and author of the standard German textbook on navigation[13] argued that Be-

---

3   Cf. Berger, 2013.
4   An impression of the impact of scientific books by German printers in Portugal can be gained by perusing Leitão, 2004.
5   Kellenbenz, 1966: 317.
6   Kellenbenz, 1990.
7   Kellenbenz, 1966: 317.
8   Werner, 1965.
9   Publisher of the *Reportorio dos tempos*, Lisboa 1518; see also: Hendrich, 2007.
10 *Regimento do estrolabio e do quadrante*, Lisboa ca. 1509.
11 Willers, 1992.
12 Humboldt, 1852: 227 and 234 under reference to Barros, *Da Asia*. Lisboa 1778: Dec. I, liv. 4, c. 2, 282.
13 Breusing, 1860; reprinted nine times until 1927.

haim introduced the cross staff in Portugal[14]. Eugen Gelcich, head of the nautical school of the Austrian navy in Lussinpiccolo after 1881, asserted in addition that Behaim introduced Regiomontan's tables in Portugal[15] thereby revolutionizing astronomical navigation and facilitating the Portuguese discoveries.

The Behaim myth was totally debunked by Ernst George Ravenstein in 1908[16], though. The assertion that he introduced the cross staff in Portugal does not fit the historical facts as the earliest date that a cross staff can positively be identified on board a Portuguese ship was in 1529[17]; and that the ephemerides of Regiomontanus cannot be the source of the Portuguese tables follows simply from the fact that the maximum solar declination of the Portuguese tables is 23° 33', the value given in the *Almanach Perpetuum* of Abraham Zacut, whereas Regiomontanus reckoned with a value of 23° 30'[18].

About 100 years ago then arose a new controversy in the history of navigation and the discoveries between Joaquim Bensaude[19] and Hermann Wagner[20] about the priority of the discovery and first application of the rhumb line. The rhumb line (Loxodrome) is the course a ship takes if it sails on a constant course, it is not a part of a great circle, but a spiral line converging towards the poles, in fact never reaching it. The important question is how this course can be represented as a straight line on a nautical chart, a question that was first raised by Pedro Nunes in his *Tratado da sphera* (1537) where he also gave a diagram showing the run of the loxodrome[21]. The practical solution was then given by Gerhard Mercator in his 1569 world chart *Ad usum navigantium*[22] and the theoretical foundation by Edward Wright in his *Certaine errors in navigation* (1599)[23]. The question of priority – probably mostly fueled by national pride – raised 100 years ago remains unsolved today. The question of German impact on Portuguese nautical science in the Renaissance and vice versa can be summed up easily, though: there wasn't any that can be identified[24].

---

14 Breusing, 1869/1982.
15 Gelcich, 1892: 77.
16 Ravenstein 1908.
17 Albuquerque, 1988 a: 12.
18 Albuquerque, 1988 b: 85 f.
19 Bensaude, 1917–1920, 19 seq.
20 Wagner, 1915 and Wagner, 1917.
21 Nunes, 1537: fol. B V.
22 Among the abundant literature the latest comprehensive works are: D'Hollander, 2005; Krücken, 2011 as well as Gaspar and Leitão, 2014.
23 Wright, 1599.
24 This corresponds to the fact that today only three copies of *De arte atque ratione navigandi libri duo* (1573) can be found in libraries in southern Germany: Staatsbibliothek

Responses to Nunes' contributions to nautical science in Northern Europe

From many sources we may deduce that Nunes – the foremost mathematician of Portugal – was well known in European scientific circles of his time. He was a friend of John Dee who even willed that

> [...] if my work cannot be finished or published while I remain alive, I have bequeathed it to that most learned and grave man who is the sole relic and ornament and prop of the mathematical arts among us, D D Pedro Nunes, of Salacia.[25]

Tycho Brahe knew him as documented in his correspondence and in Mercator's library was found: *Nonius de Arte Navigandi. 1573* and *Petrus Nonius de Erratis Orontij, manuscriptus*.[26]

It seems, though, that Nunes' influence on writers of navigational books in northern Europe was marginal if not nonexistent:

Gemma Frisius (1508–1555), the astronomer who first proposed the solution to the longitude problem by watch[27] took up the properties of the rhumb line in his edition of the Apian *Cosmographia*, but without even mentioning Nunes:

> Aliis autem omnibus directis licet secundum Magnetis ductum navigationibus; curva fiunt itinera, quæ circulis maximis non sunt similes; neque parrallelis, sed neque circuli sunt, verum lineæ curvæ tantum, omnes tandem in polorum alterum concurrentes.[28]

Michel Coignet, in his *Instruction nouvelle* (1581)[29] even reproduced Nunes' diagram[30]. Still he maintained that Nunes' explications were impractical:

> [...] toutes ses imaginaõs ne sont, pour la plus grand part, que choses peu praticables, & pour ceste raison de petite efficace pour les Pilotes.[31]

William Borough, one of the most influential persons in England in the second half of the sixteenth century regarding the nautical aspirations of the English nation and

---

Bamberg, Bayerische Staatsbibliothek München, Staats- und Stadtbibliothek Augsburg. Apart from that there is one copy of Nunes' *Tratado da sphera: com a theorica do sol e da lua* (1537) in the "Herzog-August-Bibliothek" bound together with the Wolfenbüttel manuscript. No copy can be found anywhere in libraries near the German coasts.

25 Translated from Dee, 1567: Aiiiv.
26 Watelet, 1994: 410.
27 Cf. Pogo, 1934/1935.
28 Frisius, 1553: 25 r.
29 Coignet, 1581.
30 Ibid.: 21.
31 Ibid.: 26.

author of a treatise on the variation of the compass – a salient problem at the time and for at least 200 hundred years more – wrote about Nunes in that work:

> But how that may stande with the principles of Geometry, I referre the iudgement to the expert Mathematicians, for it is like as a circle should be made of straight lines, which is impossible.[32]

Finally Simon Stevin, who dealt with the problems of loxodromes in his *Eertclootschrift*, stated in 1608:

> Waer af den vermaerden Wisconstnaer *Petrus Nonius* handelende, heeft gheschreven vande ghetalen dienende tottet formen der selve, maer sy en wierden by hem niet recht ghenouch getroffen[33].

Traces of Portuguese nautical science in the German territories in the sixteenth century

In view of the facts shown above an influence of the advanced knowledge in navigational matters amassed in Portugal since the end of the fifteenth century in general cannot be discovered in the nautical literature of the German territories of the Renaissance and later.

There are a few factors that must be considered as a cause: On one hand ships from the German coast and the ports of the Hanseatic League were mostly trading in the North Sea and the Baltic, their navigators therefore didn't need any of the advanced knowledge of astronomical navigation. For them the traditional "Seebücher" (pilot books)[34] contained everything they daily needed to sail from port to port on their usual routes.

Later on when trade expanded to Portugal, Spain and the Mediterranean German navigators and pilots relied on Dutch manuals and textbooks. The prominent Dutch nautical manuals, for example Waghenaer[35], Metius[36] or Lastman[37] don't even mention Nunes when discussing problems of the sailings or nautical instruments.

As a consequence Pedro Nunes or other Portuguese authors on navigation (Faleiro, Lavanha, Naiera, Pimentel among others) are never even mentioned in

---

32 Borough, 1581: G1 r.
33 Stevin, 1608: 153.
34 Regarding this type of rutters see: Sauer, 1996.
35 Waghenaer, 1584.
36 Metius, 1614.
37 Lastman, 1629.

German manuals although they discuss charts on Mercator projection and the so called "Krumstrecks Rekening" (course calculation by spherical trigonometry) since the middle of the seventeenth century[38].

Still there are two isolated items that show connections between Portugal and the German territories in the sixteenth century: an early Portuguese nautical manuscript and the earliest German printed manual of navigation.

The Wolfenbüttel manuscript (Cod. Guelf. 131.4 Quodl. 2°)

Among the treasures of one of the oldest and greatest baroque libraries in Europe, the "Herzog-August-Bibliothek" in Wolfenbüttel – nowadays a rather unimportant town, but formerly the capital of the dukes of Braunschweig and Lüneburg[39] – is a rather unusual manuscript. It had remained unnoticed for about 150 years after it entered the library, but then a rather unlikely person, Heinrich David Wilckens, later a professor of forestry at the Forstakademie in Schemnitz, stumbled across it and published a transcript of it along with a translation into German[40]. Regardless of the fact that Wilckens – as he confessed himself – knew no Portuguese and most probably was not acquainted with problems of navigation as well the translation is quite accurate. He only reproduced the text, though, and did not try to analyze the content.

After Wilckens the manuscript fell again into oblivion until 1960, when Luis de Albuquerque convinced his friend Helmut Peter Schwake to publish an article about it[41]. At the time the manuscript seemed to be lost[42], but that was an error due to the fact that the manuscript was bound between two printed items, the 1537 *Tratado da sphera* by Pedro Nunes[43] and the 1530 *Suma de geographia* by Martin Fernández de Enciso[44]. Schwake had a look at it as did João Alves Dias when he, Luis de Albuquerque and Marilia dos Santos Lopes visited the "Herzog-August-Bibliothek" in 1989[45].

---

38 Tangerman, 1656: 99.
39 This accounts for the fact that one of the great German minds of the time, Gotthold Ephraim Lessing, famous author of plays and literary critic, was librarian there from 1770 on.
40 Wilckens, 1793.
41 Schwake, 1967.
42 Albuquerque, 1965: 5.
43 Shelf mark A: 131.4 Quod. 2° (1).
44 Shelf mark A: 131.4 Quod. 2° (2).
45 Lopes and Albuquerque, 1993.

The manuscript has not been studied in its content so far[46]; Luis de Albuquerque only reprinted the transcript by Wilckens in his *Os Guias Náuticos de Munique e Évora*[47] with a few annotations relating the text to other Portuguese nautical texts.

It consists of 16 paper leaves numbered 1–16 measuring 30 x 19,5 cm. The original leaves seem to have been bigger as leaves 13–16 are cut right to the margin of the written text – without loss of text, though[48].

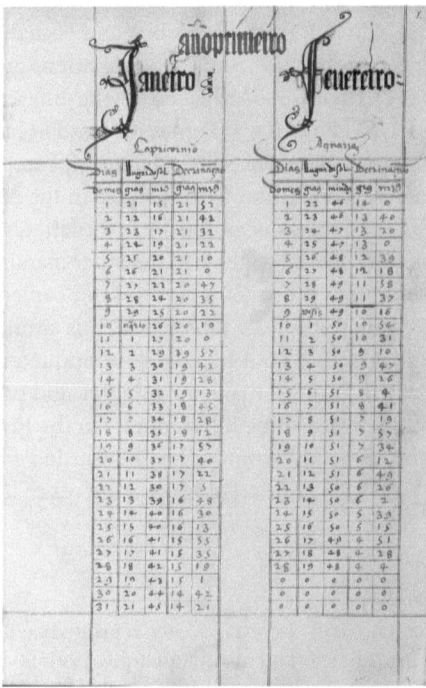

The first 12 leaves contain solar declination tables for a four year cycle starting with the first year after the leap year. These tables which were based on the *Almanach Perpetuum* by Abraham Zacuto[49] were necessary to ascertain the latitude of a place or ship by means of a quadrant or astrolabe[50]. They can be found in the

---

46 Schwake, 1967 discusses the textual appearance but not the nautical content.
47 Albuquerque, 1965: 251–279.
48 For more information about the textual appearance see Schwake, 1967: 22–25.
49 First conclusively shown by Luciano Pereira da Silva, see Pereira da Silva, 1945.
50 Cf. Albuquerque, 1988 b, 61–76.

earliest printed manual of navigation, the *Regimento do estrolabio*[51], and in many manuscripts and manuscript atlases all through the sixteenth and into the seventeenth century[52]. When comparing the tables of the Wolfenbüttel manuscript to other declination tables there is a congruence to the greatest extent with the values in the *Manual de Évora* and a subset of other manuscripts: the *Breve tratado de marinharia* by João de Lisboa[53], the *Regimento de navegação* by Andre Pires[54], the *Cosmographie* by Jean Alfonce[55] and the *Livro de marinharia* by Bernardo Fernandes[56]. Of the 366 values for the daily solar declination of the leap year only 7 differ from the Évora values, 9 from the Lisboa and Fernandes manuscripts, 17 from the (second) Pires table and 19 from the declinations given in the Saintonge manuscript. If one disregards obvious scribal errors the differences narrow down to 1 (Évora), 4 (Saintonge), 12 (Pires II), 5 (Fernandes) and 5 (Lisboa). In spite of the marginal difference between the Wolfenbüttel manuscript and the *Manual de Évora* one cannot safely draw the conclusion that the table in the manuscript is a direct copy of the *Manual de Évora*: the value of April 1 in the latter is clearly false (8° 20'), but the manuscript contains the value given in the other manuscripts (8° 26') which seems to be an indication that the manuscript was not copied from it[57].

Then follows an explanation of the tables, which is similar to sections in the Paris Pires manuscript as pointed out by Luis de Albuquerque[58], and a short paragraph noting how many minutes make up a degree and parts of a degree. This explication can be found – with some differences – in the Pires manuscript[59] and in the *Évora guide*[60], the *Livro de Marinharia* by Bernardo Fernandes[61], the *Códice Bastião Lopes*[62] and – which is remarkable – in the other item containing Por-

---

51 Anon., *Regimento do estrolabio, Lisboa ca. 1509*, reprinted in: Bensaude, 1914.
52 An – incomplete – list can be found in Albuquerque, 1991: 50 f.
53 Arquivo Nacional da Torre do Tombo, Lisboa, Casa Forte, s. inv.; published by Rebello, 1903.
54 Bibliothèque Nationale Paris, Ms. portugais 40; published by Albuquerque, 1963.
55 Bibliothèque Nationale Paris, Ms. francais 676; published by Musset, 1904.
56 Biblioteca Vaticana Roma, Borg. lat. 153; published by Costa, 1940.
57 There is, of course the possibility that the copyist corrected the mistake while copying the Évora table but that does not square with the fact that he committed quite a number of other mistakes himself.
58 Albuquerque, 1965: 275.
59 Albuquerque, 1963: 203.
60 Albuquerque, 1965: 192.
61 Costa, 1940: 3.
62 Albuquerque, 1987: 4.

tuguese nautical science in the German territories, the *Instrument unde Declinatie der Sünnen* (1578) by Jakob Alday[63] of which more later.

Following is the chapter for calculating the latitude from an observation of the culmination of the sun at noon. It consists of 13 rules: five rules for southern declinations of the sun, five rules for northern declinations of the sun, one rule for cases when the declination is zero and finally a rule that makes no sense at all. Whereas today the latitude can be calculated from the noon altitude of the sun by a single simple formula[64] the pilots at that time could not handle negative numbers and therefore had to distinguish the different observational cases (northern and southern solar declination, sun north or south of the observer, sun overhead) and apply different rules for each case.

The rules are mostly similar – but not identical – to the rules given in the *Regimento do estrolabio e do quadrante*[65] which may indicate that they were written early in the sixteenth century – or taken from an early copy of the rules contained in the *Regimento*.

As was usual in the Portuguese manuals the Wolfenbüttel manuscript then gives the rule for ascertaining the latitude by observation of the height of the North Star. As the North Star is not exactly at the Celestial Pole but circles it a correction[66] must be applied to the observed height depending on the position of the North Star with regard to the Pole. This position was indicated in turn by the position of the "Guards", a pair of stars in the "Great Bear". The text then gives the values for correction with reference to the position of the "Guards"; they are – among others – identical with the values in the *Regimento do estrolabio e do quadrante* and the Évora manual, also in the *Livro de Marinharia* of João de Lisboa[67], and in foreign works like the *Boke of Idrography* by John Rotz[68] and the *Instrument unde Declinatie der Sünnen*[69] as well as in Portuguese manuscript atlases[70].

---

63 Alday, 1578: A IV v.
64 Lat = Dec + Zenith distance.
65 See Albuquerque, 1965: 275–277.
66 At that time the Pole Star was about 3 ½ degrees from the Celestial Pole, so this had to be subtracted when the Pole Star was right above or added if it was right below the Pole – with appropriate intermediate values for the positions in between. For a comprehensive explanation of the procedure and its development in the sixteenth century, see Albuquerque, 1988 b, 29–57.
67 Rebello, 1903: 46.
68 Wallis, 1981: Fol. 4 r.
69 Alday, 1578: fol. E.
70 See for instance the Vaz Dourado Atlas in the Biblioteca Nacional Lisboa (Cod. 171), fol. 1.

This "Regiment of the North Star" is accompanied by a circular drawing which shows the correction values graphically. This is also a common feature of almost all the manuals and manuscripts of the time[71]. The drawing in the Wolfenbüttel manuscript most closely resembles the diagram in the *Suma de Geographia* (1519) by Fernández de Enciso[72]. It may therefore be later than the Évora manual – which also has a more primitive diagram.

Finally the manuscript explains how to find the time of midnight by the North Star and the position of the guards. This description is also fairly common in the Portuguese nautical texts of the sixteenth century. A similar – almost verbatim – explanation is contained in the *Livro de Marinharia*[73] and in the Pires manuscript[74] which itself is a verbatim copy of the text in the Évora manual[75].

This leaves the questions when the manuscript was produced and how it came to be in the Herzog-August-Bibliothek. These can only be answered tentatively: Several participants in the colloquium guessed that the manuscript is in a hand of about 1530 – by an official scribe. Prof. Marília dos Santos Lopes was so kind to have the curator of manuscripts at the Herzog-August-Bibliothek look at the watermark when she was visiting Wolfenbüttel in early 2015. The result was that the watermark points to a paper produced at Genoa in 1516[76]. From this information and the clues from the text and diagram one may infer that it was written in the first half of the sixteenth century – post 1519.

The manuscript was entered in the library catalogue between 1658 and 1659 by Duke August himself; it was apparently not bound after it entered the Herzog-August-Bibliothek because the bindings that the Herzog used for his books were different. It is therefore assumed that the manuscript was bound with the Nunes book dealing with the same subject matter before entering the library; this seems to be indicated by the fact that the leaves are cut right to the margin[77].

How it entered the library can only be guessed: Although the Duke had agents in Southern Europe he received large shipments from his agents in South Germany, so

---

71 See for instance Alday, 1578: fol. E; Rebello, 1903: 36; Albuquerque, 1987: 5; Wallis, 1981: Fol. 4 r.
72 Fernández de Enciso, 1519: fol. B VIII v.
73 Rebello, 1903: 47.
74 Albuquerque, 1963: 206.
75 Albuquerque, 1991: 41 (of facsimile text).
76 According to the information given the watermark resembles Briquet, 1968: Nr. 5258, i. e. a paper produced in Genoa in 1516.
77 A clue to the provenance may be derived from the binding, but my inquiries so far have been fruitless.

we may speculate that it came from Augsburg[78]. It could have arrived there through the same connections that brought the *Regimento do estrolabio* to the southern territories of Germany and finally to the Bayerische Staatsbibliothek in Munich.

The *Instrument unde Declinatie der Sünnen*

The second link between Portugal and the German territories is contained in the little booklet already mentioned with the title *Instrument unde Declinatie der Sünnen* published in 1578 in Lübeck[79].

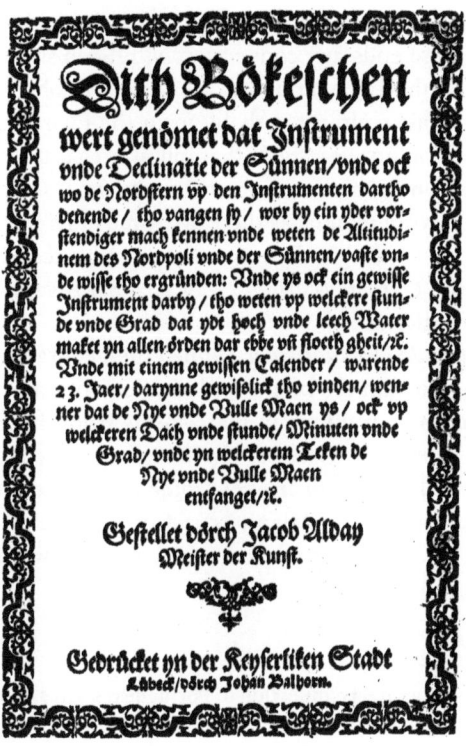

---

78 Bepler and Härtel, 2004.
79 The only known complete copy is now in the "Deutsches Schiffahrtsmuseum" in Bremerhaven; shelf mark 97–0459. It was first recognized by me as the earliest German manual of navigation in 1982. I analyzed its content in Köberer, 1983, this supplements the information given here.

The book merges two different sources of navigational information: in the first part it reproduces information about astronomical navigation based mostly on Portuguese sources, the rest of it deals with the traditional procedures necessary for navigation in the northern tidal waters[80].

It starts – after the dedication to a Lübeck merchant of English origin – with the declaration of the usefulness of astronomical navigation; this might have been necessary with regard to the shipping routes plied by ships of the German ports: the Baltic and the North Sea, but hardly further which did not require astronomical navigation.

This paragraph is taken almost verbatim from Pedro de Medina's *Arte de navegar* (1545), but there ends any connection with Spanish texts. The following explanation of the method of ascertaining the latitude by observation of the noon height of the sun is taken almost verbally from the Évora manual[81]. The solar declination tables for a four year cycle that follow are clearly similar to the Portuguese tables based on Zacuto's *Almanach Perpetuum*. It seems that they are not a direct copy of the Évora manual as there are some different values that cannot be viewed as errors in copying[82]; there is rather a semblance with the Bastião Lopes manuscript[83], the *Livro de Marinharia* by Manuel Álvares[84] and the *Voyages Avantureux* of Jean Alfonce[85].

The *Instrument unde Declinatie der Sünnen* then contains another standard information of the Portuguese books and manuscripts: a table of latitudes of ports, capes and islands. It differs in two aspects from the Évora manual and the *Regimento do estrolabio*: Whereas these start at the equator and go up to the Biskaya[86], the table in the *Instrument unde Declinatie der Sünnen* starts at Cape Bojador and follows the Atlantic coast of Africa and Europe up to the North Cape. This is a clear sign of what Alday thought would be the range of the voyages of his readers.

---

80 As these do not have a direct connection to Portuguese sources and are analyzed in the commentary to the facsimile edition in Alday, 1578/2009 they will not be discussed here in detail.
81 Alday, 1578/2009 (Kommentarband): 16–19.
82 There is a clear mistake in the values for April of the 3rd year in the Évora manual, with the correct value in the Alday table, see ibid., 86.
83 Albuquerque, 1987.
84 Albuquerque, 1969.
85 Alfonce, 1559.
86 The Évora manual even contains latitudes of places in Brazil, all around Africa and in the Red Sea.

Most of the latitudes in the area overlapping in the Portuguese manuals and in the *Instrument unde Declinatie der Sünnen* are identical which is an indication that Alday took them from Portuguese sources. The latitudes of places in Northern Europe pose a question, though, because no such tables had been published prior to 1578, some of the latitude values of islands on the Norwegian coast were only published a decade later in Hakluyt's *Principal navigations*, relating the attempts by English sailors of the "Muscovy Company" to find a North-East Passage to China[87]. It can therefore be assumed that Alday compiled the list himself, probably based on information gained during these voyages.

Then follows a description of the use of the astrolabe and the cross staff which is quite confused and cannot be traced to a published Portuguese text, and another standard instruction: the determination of latitude by the Pole Star which has been discussed in connection with the Wolfenbüttel manual. The explanation in the *Instrument unde Declinatie der Sünnen* is almost identical with the explications in the Évora manual, but that does not necessarily mean that it was copied from it as such a text can be found in many manuscript sources and atlases of the sixteenth century[88]. As usual in the Portuguese texts the *Instrument unde Declinatie der Sünnen* also contains the circular diagram showing the values for correcting the measured height of the Pole Star according to the position of the guards; these are identical with the early Portuguese values[89].

At this point the explanations of astronomical navigation – and the traces of Portuguese sources – end. Following is a treatise on the calculation of the tides with an instrument for "shifting the tides", i.e. the deduction of the times of low and high tides at a given time based on the "establishment of the port" (local time of high and low tide at full and new moon)[90]. Finally the book provides a table of differences in longitude between Antwerp and mostly Northern European ports – with surprisingly accurate values for ports in the Eastern Baltic[91] and lunar tables for 1578–1600 taken out of Stadius followed by a confused explanation how to calculate the phases of the moon.

What can be said about the author, Jakob Alday? He was an Englishman who – as can be taken from the dedication – was well known to an English merchant – John Chappell – residing in Lübeck and trading with Russia. He may be identified with John Alday, an English sailor who prided himself to

---

87 Hakluyt, 1589: 268
88 See Alday, 1578/2009: 35.
89 Ibid.
90 For a description of this part of the book, see Alday, 1578/2009: 38–45.
91 Ibid.: 47.

be a pupil and assistant of Sebastian Cabot[92] who had learned the principles of the new, astronomical navigation in Spain[93]. Alday took part in voyages to the Lapland coast which would explain the knowledge of the latitudes there and he later led an unsuccessful Danish expedition to Greenland. This last fact may be taken as another clue to his identity as another copy of the first edition of the *Instrument unde Declinatie der Sünnen* is in the Royal library in Copenhagen: it may be a gift to King Frederik II of Denmark to recommend himself for the task.

So the oldest German manual of navigation may be written by an Englishman, but the vital information in this booklet is based on Portuguese nautical science.

Literature

Albuquerque, Luís de. 1963. *O Livro de Marinharia de André Pires*. Lisboa: Junta de Investigações do Ultramar.

Albuquerque, Luís de. 1965. *Os Guias Náuticos de Munique e Évora*. Lisboa: Junta de Investigações do Ultramar.

Albuquerque, Luís de. 1969. *O Livro de Marinharia de Manuel Álvares*. Lisboa: Junta de Investigações do Ultramar.

Albuquerque, Luís de (ed.). 1987. *Códice Bastiao Lopez (De Autor Anónimo)*. Lisboa: Imprensa Nacional-Casa da Moeda.

Albuquerque, Luís de. 1988 a. *Instrumentos de navegação*. Lisboa: Com. Nacional para as Comemorações dos Descobrimentos Portugueses.

Albuquerque, Luís de. 1988 b. *Astronomia náutica*. Lisboa: Com. Nacional para as Comemorações dos Descobrimentos Portugueses.

Albuquerque, Luís de. 1991. *Guía Náutico de Munique e Guía Náutico de Évora*. Lisboa: Com. Nacional para as Comemorações dos Descobrimentos Portugueses.

Alday, Jakob. 1578/2009. *Dith Bökeschen wert genömet dat Instrument vnde Declinatie der Suennen ....* Lübeck 1578; reedited with a transcription and commentary by Wolfgang Köberer. Wiefelstede: Oceanum-Verlag.

Alfonce, Jean. 1559. *Les Voyages auantureux du Capitaine Ian Alphonce, Sainctongeois. Contenant les Reigles et enseignemens necessaires à la bonne et seure Navigation*. Poitiers: J. de Marnef.

---

92 See Taylor, 1954: 168.
93 See Sandman and Ash, 2004.

Bensaude, Joaquim. 1914. *Histoire de la science Portugaise nautique a l'époque des Grandes Découvertes*, Vol. 1. Munich: Carl Kuhn.

Bensaude, Joaquim. 1917–1920. *Les légendes allemandes sur l'histoire des découvertes maritimes portugaises*. Genéve: A. Kundis.

Bepler, Jill and Härtel, Helmar. 2004. „Das Netzwerk des Herzogs". *Vernissage* 12, Nr. 136, issue 14/04, 24–31.

Berger, Frank (ed.). 2013. *Der Erdglobus des Johann Schöner von 1515*. Frankfurt am Main: Henrich Editionen.

Borough, William. 1581. *A discourse of the Variation of the cumpas, or Magneticall Needle*. London: Richard Ballard.

Breusing, Arthur. 1860. *Steuermannskunst*. Bremen: Strack.

Breusing, Arthur. 1869. „Regiomontanus, Martin Behaim und der Jakobstab". In: Köberer, Wolfgang (ed.). 1982. *Das Rechte Fundament der Seefahrt. Deutsche Beiträge zur Geschichte der Navigation*. Hamburg: Hoffmann und Campe, 186–193.

Briquet, Charles-Moïse. 1968. *Les filigranes: dictionnaire des marques du papier*. Amsterdam: The Paper Publ. Soc.

Coignet, Michel. 1581. *Instruction nouvelle des poincts plus excellents & necessaires, touchant l'art de nauiguer*. Anvers: Henry Hendrix.

Costa, A. Fontoura da (ed.). 1940. *Livro de marinharia de Bernardo Fernandes (cêrca de 1548)*. Lisboa: Agência Geral das Colónias.

Dee, John. 1567. *Propaedeumata aphoristica*. London: Apud Reginaldum Vuolfium.

D'Hollander, Raymond. 2005. *Loxodromie et projection de Mercator*. Paris/Monaco: Institut Océanographique.

Fernández de Enciso, Martín. 1519. *Suma de geographia q̃ trata de todas las partidas y prouincias del mundo*. Sevilla: Jacob Cromberger.

Frisius, Gemma. 1553. *Cosmographia Petri Apiani per Gemmam Frisium apud Lovanienses medicum et mathematicum insignem....* Paris: Gaultherot.

Gaspar, Joaquim Alves and Leitão, Henrique. "Squaring the Circle: How Mercator Constructed His Projection in 1569", in: *Imago Mundi* 66/1 (2014), 1–24.

Gelcich, Eugen. 1892. „Die Instrumente und die wissenschaftlichen Hülfsmittel der Nautik zur Zeit der grossen Länder-Entdeckung". In: *Hamburgische Festschrift zur Erinnerung an die Entdeckung Amerika's*, Vol. I. Hamburg: Friederichsen, 1–90.

Hakluyt, Richard. 1589. *The Principall Navigations, Voiages & Discoveries of the English Nation*. London: Georg Bishop (Facsimile 1965. Cambridge: AMS Press).

Hendrich, Yvonne. 2007. *Valentim Fernandes: ein deutscher Buchdrucker in Portugal um die Wende vom 15. zum 16. Jahrhundert und sein Umkreis.* Frankfurt am Main: Peter Lang.

Humboldt, Alexander von. 1852. *Kritische Untersuchungen über die historische Entwickelung der geographischen Kenntnisse von der Neuen Welt und die Fortschritte der nautischen Astronomie in dem 15ten und 16ten Jahrhundert,* Vol. 1. Berlin: Nicolai'sche Buchhandlung.

Kellenbenz, Hermann. 1990. *Die Fugger in Spanien und Portugal bis 1560. Ein Großunternehmen des 16. Jahrhunderts.* München: Vögel.

Kellenbenz, Hermann. 1966. „La participation des capitaux de l'Allemagne meridionale aux entreprises Portugaises d'outre-mer au tournant du XVI siècle". In: Mollat, Michel and Adam, Paul (eds.). *Actes du 5e Colloque International d'Histoire Maritime (Lisbonne, 14.-16. Septembre 1960).* Paris: S.E.V.P.E.N., 309–317.

Köberer, Wolfgang. 1983. „Ein niederdeutsches Navigationshandbuch aus dem 16. Jahrhundert". *Deutsches Schiffahrtsarchiv* 6, 151–173; reprinted in: *Mare Liberum* 1 (1990), 153–170.

Krücken, Friedrich Wilhelm. 2011. *Ad Maiorem Gerardi Mercatoris Gloriam.* Vol. V. Münster: MV-Wissenschaft.

Lastman, Cornelis Jansz. 1629. *De Schatkamer des Grooten See-vaerts-Kunst.* Amsterdam: Lastman.

Leitão, Henrique. 2004. *O Livro Científico dos Séculos XV e XVI, Ciências Físico-Mathemáticas na Biblioteca Nacional.* Lisboa: Biblioteca Nacional.

Lopes, Marília dos Santos and Albuquerque, Luis de. 1993. "Heinrich David Wilckens, primeiro historiador de náutica portuguesa do século XV". *Mare Liberum* 6 (Dezembro 1993), 25–29.

Metius, Adriaan. 1614. *Institutiones Astronomicae et Geographicae.* Franeker: Jansz.

Musset, Georges (Ed.). 1904. *La Cosmographie avec l'espère et régime du soleil et du nord par Jean Fonteneau dit Alfonse de Saintonge (Recueil de voyages et de documents pour servir à l'histoire de la géographie 20).* Paris: Ernest Leroux.

Nunes, Pedro. 1537. *Tratado da sphera com a Theorica do Sol e da Lua.* Lisboa: per Germão Galharde.

Silva, Luciano Pereira da. 1945. As tábuas náuticas portuguesas e o Almanach perpetuum de Zacuto, in: Silva, Luciano Pereira da, *Obras Completas,* Vol. II. Lisboa: Agência Geral dos Colónias, 3–19.

Pogo, Alexander. 1934/1935. "Gemma Frisius, his method of determining differences of longitude by transporting timepieces (1530), and his treatise on triangulation (1533)". *Isis* 22, 469–505.

Ravenstein, Ernest. 1908. *Martin Behaim. His life and his globe.* London: Philip.

Rebello, Jacinto Ignacio de Brito (ed.). 1903. *Livro de marinharia: tratado da agulha de marear de João de Lisboa. Roteiros, sondas e outros conhecimentos relativos à navegação.* Lisboa: Imprensa de Libanio da Silva.

Sandman, Alison and Ash, Eric H. 2004. "Trading Expertise: Sebastian Cabot between Spain and England". *Renaissance Quarterly* 57, 813–846.

Sauer, Albrecht. 1996. *Das >Seebuch<. Das älteste erhaltene Seehandbuch und die spätmittelalterliche Navigation in Nordwesteuropa.* Hamburg: Kabel.

Schwake, Helmut Peter. 1967. „Heinrich David Wil(c)ken(s) und eine portugiesische Handschrift der Wolfenbütteler Bibliothek". *Revista da Faculdade de Ciências Coimbra* 39, 339–361.

Stevin, Simon. 1608. *Wisconstige gedachtenissen, Deel 5, Tweede deel des Weereltschrifts, vant Eertclootschrift.* Leiden: Ian Bouwensz.

Tangerman, Hans. 1656. *Wechwyser Tho de Kunst der Seevaert Allen Seevaerenden sehr nütte und Deenstlick.* Hamburg: Pfeiffer.

Taylor, E.G.R. 1954. *The Mathematical Practitioners of Tudor and Stuart England.* London: Hollis & Carter.

Waghenaer, Lucas Janszoon. 1584. *T'eerste deel vande Spieghel der zeevaerdt, van de navigatie der Westersche zee, innehoudende alle de custen van Vranckrijck, Spaingen ende 't principaelste deel van Engelandt, in diverse zee caerten begrepen.* Leiden: Plantijn.

Wagner, Hermann. 1915. „Gerhard Mercator und die ersten Loxodromen auf Karten". *Annalen der Hydrographie und maritimen Meteorologie* 43, 299–311 and 343–352.

Wagner, Hermann. 1917. „Die loxodromische Kurve bei G. Mercator. Eine Abwehr gegenüber Senhor Joaquim Bensaude (1917)". *Nachrichten der Königlichen Gesellschaft der Wissenschaften zu Göttingen, Philologisch-historische Klasse 1917*, 254–267.

Wallis, Helen (ed). 1981. *The maps and text of the Boke of idrography presented by Jean Rotz to Henry VIII.* Oxford: The Roxburgh Club.

Watelet, Michel (ed.). 1994. *Gérard Mercator cosmographe: le temps et l'éspace.* Anvers: Fonds Mercator Paribas.

Werner, Theodor Gustav. 1965. „Nürnbergs Erzeugung und Ausfuhr wissenschaftlicher Geräte im Zeitalter der Entdeckungen. Das Behaim-Problem in wirtschaftsgeschichtlicher Betrachtung". *Mitteilungen des Vereins für Geschichte der Stadt Nürnberg* 53, 69–149.

Wilckens, Heinrich David. 1793. *Ueber eine portugiesische Handschrift der Wolfenbütteler Bibliothek.* Wolfenbüttel: H. G. Albrecht.

Willers, Johann. 1992. „Leben und Werk des Martin Behaim". In: Ibid. (ed.). *Focus Behaim-Globus.* Teil 1. Nürnberg: Verlag des Germanischen Nationalmuseums, 173–188.

Wright, Edward. 1599. *Certaine Errors in Navigation, arising either of the Ordinarie Erroneous Making or Vsing of the Sea Chart, Compasse, Crosse Staffe, and Tables of Declination of the Sunne, and Fixed Starres Detected and Corrected.* London: Valentine Sims.

Notes on Contributors

Torsten dos Santos Arnold is a Doctoral Researcher within the DFG Project "The Globalized Periphery: Atlantic Commerce, Socioeconomic and Cultural Change in Central Europe (1680–1850)" at the European University Viadrina at Frankfurt (Oder), Germany. He graduated with his MA in Maritime History from the Faculty of Letters, Lisbon University (FLUL) and the Portuguese Naval College in 2014. In 2013, his research of the Indo-Portuguese copper trade during the first half of the sixteenth century was awarded with the annual price of the Portuguese Association of Economic and Social History (APHES). The same year, he organized the exhibition "Hermann Kellenbenz (1913–1990): ao Serviço da História" (Hermann Kellenbenz [1913–1990]: in the service of History) in collaboration with the National Library of Portugal.

After a degree in physics **Samuel Gessner** became historian of science obtaining the degree of PhD in that field at university Paris 7, France. His first post-doc project *Instruments in texts and in the practitioners' hands* was financed by the Portuguese Science Foundation FCT (2007–2013). As an associate of the CIUHCT (Centro Interuniversitário de História das Ciências e da Tecnologia) at the University of Lisbon, he focuses on the diverse mathematical cultures in early modern Europe, and the role of mathematical instruments as conceived of by both theoreticians and practitioners. He insists on using artefacts of material culture as primary sources alongside textual and iconographic documents. Samuel Gessner won a grant from the cogito foundation in 2007, and a grant from the Scientific Instrument Society in 2010 which allowed him to visit and study instruments at several important collections in Paris, Munich, Florence, London, Oxford, Cambridge, Edinburgh and Krakow. In 2014 he was chosen to conduct research on the Eisinga planetarium (Netherlands) to underpin its UNESCO World Heritage status. Since 2015 he investigates the cognitive and technological context of Renaissance planetary clocks. He won the prestigious fellowship of the Kulturstiftung des Bundes KSB in the program *International Museum* for 2016 to 2017 to work at the Mathematisch Physikalischer Salon in Dresden (Germany).

Annemarie Jordan Gschwend received her Ph. D. in Art History at Brown University, Providence, R. I. in 1994 with her dissertation, *The Development of Catherine of Austria's Collection in the Queen's Household: Its Character and Cost*. She is specialized in royal patronage in Portugal, Spain, Austria and the Netherlands (1500–1700); the history of collecting at Renaissance Habsburg and Portuguese courts; art, architecture, court portraiture, crown jewels and treasuries, menageries and Flemish tapestries in Portugal and Spain. One of her main interests are the cultural transfers between the Habsburg courts, Africa, and Asia in the Renaissance. She was director and principal coordinator of the research project *Statesman, Art Agent and Connoisseur: Hans Khevenhüller, Imperial Ambassador at the Court of Philip II of Spain* (2008–2013), funded by the Getty Foundation, Los Angeles, and which publication is expected in 2017.

Achim Thomas Hack was born and raised in Stuttgart. After almost two years of civil service in an institution for the disabled, he studied medieval history and comparative religion at the Universities of Tübingen and Rome (La Sapienza). He completed both his Master's Degree (1994) and his "Promotion" (1996) at Tübingen's Eberhard-Karls-Universität. He joined the scientific staff of the Regesta imperii (papal regests of the 9th century) and collaborated in the same capacity on the MGH edition of Thomas Ebendorfer's "*Chronica regum Romanorum*". In 1998, he became Scientific Assistant at the University of Regensburg, where he defended his "Habilitation" and was made "Privatdozent" in 2005. He taught at the Universities of Regensburg, Constance and Munich and in 2010 was appointed to the chair of Medieval History at Jena's Friedrich-Schiller-Universität.

Yvonne Hendrich, born in 1977 in Worms (Germany), studied History, German Studies, and Portuguese Studies at the Johannes Gutenberg University of Mainz and the Universidade Nova of Lisbon. She earned her Ph.D. in History in 2006 with a dissertation on Valentim Fernandes, typographer of Moravian-German descent who worked in Portugal at the turn of the 16th century. Since April 2009 she has been a continuing lecturer in the Department of Romance Languages at the Johannes Gutenberg University of Mainz. Her teaching responsibilities include courses on Portuguese language acquisition, lusophone literature, and cultural studies. Her research areas cover issues of migration and identity, as well as fictional discourse in history and the German-Portuguese relations since the age of discovery.

Thomas Horst, born in Munich in 1980, studied history and anthropology at the universities of Munich and Vienna. In 2003 and 2005 he carried out twice an ethnological field research on the descendants of the Munduruků-Indians in the Amazon region (Brazil). After his PhD in 2008 (on the development of manuscript maps of Bavaria as sources for the history of climatology) he specialized in the analysis of old globes and won the prestigious "Fiorini-Haardt-Prize" of the International Coronelli-Society for the Study of Globes in 2010. His book about Gerhard Mercator and his atlas of 1595, translated into French and Dutch, has been distinguished by the "Société de Géographie" with a special award in Paris in 2012. Since September 2013 he is working on the postdoc-project *Maps, Globes and Texts: Cosmographical knowledge in early Modern Europe* at the Centro Interuniversitário de História das Ciências e da Tecnologia (CIUHCT), University of Lisbon, which is financed by the FCT (Fundação para a Ciência e a Tecnologia), the Portuguese Foundation for Science and Technology (FCT SFRH/ BPD/85102/2012). His main areas of interest include the history of early modern cosmography, the history of climate and the history of discovery, the study of globes and historical visual culture.

Gabriele Kaiser holds a MA degree in Library Sciences and History. She obtained her PhD with a work about Leonhard Thurneysser, who was a doctor, alchemist and the first printer in 16th century Berlin. Commemorating the 400th anniversary of Thurneysser's death, she curated exhibitions in Berlin and in Basel. In 1991 she started her engagement for the Staatsbibliothek zu Berlin (Preußischer Kulturbesitz).

Gabriele Kaiser held various positions, such as the planning director for the Library's historical building "Unter den Linden". She was a fundraiser for the restoration of the Library's J. S. Bach autographs ("Bachpatronat"). Now she works as a specialist for autographs, legacies and art collection in the Manuscript Department of the Staatsbibliothek zu Berlin. Gabriele Kaiser consistently devoted herself to the legacy of Leonhard Thurneysser. In 2012 she published together with Diethelm Eikermann *"Die Pest in Berlin 1576"* – an edition of the medical recommendations of Thurneysser during the Berlin plague epidemic of 1576.

Wolfgang Köberer grew up in the middle of Germany – far from the sea. After graduating from High School he lived in the United States doing social work in Washington, D.C. and Philadelphia, Pa. from 1969–1970. On return to Germany he went to law school in Frankfurt am Main. Later he worked a couple of years as assistant teacher at the Johann-Wolfgang-Goethe-University, then joined a Frankfurt law firm specialized in criminal defence in 1984. He earned his doctorate in jurisprudence with a thesis about the feasibility of mathematically formalizing criminal sentencing. He has been interested in the history of navigation since the mid-70s. Published among other things a compilation of articles by German authors on the history of navigation: *Das Rechte Fundament der Seefahrt* (1982), edited and annotated the facsimile of the earliest German manual of navigation: *Dith boekeschen werd genoemet...,1578* (2010) and lately a bibliography dealing with the German literature about the history of navigation: *Bibliographie zur Geschichte der Navigation in deutscher Sprache* (2011). He was elected an associated member of the "Academia de Marinha" of the Portuguese Navy in November 2012.

Henrique Leitão is Senior Researcher (*Investigator Principal*) at the Interuniversity Center for the History of Science and Technology, CIUHCT, at the University of Lisbon, and professor at the Faculty of Sciences, University of Lisbon. He is mostly interested in the history of exact sciences in the 15th to 17th centuries (mathematics, astronomy, navigation science, physics, etc.), with a special, but not exclusive, focus on Portugal. His main project in the past years has been the edition of the complete works of the Portuguese mathematician and cosmographer Pedro Nunes (1502–1578) beside research on the history of cartography, the history of libraries, the practice of science in Jesuit colleges, and other topics. In 2014 he received the prestigious Pessoa Prize (Prémio Pessoa).

Marília dos Santos Lopes teaches History and Culture Studies at the School of Human Sciences, Universidade Católica Portuguesa. She is member of the Research Centre for Communication and Culture (CECC), Portugal, and Senior Fellow at the Herzog August Library in Wolfenbüttel, Germany. Her research focuses on intercultural processes, on visual culture, and on the history of knowledge exchange in early modern Europe. Her publications include the following single authored books: *Writing New Worlds. The Cultural Dynamics of Curiosity in Early Modern Europe* (2016), *Identidade em viagem. Para uma história da cultura portuguesa* (2015), *Ao cheiro desta canela. Notas para a história de uma especiaria rara* (2002), *Wonderful things never yet seen. Iconography of the Discoveries* (1998). *Afrika. Eine neue Welt in deutschen Schriften des 16. und 17. Jahrhunderts* (1992).

Jürgen Pohle, born in 1965 in Trier (Germany), studied history and geography at the Albertus-Magnus-University in Cologne and in Lisbon. His Ph.D. (finished in 1999/2000) deals with Germany and the overseas expansion of Portugal in the 15th and 16th centuries (*Deutschland und die überseeische Expansion Portugals im 15. und 16. Jahrhundert*). With this standard-work for the German-Portuguese relationship he became Assistant Professor for Social and Economic History on various Portuguese Universities from 2000 to 2014. Since 2009/2010 he is "integrated researcher" at the research centre for Global History (CHAM – Universidade Nova de Lisboa/ Universidade dos Açores) and research fellow of the FCT (Fundação para a Ciência e a Tecnologia), the Portuguese Foundation for Science and Technology.

Yves Schumacher was born in 1946 in Zurich, Switzerland. After a commercial education he devoted himself to European ethnology and attended lectures by Professors Arnold Niederer and Rudolf Schenda. Initially he earned his living in marketing and advertising. Following a two-year stay (1980–1982) in Beijing as Regional Manager of a trading company, he worked for a French advertising agency group. Besides this public work he was always active in publishing und curating exhibitions, focusing on cultural and historical topics. During the last 15 years, Yves Schumacher worked as CEO of the Association of Zurich Museums und runs today his own communication agency.

Dieser Band wurde mit zahlreichen Abbildungen illustriert. Die jeweiligen Autoren waren bemüht, die Urheberrechte nach bestem Wissen hierfür einzuholen. Falls einzelne urheberrechtliche Ansprüche unberücksichtigt blieben, ersuchen die Herausgeber um nachträgliche Mitteilung.

passagem
Estudos em Ciências Culturais
Studies in Cultural Sciences / Kulturwissenschaftliche Studien

Ed. Marília dos Santos Lopes & Peter Hanenberg

Band 1 Adriana Alves de Paula Martins: A construção da memória da nação em José Saramago e Gore Vidal. 2006.

Band 2 Fernando Clara: Mundos de Palavras. Viagem, História, Ciência, Literatura: Portugal no Espaço de Lingua Alemã (1770–1810). 2007.

Band 3 Ana Margarida Abrantes: Meaning and Mind. A Cognitive Approach to Peter Weiss' Prose Work. 2010.

Band 4 Peter Hanenberg / Isabel Capeloa Gil / Filomena Viana Guarda / Fernando Clara (Hrsg.): Kulturbau. Aufräumen, Ausräumen, Einräumen. 2010.

Band 5 Ana Margarida Abrantes / Peter Hanenberg (eds.): Cognition and Culture. An Interdisciplinary Dialogue. 2011.

Band 6 Gerald Bär / Howard Gaskill (eds.): Ossian and National Epic. 2012.

Band 7 Fernando Clara / Cláudia Ninhos (eds.): A Angústia da Influência. Política, Cultura e Ciência nas relações da Alemanha com a Europa do Sul, 1933–1945. 2014.

Band 8 Eduardo Cintra Torres / Samuel Mateus (eds.): From Multitude to Crowds: Collective Action and Media. 2015.

Band 9 Lydia Schmuck: Mio Cid e D. Sebastião. Construções de unidade e diferença nas literaturas ibéricas do século XX. 2016.

Band 10 Thomas Horst / Marília dos Santos Lopes / Henrique Leitão (eds.): Renaissance Craftsmen and Humanistic Scholars. Circulation of Knowledge between Portugal and Germany. 2017.

www.peterlang.com

www.ingramcontent.com/pod-product-compliance
Ingram Content Group UK Ltd.
Pitfield, Milton Keynes, MK11 3LW, UK
UKHW041923210426
5322IPUK00002B/29